Communications
in Computer and Information Science 272

Ana Fred   Jan L.G. Dietz   Kecheng Liu
Joaquim Filipe (Eds.)

Knowledge Discovery, Knowledge Engineering and Knowledge Management

Second International Joint Conference, IC3K 2010
Valencia, Spain, October 25-28, 2010
Revised Selected Papers

Springer

Volume Editors

Ana Fred
IST - Technical University of Lisbon, Portugal
E-mail: afred@lx.it.pt

Jan L.G. Dietz
Delft University of Technology, The Netherlands
E-mail: j.l.g.dietz@tudelft.nl

Kecheng Liu
University of Reading, UK
E-mail: k.liu@henley.reading.ac.uk

Joaquim Filipe
INSTICC and IPS
Estefanilha, Setúbal, Portugal
E-mail: joaquim.filipe@estsetubal.ips.pt

ISSN 1865-0929   e-ISSN 1865-0937
ISBN 978-3-642-29763-2   e-ISBN 978-3-642-29764-9
DOI 10.1007/978-3-642-29764-9
Springer Heidelberg Dordrecht London New York

Library of Congress Control Number: 2012954532

CR Subject Classification (1998): H.4, H.3, C.2, H.5, D.2, J.1

Typesetting: Camera-ready by author, data conversion by Scientific Publishing Services, Chennai, India

Printed on acid-free paper

Springer is part of Springer Science+Business Media (www.springer.com)

Preface

The present book includes extended and revised versions of a set of selected papers from the Second International Joint Conference on Knowledge Discovery, Knowledge Engineering and Knowledge Management (IC3K 2010), held in Valencia, Spain, during October 25–28, 2010. IC3K was sponsored by the Institute for Systems and Technologies of Information Control and Communication (INSTICC) in cooperation with the Association for the Advancement of Artificial Intelligence (AAAI) and the Workflow Management Coalition (WfMC).

The purpose of IC3K is to bring together researchers, engineers and practitioners in the areas of Knowledge Discovery, Knowledge Engineering and Knowledge Management, fostering scientific and technical advances in these areas.

IC3K is composed of three concurrent and co-located conferences, each specialized in at least one of the afore-mentioned main knowledge areas, namely:

- KDIR (International Conference on Knowledge Discovery and Information Retrieval). Knowledge discovery is an interdisciplinary area focusing on methodologies for identifying valid, novel, potentially useful and meaningful patterns from data, often based on underlying large data sets. A major aspect of knowledge discovery is data mining, i.e., applying data analysis and discovery algorithms that produce a particular enumeration of patterns (or models) over the data. Knowledge discovery also includes the evaluation of patterns and identification of which add to knowledge. Information retrieval (IR) is concerned with gathering relevant information from unstructured and semantically fuzzy data in texts and other media, searching for information within documents and for metadata about documents, as well as searching relational databases and the Web. Information retrieval can be combined with knowledge discovery to create software tools that empower users of decision support systems to better understand and use the knowledge underlying large data sets.

- KEOD (International Conference on Knowledge Engineering and Ontology Development). Knowledge engineering (KE) refers to all technical, scientific and social aspects involved in building, maintaining and using knowledge-based systems. KE is a multidisciplinary field, bringing in concepts and methods from several computer science domains such as artificial intelligence, databases, expert systems, decision support systems and geographic information systems. Currently, KE is strongly related to the construction of shared knowledge bases or conceptual frameworks, often designated as ontologies. Ontology development aims at building reusable semantic structures that can be informal vocabularies, catalogs, glossaries as well as more complex finite formal structures representing the entities within a domain and the relationships between those entities. A wide range of applications is emerging, especially given the current Web emphasis, including library science, ontology-enhanced search, e-commerce and configuration.

- KMIS (International Conference on Knowledge Management and Information Sharing). Knowledge management (KM) is a discipline concerned with the analysis and technical support of practices used in an organization to identify, create, represent, distribute and enable the adoption and leveraging of good practices embedded in collaborative settings and, in particular, in organizational processes. Effective knowledge management is an increasingly important source of competitive advantage, and a key to the success of contemporary organizations, bolstering the collective expertise of its employees and partners. Information sharing (IS) is a term used for a long time in the information technology (IT) lexicon, related to data exchange, communication protocols and technological infrastructures.

IC3K received 369 paper submissions from 56 countries in all continents. In all, 38 papers were published and presented as full papers, i.e., completed work, 89 papers reflecting work-in-progress or position papers were accepted for short presentation, and another 86 contributions were accepted for poster presentation. These numbers, leading to a "full-paper" acceptance ratio of about 10% and a total oral paper presentation acceptance ratio close to 34%, show the intention of preserving a high-quality forum for the next editions of this conference. This book includes revised and extended versions of a strict selection of the best papers presented at the conference.

On behalf of the conference Organizing Committee, we would like to thank all participants. First of all the authors, whose quality work is the essence of the conference, and the members of the Program Committee, who helped us with their expertise and diligence in reviewing the papers. As we all know, producing a conference requires the effort of many individuals. We wish to also thank all the members of our Organizing Committee, whose work and commitment were invaluable.

September 2011

Ana Fred
Jan L.G. Dietz
Kecheng Liu
Joaquim Filipe

Organization

Conference Chair

Joaquim Filipe — Polytechnic Institute of Setúbal / INSTICC, Portugal

Program Co-chairs

KDIR

Ana Fred — Technical University of Lisbon / IT, Portugal

KEOD

Jan L.G. Dietz — Delft University of Technology, The Netherlands

KMIS

Kecheng Liu — University of Reading, UK

Organizing Committee

Sérgio Brissos	INSTICC, Portugal
Helder Coelhas	INSTICC, Portugal
Vera Coelho	INSTICC, Portugal
Andreia Costa	INSTICC, Portugal
Patrícia Duarte	INSTICC, Portugal
Bruno Encarnação	INSTICC, Portugal
Liliana Medina	INSTICC, Portugal
Elton Mendes	INSTICC, Portugal
Carla Mota	INSTICC, Portugal
Raquel Pedrosa	INSTICC, Portugal
Vitor Pedrosa	INSTICC, Portugal
Daniel Pereira	INSTICC, Portugal
Filipa Rosa	INSTICC, Portugal
José Varela	INSTICC, Portugal
Pedro Varela	INSTICC, Portugal

KDIR Program Committee

Eduarda Mendes Rodrigues, Portugal
Marcos Gonçalves Quiles, Brazil
Luigi Pontieri, Italy
Eva Armengol, Spain
Pedro Ballester, UK
Roberto Basili, Italy
M.M. Sufyan Beg, India
Florian Boudin, Canada
Marc Boullé, France
Maria Jose Aramburu Cabo, Spain
Rui Camacho, Portugal
Luis M. de Campos, Spain
Keith C.C. Chan, China
Shu-Ching Chen, USA
Koh Hian Chye, Singapore
Jerome Darmont, France
Spiros Denaxas, UK
Dejing Dou, USA
Antoine Doucet, France
Floriana Esposito, Italy
Juan M. Fernández-luna, Spain
Daan Fierens, Belgium
Ana Fred, Portugal
Gautam Garai, India
Susan Gauch, USA
Gianluigi Greco, Italy
José Hernández-Orallo, Spain
Kaizhu Huang, China
Yo-Ping Huang, Taiwan
Beatriz de la Iglesia, UK
Szymon Jaroszewicz, Poland
Piotr Jedrzejowicz, Poland
Liu Jing, China
Estevam Hruschka Jr., Brazil
Wai Lam, China
Anne Laurent, France
Carson K. Leung, Canada
Changqing Li, USA
Chun Hung Li, China
Kin Fun Li, Canada
Wenyuan Li, USA

Xia Lin, USA
Berenike Litz, Germany
Devignes Marie-Dominique, France
Ben Markines, USA
Paul McNamee, USA
Rosa Meo, Italy
Stefania Montani, Italy
Pierre Morizet-Mahoudeaux, France
Henning Müller, Switzerland
Giorgio Maria Di Nunzio, Italy
Chen-Sen Ouyang, Taiwan
Ari Pirkola, Finland
Pascal Poncelet, France
Ronaldo Prati, Brazil
Huzefa Rangwala, USA
Zbigniew W. Ras, USA
Jan Rauch, Czech Republic
Arun Ross, USA
Leander Schietgat, Belgium
Filippo Sciarrone, Italy
Dou Shen, USA
Fabricio Silva, Portugal
Andrzej Skowron, Poland
Dominik Slezak, Poland
Amanda Spink, UK
Marcin Sydow, Poland
Kosuke Takano, Japan
James Tan, Singapore
Andrew Beng Jin Teoh, Korea,
   Republic of
Dian Tjondronegoro, Australia
Christos Tjortjis, Greece
Kar Ann Toh, Korea, Republic of
Volker Tresp, Germany
Joaquin Vanschoren, Belgium
Michal Wozniak, Poland
Qing Xu, USA
Kiduk Yang, USA
Xiao-Jun Zeng, UK
Chengcui Zhang, USA
Dengsheng Zhang, Australia

KDIR Auxiliary Reviewers

Laurence Amaral, Brazil
Maria Angelica Camargo-Brunetto,
    Brazil
Daniele Codetta-Raiteri, Italy
Bindu Garg, India
Danny Johnson, Canada
Ann Joseph, USA

Hiep Luong, USA
Ruggero Pensa, Italy
Achim Rettinger, Germany
Greg Stafford, USA
Alessia Visconti, Italy
Qiang Wang, USA

KEOD Program Committee

Annette Ten Teije, The Netherlands
Alia Abdelmoty, UK
Andreas Abecker, Germany
Yuan An, USA
Francisco Antunes, Portugal
Choorackulam Alexander Augustine,
    India
Teresa M.A. Basile, Italy
Maroua Bouzid, France
Rafik Braham, Tunisia
Patrick Brezillon, France
Vladimír Bureš, Czech Republic
Núria Casellas, Spain
Yixin Chen, USA
Samuel Cruz-Lara, France
Juergen Dix, Germany
Pawel Drozda, Poland
Magdalini Eirinaki, USA
Stefan van der Elst, The Netherlands
Simon Fong, Macau
Johannes Fuernkranz, Germany
Fabien Gandon, France
Raul Garcia-Castro, Spain
Maria Paula Gonzalez, Argentina
Stephan Grimm, Germany
Sven Groppe, Germany
Katerina Kabassi, Greece
C. Maria Keet, Italy
Katia Lida Kermanidis, Greece
Tetsuo Kinoshita, Japan
Mare Koit, Estonia
Zora Konjovic, Serbia
Pavel Kordik, Czech Republic

Patrick Lambrix, Sweden
Jiang Lei, Canada
Ming Li, China
Antoni Ligeza, Poland
Rocio Abascal Mena, Mexico
Munir Merdan, Austria
Malgorzata Mochol, Germany
Claude Moulin, France
Kazumi Nakamatsu, Japan
Keiichi Nakata, UK
Roberto Navigli, Italy
Nan Niu, USA
Yanuar Nugroho, UK
Emilia Pecheanu, Romania
Mihail Popescu, USA
Juha Puustjärvi, Finland
Rong Qu, UK
Amar Ramdane-Cherif, France
Domenico Redavid, Italy
Thomas Reineking, Germany
Dominique Renaud, France
François Rousselot, France
Maria Theresia Semmelrock-Picej,
    Austria
Nuno Silva, Portugal
Anna Stavrianou, France
Mari Carmen Suárez-Figueroa, Spain
Christos Tatsiopoulos, Greece
Orazio Tomarchio, Italy
Rafael Valencia-Garcia, Spain
Iraklis Varlamis, Greece
Cristina Vicente-Chicote, Spain
Bruno Volckaert, Belgium

Sebastian Wandelt, Germany
Franz Wotawa, Austria
Yue Xu, Australia

Gian Piero Zarri, France
Jinglan Zhang, Australia
Catherine Faron Zucker, France

KEOD Auxiliary Reviewers

Alexandre Gouveia, Portugal
Willem van Hage, The Netherlands
Lina Lubyte, Italy
Paulo Maio, Portugal
Hélio Martins, Portugal

Slawomir Nowaczyk, Poland
Heiko Paulheim, Germany
Daniel Rodriguez, Spain
Olga Streibel, Germany
Julien Velcin, France

KMIS Program Committee

Marie-Helene Abel, France
Miriam C. Bergue Alves, Brazil
Bernd Amann, France
Ioannis Anagnostopoulos, Greece
Timothy Arndt, USA
Aurelie Aurilla Bechina Arnzten,
    Norway
Sandra Begley, UK
Eva Blomqvist, Italy
Nunzia Carbonara, Italy
Badrish Chandramouli, USA
Ying Chen, USA
Reynold Cheng, China
David Cheung, China
Dominique Decouchant, Mexico
Mariagrazia Dotoli, Italy
Zamira Dzhusupova, Macau
Alptekin Erkollar, Austria
Vadim Ermolayev, Ukraine
Marco Falagario, Italy
Joan-Francesc Fondevila-Gascón,
    Spain
Anna Goy, Italy
Felix Hamza-Lup, USA
Jan Hidders, The Netherlands
Carlos Arturo Raymundo Ibañez, Peru
Anca Daniela Ionita, Romania

Mounir Kehal, France
Elise Lavoué, France
Chengkai Li, USA
Peter Maher, USA
Sonia Mendoza, Mexico
Christine Michel, France
Taneli Mielikainen, USA
Owen Molloy, Ireland
Ana Marlene Freitas de Morais, Brazil
Fei Nan, USA
Odysseas Papapetrou, Germany
Marina Ribaudo, Italy
Luis Anido Rifón, Spain
Waltraut Ritter, China
John Rohrbaugh, USA
Manas Somaiya, USA
Hiroyuki Tarumi, Japan
Paul Thompson, USA
Ricardo da S. Torres, Brazil
Christopher Turner, UK
Maggie Minhong Wang, China
Leandro Wives, Brazil
Lugkana Worasinchai, Thailand
Nan Zhang, USA
Ying Zhao, USA
Yongluan Zhou, Denmark

KMIS Auxiliary Reviewers

Maxim Davidovsky, Ukraine
Kimberly García, Mexico
Xin Jin, USA
Wenjing Ma, USA

Avila Mora Ivonne Maricela, Mexico
Aniket Pingley, USA
Di Wang, USA

Invited Speakers

Edwin Diday	CEREMADE, Université Paris Dauphine, France
Rudi Studer	Karlsruhe Institute of Technology (KIT), Germany
Joydeep Ghosh	University of Texas at Austin, USA
Oscar Pastor	Universidad Politécnica de Valencia, Spain
Alain Bernard	Ecole Centrale de Nantes, France

Table of Contents

Invited Papers

Part I: Knowledge Discovery and Information Retrieval

Part II: Knowledge Engineering and Ontology Development

Part III: Knowledge Management and Information Sharing

Invited Papers

Actionable Mining of Large, Multi-relational Data Using Localized Predictive Models

Joydeep Ghosh and Aayush Sharma

Department of Electrical and Computer Engineering
The University of Texas at Austin, Austin, Texas 78712, U.S.A.

Abstract. Many large datasets associated with modern predictive data mining applications are quite complex and heterogeneous, possibly involving multiple relations, or exhibiting a dyadic nature with associated side-information. For example, one may be interested in predicting the preferences of a large set of customers for a variety of products, given various properties of both customers and products, as well as past purchase history, a social network on the customers, and a conceptual hierarchy on the products. This article provides an overview of recent innovative approaches to predictive modeling for such types of data, and also provides some concrete application scenarios to highlight the issues involved. The common philosophy in all the approaches described is to pursue a simultaneous problem decomposition and modeling strategy that can exploit heterogeneity in behavior, use the wide variety of information available and also yield relatively more interpretable solutions as compared to global "one-shot" approaches. Since both the problem domains and approaches considered are quite new, we also highlight the potential for further investigations on several occasions throughout this article.

1 Introduction

Classical methods for predictive modeling (regression, classification, imputation, etc.) typically assume that the available data is in the form of a single "flat" file that provides, for each record or entity, a list of the values for the independent and (when available) dependent variables associated with it. However many modern data driven applications involve much more *complex* data structures such as multi-modal tensors, sets of interlinked relational tables, networks of objects where both nodes and links have properties and relationships, as well as other dependencies/constraints such as hierarchical or spatial orderings [17]. Drastic problems, including entry duplication and skewing of counts, that can occur when such data forms are forced into a single "flat" format are well known [23,17]. Therefore, researchers have increasingly concentrated on ways to directly analyze datasets in their natural format, including notable efforts on tensor [49,29,30] and multi-relational [19] data mining.

This article focuses on predictive modeling of *dyadic (bi-modal)* data that consist of measurements on dyads, which are pairs of entities from two different sets (modes). The measurements can be represented as the entries of a matrix, whose rows and columns are the two sets of entities. Moreover, independent variables (attributes or *covariates*) are associated with the entities along the two modes. For concreteness, consider a movie

A. Fred et al. (Eds.): IC3K 2010, CCIS 272, pp. 3–22, 2013.

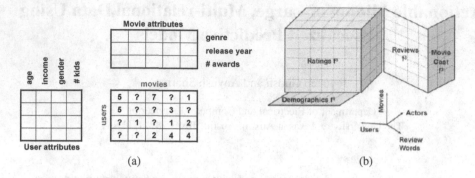

Fig. 1. (a) Conceptual representation of a "Dyadic data with Covariates" (DyaC) dataset on user-movie ratings. (b) Example of multiple relations with some shared modes [8].

recommendation system to predict user ratings for movies, for which there are additional covariates associated with each user (age, gender, etc) and each movie (genre, release year etc), in addition to ratings data. Attributes may also be associated with a user-movie pair, e.g., whether a user's favorite actor is in the movie. Such data structures, which we shall refer to as "Dyadic data with Covariates" or **DyaC**, can be conceptually visualized as in Fig. 1(a). From the figure, it is clear that the data involves multiple tables and cannot be naturally represented as a single flat file.

The characteristic problem to be solved with such data is one of estimating the affinity (e.g. rating) between the modes (e.g. users and movies) given the values of a small number of such affinities. Indeed, recommender systems, which are a special case of this setting, have proved to be very successful at identifying affinities between users and items. Identifying personalized content of interest can greatly enrich the user experience and help institutions offering the recommendations to effectively target different kinds of users by predicting the propensity of users towards a set of items. Marketing data also lends itself perfectly for an affinity estimation problem wherein effective marketing strategies can be formulated based on the predicted affinities. Additionally, there are useful applications in estimating affinities as clickthrough rates for online ads associated with users, search queries, or web pages.

Many current approaches for affinity estimation have concentrated only on a small number of known affinities to infer the missing ones [41]. However, there are often available, many auxiliary sources of information associated with the entities that can aid the estimation of affinities between them. For example, in a movie recommendation engine, the attributes associated with a user might consist of demographic information such as age, gender, geo-location etc. that are often collected as profile information at the time of user registration. Similarly, movie attributes consist of readily available features such as genre, release date, running time, MPAA rating etc. The attributes associated with entities can have a strong bearing on the affinity relationships. For example, it may be common for adolescent males to enjoy movies about comic book characters. In this case, it could be very helpful to have the age and gender information of the user when attempting to predict the affinity of that user for such a movie. Another important source of auxiliary information about entities is a neighborhood structure such as a social network represented by a user-user directed graph. The linkage structure can have an impact on a user's affinities, since preferences are often influenced by preferences of one's friends.

Another problem associated with the methods relying only on past affinities is their inability to intelligently cater to affinity estimation for new entities with no prior history of affinities. This is referred to as a *cold-start* problem. The best one can do with these methods is to utilize a global average model, which however, fails to capture subtle correlations that might exist between a few existing and the new entities. Accurate prediction of affinities for new entities is very crucial for many applications. In the recommender system example, predicting affinities for a new product before its launch could help companies to use more targeted marketing techniques, and could help users recently introduced to the system to quickly find products that they will find useful.

A third aspect of large affinity datasets is that they typically exhibit substantial heterogeneity. Here heterogeneity implies that the relationship between the independent and dependent attributes varies substantially in different parts of the problem space. In such scenarios, it is more appropriate to develop multiple predictive models, one for each relatively homogeneous part of this space, rather than building a single global model. A multi-model approach has several advantages in terms of accuracy, interpretability, reliability and computation [37]. It also provides alternate, more effective ways of active learning and incremental learning as well [16]. However such approaches also raise additional issues such as how to best determine the nature and scope of each model as well as the total number of models developed. The issue of hard vs. soft decomposition of the problem space is also intimately related to interpretability and actionability of the overall solution, leading to non-trivial tradeoffs.

The vast majority of the substantial recent literature on recommender systems (many motivated by Netflix!), including collaborative filtering and matrix factorization methods [26,48,41] simply ignore the covariate information if present, and concentrate solely on the ratings matrix. This is also true of co-clustering or stochastic blockmodel based approaches [22,38] that group elements along both modes, and then model the responses to be homogeneous within each (user, movie) group in the cartesian-product space. A few works do incorporate "side-information" provided by covariates indirectly, typically through a kernel, or as a regularizer to matrix factorization [1,33,9], but they do not exploit heterogeneity. On the other hand, while the use of multiple predictive models to deal with heterogeneity is encountered in a wide range of disciplines, from statistics to econometrics to control and marketing [37,39,32,21,24,40,35], these approaches typically apply only to single flat-file data representations.

In KDD'07, two related approaches were introduced: SCOAL (Simultaneous Co-clustering and Learning) [12] and PDLF (Predictive Discrete Latent Factor Modeling) [5], the first work being nominated for and the second one receiving the Best Research Paper Award. Both papers proposed ways to address DyaC data using localized models that could exploit data heterogeneity, and these approaches have been subsequently refined and expanded. Considering the user-movie recommendation problem again for concreteness, both SCOAL and PDLF will iteratively partition the user-movie matrix into a grid of blocks (co-clusters) of related users and movies, while simultaneously learning a predictive model on each formative co-cluster. The predictors *directly* use covariate information as opposed to the indirect usage of "side-information" through a soft similarity constraint [1,33,9]. The organic emergence of predictive models together with the co-clusters that determine their domains, improves interpretability as well as accuracy in modeling several heterogeneous, DyaC datasets, as this mechanism is able to exploit both local neighborhood information and the available attributes effectively [12,5]. We call the strategies taken by these two methods *Simultaneous Decomposition and Prediction (SDaP)* approaches.

This article first motivates the need for addressing DyaC data through real-world application scenarios outlined in Sec 2. Then it provides a summary of the key ideas behind the SCOAL and PDLF approaches (Sec 3), and sets the context for a novel approach based on a generative (probabilistic) model of DyaC data, which is presented in Sec 4 in its simplest form. More advanced formulations and future work are suggested in the concluding section.

2   Illustrative Applications

In this subsection, we briefly describe three areas that can greatly benefit from SDaP. For all three examples, we show why DyaC based approaches are suitable and also highlight certain unresolved challenges that motivate further work.

Ecology. The analysis of population dynamics and their interaction with the environment is a central problem in the field of ecology. A typical data setup for this problem consists of population count data for different species across varying environmental conditions. The objective then is to predict the population counts for species of interest in certain locations/environments at current or future times. Typical datasets of this nature are highly heterogeneous with differing population patterns across different species, environments and seasons, and is also extremely sparse with unavailable counts for many species and environments. For example, the recently compiled and NSF funded *eBird Reference Dataset* [36] (avianknowledge.net, ebird.org) contains over 27 million observations of birds count for over 4000 species collected from more than 250,000 locations. Each location is annotated with over 50 covariates such as habitat, climatic conditions, elevation etc. Bird species are described by about 25 attributes such as preferred habitat, global population size, breeding information etc. The data is collected by human observers over different times (sampling events which are further associated with 15 covariates) adding a time dimension to the dataset as well, which makes the DyaC even more rich and complex.

Current approaches typically model each species separately using poisson regression over the independent environmental variables [11,43], and thus flatten the data. Generalization ability is inadequate [11]. We note that the problem domain is inherently dyadic in nature, with the species and the environments forming the two modes of variation, each with associated covariates. Considering time adds a third mode with strong seasonality properties. Current methods fail to leverage the dyadic or tensor nature of the data: each species is treated independently of others and their attributes are ignored. Learning multiple models using SDaP can greatly improve the performance by efficiently capturing the data heterogeneity as well as relations among sites and among species. Equally important, it can significantly enhance interpretability and actionablity as closely related subsets of species (in terms of the influence of environment on their count) and associated locations will automatically emerge from the model. This domain also imposes (soft) spatial constraints and partial orderings via factors such as geographical location, altitude etc. A hierarchy defined on the birds (ebird.org) adds a different type of soft constraint on a different mode! Such constraints can also be exploited to influence the decomposition process, for example via an efficient markov random field (MRF) based latent dyadic model.

Customer Product Preference Analysis. The problem of analyzing customer purchase or preference behavior involves multi-modal, inter-connected relational data and forms a suitable and broad application domain for SDaP. Consider the publicly available ERIM household panel dataset that has been widely studied in the marketing research community [27,28,44]. This dataset has purchase information over a period of 3 years and covers 6 product categories (ketchup, tuna, sugar, tissue, margarine and peanut butter) with a total of 121 products. Each household is annotated with demographic information including income, number of residents, male head employed, female head employed, etc. Products are described by attributes such as market share, price and advertising information. Details of each shopping visit of each household over the 3 year period are recorded, adding a third time dimension to the dataset.

SCOAL has been applied to ERIM for predicting the number of units of specified products purchased by households, given household and product covariate information. For this problem, SCOAL substantially improves accuracy over alternative predictive techniques [12,14], including sophisticated Hierarchical Bayesian approaches on a flattened data representation, thereby pointing to the utility of the dyadic viewpoint. In addition, SCOAL also provides interpretable and actionable results by indicating what factors influence purchases in different household-product groups [13]. However, the current approach needs further extension to cater well to additional related information that is available, including attributes of the shops and of the city of residence. Moreover, the increasing popularity of customer interaction and feedback channels, including social networks and ratings sites, is bound to lead to additional acquired data that add new dimensions to the customer purchase behavior modeling problem, which also need to be leveraged.

Click-Through-Rate Prediction. A key goal of content providers and search engines such as Yahoo! and Google is to get as high a click-through-rate (CTR) as possible by serving users the content and ads they are most likely to click on. The massive scale of the ads targeting problem and its obvious business relevance has started attracting attention from the data mining community [3,2]. A typical data setup for the problem is as follows: Ads are categorized into a hierarchical taxonomy. Each category in the taxonomy is a specific topic of interest, e.g., loans, travel, parenting and children. The categories are annotated by attributes such as descriptive keywords, historic CTR rates and volume. Users are also described by features such as demographics, geographical location and metrics computed based on previous browsing behavior. For some (ad category, user) pairs, the target CTR value is known or easily estimated, and these form the training data. Given a user, the objective is then to select the categories to be served based on the highest predicted CTRs, among other criteria.

Once again we have DyaC data, with users and ad categories representing two sets of entities. Also, such data is very large (typically several hundred million users per week and several thousand categories), very sparse and noisy, with little activity in some low traffic categories. Moreover, the data is very heterogeneous, with widely varying patterns of user behavior across different user and category groups.

Initial results on applying SCOAL and PDLF approaches to an internal Yahoo! dataset showed substantial promise in terms of both accuracy and speed as compared to traditional predictive models. Yet, they are wanting in several important aspects: (a) they don't have an effective mechanism for exploiting the taxonomy available on the category mode, (b) user behavior is not static but even differs by the day, hence distinct (though related) models for each day, or for weekdays vs. weekends, are desired; and (c) the cold-start problem of determining what ads to serve to new users and predicting user propensity towards a new category, needs to be robustly addressed. Mechanisms for incorporating constraints among entities, for multitask learning and for cold-start can however be added to the generative approach presented in Section 4 to address these challenges. Finally, the ad targeting problem is essentially dynamic. Ad views and clicks are recorded and CTR values are updated in near real time. So SDaP needs to have a scalable, incremental version that is capable of effectively modeling streaming data.

Other application domains that can benefit from SDaP include (i) microarray data annotated with regulatory network information, gene/condition metadata, colocation of names in medical abstracts, etc., (ii) cross-language information retrieval, and (iii) scene analysis of a large number of images, each containing a variety of objects with geometric and co-locational constraints. It is also suitable for a large class of problems that can be represented as directed graphs where both nodes and edges have associated attributes. Email data (4 mode tensor with sender, recepient, time and content/topic; attributes of the people are also provided), and 3-mode web data with source, destination and anchor text, etc, fall in this category, and corresponding large datasets - Enron email and the substantial TREC WT10g Web Corpus - are already available.

3   Latent Factor Modeling of Dyadic Data with Covariates

In this section, we consider the two aforementioned SDaP approaches [12,5] in a bit more detail.

Simultaneous Co-clustering and Learning (SCOAL). In complex classification and regression problem domains, a single predictive model is often inadequate to represent the inherent heterogeneity in the problem space. The traditional "divide and conquer" solution that partitions the input space *a priori* and then learns models in each "homogeneous" segment is inherently sub-optimal since the partitioning is done independent of the modeling. The key idea behind SCOAL is to partition the entities along each mode, thus leading to blocks or co-clusters representing the cartesian-product of such partitions across different modes. If a mode has innate ordering, e.g. time, then the partitions need to be contiguous along that axis [14]. For example, a 3-D block in a user x movie x time tensor would be formed by subset of users, rating a subset of movies over a contiguous time period. For each block, a predictive model that relates the independent variables to the response variable in the co-cluster, is learnt. Note that *within a block, the responses themselves do not need to be similar, as distinct from the blocks formed in partitional co-clustering* [34,7]. A key property of SCOAL is that the fitting of the "local" predictive models in each block is done simultaneously with block formation. The overall goal is to obtain a partitioning such that each co-cluster can be well characterized by a single predictive model.

Specifically, SCOAL aims at finding a co-cluster assignment and corresponding set of co-cluster specific models that minimize a global objective function, e.g. the prediction error summed over all the the known entries in the data matrix. For instance, with linear regression models, the objective function is a suitably regularized sum squared error. A simple iterative algorithm that alternately updates the co-cluster models and the row and column cluster assignments can be applied to obtain a local minimum of the objective function. A variety of regression models can be used for the predictive learning. For example, the data can be modeled by a collection of neural network models or regression models with the L_1 norm (Lasso [25]). The mathematics also carries through for all generalized linear models (GLMs), thus covering binary (classification) and count responses as well. It also generalizes to tensor data, and for any noise term belonging to the exponential family. Recent advances in the SCOAL approach include a dynamic programming solution to segment ordered modes, an incremental way to increase the number of models, active learning methods that exploit the multiple-model nature, and novel ways of determining the most reliable predictions that also exploit the presence of multiple local models [16,15,14].

Predictive Discrete Latent Factor Modeling (PDLF). While SCOAL learns multiple, independent local models, the Predictive Discrete Latent Factor (PDLF) model simultaneously incorporates the effect of the covariates via a *global* model, as well as any local structure that may be present in the data through a block (co-cluster) specific constant. Similar to SCOAL, the dyadic data matrix is partitioned into a grid of co-clusters, each one representing a local region. The mean of the response variable is modeled as a sum of a function of the covariates (representing global structure) and a co-cluster specific constant (representing local structure). The co-cluster specific constant can also be thought of as part of the noise model, teased out of the global model residues. The authors also formulate scalable, generalized EM based algorithms to estimate the parameters of hard or soft versions of the proposed model.

SCOAL and PDLF show benefit in complementary situations; SCOAL works well in domains with very high heterogeneity where sufficient data is available to learn multiple models, while PDLF shows better value in situations where the training data is limited and several outliers are present. Also note that the standard approach in the statistics community for such problems would be to develop a semi-parametric hierarchical model, which at first blush is structurally similar to PDLF [21]. However, as discussed in great detail by [5,2], the assumption made in PDLF that block membership can be completely specified in terms of row and column memberships, is a key feature that makes it much more scalable. The corresponding factorization of the joint space also leads to vastly simpler and efficient inference. Similarly, the smoothing effect of a soft partitioning within the block-structure is found to be more effective than the widely used stastical approach of using a hierarchical random effects model.

Agarwal and Chen [2] recently generalized the PDLF as well as the Probabilistic Matrix Factorization [41] approach to regression based latent factor models (RLFM), which provides a unified framework to smoothly handle both cold-start and warm-start scenarios. RLFM is a two stage hierarchical model with regression based priors used to regularize the latent factors. A scalable Monte Carlo EM approach or an Iterated Conditional Mode technique can be used for model fitting. RLFM has shown substantially better results than competing techniques in a challenging content recommendation application that arises in the context of Yahoo! Front Page.

4 Latent Dirichlet Attribute Aware Bayesian Affinity Estimation (LD-BAE)

Several Bayesian formulations have been proposed in the context of affinity estimation problems. Mixed Membership stochastic Blockmodels (MMBs) [20] is one such method that utilizes affinity values to group the two entity sets via a soft co-clustering. A weighted average of the pertinent co-cluster means is then used to estimate missing affinities. The model is shown to be quite efficient in scaling to large datasets, however it fails to utilize any available side information. Other efforts include fully Bayesian frameworks for PMF([31], [42]) with differing inference techniques - ranging from Variational approximations to sampling based MCMC methods. However, the stress again is only on utilizing the available affinities. Recently, Bayesian models based on topic models for document clustering [18] have been utilized for estimating affinities between users and News articles [4]. Two sided generalizations of topic models have also been utilized for co-clustering and matrix approximation problems ([46], [45]) without taking into account auxiliary sources of information.

This section introduces a Side Information Aware Bayesian Affinity Estimation approach that is related to Latent Dirichlet Allocation [18], as explained shortly, and is hence called the Latent Dirichlet Attribute Aware Bayesian Affinity Estimation (LD-BAE) model. For simplicity, we consider the available side information to be only a set of attributes (covariates) associated with each entity. Additional sources of side information such as network structures over entities, evolution over time or knowledge that known ratings are not given at random, can be accommodated by extending this basic framework [47].

Notation. Before describing the LD-BAE framework, a quick word on the notation. We use capital script letters for sets, $\{\cdot\}$ denote a collection of variables for unnamed sets and $\dagger$ represents transpose of a matrix. Let $\mathcal{E}_1 = \{e_{1m}\}, [m]_1^M$ and $\mathcal{E}_2 = \{e_{2n}\}, [n]_1^N$ represent the sets of entities between which affinities need to be estimated. $\mathcal{Y} = \{y_{mn}\}$ is a set of $M \times N$ affinities between pairs of entities of the form $(e_{1m}, e_{2n}), e_{1m} \in \mathcal{E}_1$ and $e_{2n} \in \mathcal{E}_2$. The subset $\mathcal{Y}_{obs} \subseteq \mathcal{Y}$ is a set of observed affinities while $\mathcal{Y}_{unobs} = \mathcal{Y} \backslash \mathcal{Y}_{obs}$ denotes a set of missing affinities. A weight w_{mn} is associated with each affinity y_{mn} (affinity between a pair of entities e_{1m} and e_{2n}) such that $w_{mn} = 1$ if $y_{mn} \in \mathcal{Y}_{obs}$ and $w_{mn} = 0$ if $y_{mn} \in \mathcal{Y}_{unobs}$. The set of all $M \times N$ weights is denoted by $\mathcal{W}$. The set of entity attributes associated with $\mathcal{E}_1$ and $\mathcal{E}_2$ are respectively described by the sets $\mathcal{X}_1 = \{x_{1m}\}$

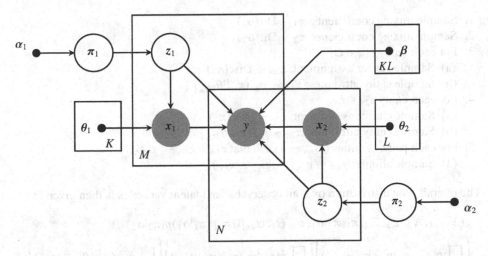

Fig. 2. Graphical model for Latent Dirichlet Attribute Aware Bayesian Affinity Estimation

and $\mathcal{X}_2 = \{x_{2n}\}$. The notation $x_{mn} = [x_{1m}^\dagger x_{2n}^\dagger]^\dagger$ is used to denote the attributes associated with the entity pair (e_{1m}, e_{2n}).

Figure 2 shows the graphical model for LD-BAE - a mixture model of KL clusters obtained as a cross-product of clustering the two sets of entities into K and L clusters respectively. First, the mixing coefficients $\pi_1(\pi_2)$ are sampled (only once) from the corresponding Dirichlet distributions parameterized by $\mathrm{Dir}(\alpha_1)$ and $\mathrm{Dir}(\alpha_2)$ for entity set $\mathcal{E}_1$ and $(\mathcal{E}_2)$ respectively. Hence, all the entities in a particular set share the same mixing coefficients, thereby inducing statistical dependency between them. Then each entity $e_{1m} \in \mathcal{E}_1$ is assigned to one of K clusters by sampling cluster assignments $z_{1m} \in \{1, \ldots, K\}$ from a discrete distribution $\mathrm{Disc}(\pi_{1m})$. Similarly, the entities $e_{2n} \in \mathcal{E}_2$, are clustered into L clusters by sampling cluster assignments $z_{2n} \in \{1, \ldots, L\}$ from a discrete distribution $\mathrm{Disc}(\pi_{2n})$. $\mathcal{Z}_1$ and $\mathcal{Z}_2$ respectively denote the sets of cluster assignments for the two entity sets. It is easy to see that by sharing mixing coefficients across entities in a set, the model is an attribute sensitive two sided generalization of the Latent Dirichlet Allocation (LDA) [18] model.

The attributes x_{1m} associated with the entity e_{1m} are drawn from one of K possible exponential family distributions of the form $p_{\psi_1}(x_{1m}|\theta_{1z_{1m}})$[1], such that the parameter $\theta_{1z_{1m}}$ of the family, is chosen according the entity cluster assignment z_{1m}. Likewise, attributes x_{2n} for an entity e_{2n} are generated from one of L possible exponential family distributions $p_{\psi_2}(x_{2n}|\theta_{2z_{2n}})$. The cluster assignments z_{1m} and z_{2n} over the two entities together determine a co-cluster (z_{1m}, z_{2n}), which then selects an exponential family distribution, $p_{\psi_y}(y_{mn}|\beta_{z_{1m}z_{2n}}^\dagger x_{mn})$ (out of KL such distributions), to generate the affinity y_{mn} associated with the entity pair (e_{1m}, e_{2n}). The parameters $\beta_{z_{1m}z_{2n}}$ of the distribution are specific to the co-cluster (z_{1m}, z_{2n}). In summary, the generative process for the attributes and the affinities between each pair of entities is as follows:

---

[1] We use the canonical form of exponential family distributions: $p_\psi(x|\theta) = p_0(x)\exp(\langle x, \theta \rangle - \psi(\theta))$.

1. Sample mixing coefficients: $\pi_1 \sim \text{Dir}(\alpha_1)$
2. Sample mixing coefficients: $\pi_2 \sim \text{Dir}(\alpha_2)$
3. For each entity $e_{1m} \in \mathcal{E}_1$
   (a) Sample cluster assignment: $z_{1m} \sim \text{Disc}(\pi_1)$
   (b) Sample entity attributes: $x_{1m} \sim p_{\psi_1}(x_{1m}|\theta_{1z_{1m}})$
4. For each entity $e_{2n} \in \mathcal{E}_2$
   (a) Sample cluster assignment: $z_{2n} \sim \text{Disc}(\pi_2)$
   (b) Sample entity attributes: $x_{2n} \sim p_{\psi_2}(x_{2n}|\theta_{2z_{2n}})$
5. For each pair of entities (e_{1m}, e_{2n}) such that $e_{1m} \in \mathcal{E}_1, e_{2n} \in \mathcal{E}_2$
   (a) Sample affinity: $y_{mn} \sim p_{\psi_y}(y_{mn}|\beta_{z_{1m}z_{2n}}^{\dagger} x_{mn})$

The overall joint distribution over all observable and latent variables is then given by:

$$p(\mathcal{Y}, \mathcal{X}_1, \mathcal{X}_2, \mathcal{Z}_1, \mathcal{Z}_2, \pi_1, \pi_2 | \alpha_1, \alpha_2, \Theta_1, \Theta_2, \beta) = p(\pi_1|\alpha_1)p(\pi_2|\alpha_2)$$

$$\left(\prod_m p(z_{1m}|\pi_1)p_{\psi_1}(x_{1m}|\theta_{1z_{1m}}) \right)\left(\prod_n p(z_{2n}|\pi_2)p_{\psi_2}(x_{2n}|\theta_{2z_{2n}}) \right)\left(\prod_{m,n} p_{\psi_y}(y_{mn}|\beta_{z_{1m}z_{2n}}^{\dagger} x_{mn}) \right)$$

Marginalizing out the latent variables, the probability of observing the known affinities and the attributes is:

$$p(\mathcal{Y}_{\text{obs}}, \mathcal{X}_1, \mathcal{X}_2 | \alpha_1, \alpha_2, \Theta_1, \Theta_2, \beta) = \int_{\mathcal{Y}_{\text{unobs}}} \int_{\pi_1} \int_{\pi_2} (p(\pi_1|\alpha_1))(p(\pi_2|\alpha_2))$$

$$\sum_{\mathcal{Z}_1} \sum_{\mathcal{Z}_2} \left(\prod_m p(z_{1m}|\pi_1)p_{\psi_1}(x_{1m}|\theta_{1z_{1m}}) \right)\left(\prod_n p(z_{2n}|\pi_2)p_{\psi_2}(x_{2n}|\theta_{2z_{2n}}) \right)$$

$$\left(\prod_{m,n} p_{\psi_y}(y_{mn}|\beta_{z_{1m}z_{2n}}^{\dagger} x_{mn}) \right) d\mathcal{Y}_{\text{unobs}} d\pi_1 d\pi_2$$

Note that even marginalization of only the mixing coefficients π_1 and π_2 induces dependencies between the clustering assignments $\mathcal{Z}_1$ and $\mathcal{Z}_2$.

Inference and Learning. As a result of the induced dependencies, direct maximization of the observed log-likelihood is intractable using an EM algorithm. One instead needs to resort to Gibbs sampling or related approaches, or use variational methods. In this subsection we take the latter route by constructing tractable lower bounds using a fully factorized mean field approximation to the true posterior distribution over the latent variables. The optimal factorized distribution over the latent variables ($\mathcal{Y}_{\text{unobs}}, \mathcal{Z}_1, \mathcal{Z}_2, \pi_1, \pi_2$) that corresponds to the tightest lower bound on the observed likelihood is then given by:

$$q^*(\mathcal{Y}_{\text{unobs}}, \mathcal{Z}_1, \mathcal{Z}_2, \pi_1, \pi_2) \tag{1}$$

$$= q^*(\pi_1|\gamma_1)q^*(\pi_2|\gamma_2)\left(\prod_{\substack{m,n \\ y_{mn} \in \mathcal{Y}_{\text{unobs}}}} q^*(y_{mn}|\phi_{mn}) \right)\left(\prod_m q^*(z_{1m}|r_{1m}) \right)\left(\prod_n q^*(z_{2n}|r_{2n}) \right)$$

Note that, since the mixing coefficients are shared across entities from the same set, we only have two variational factors corresponding to the mixing coefficients π_1 and π_2. $q_{\psi_y}(y_{mn}|\phi_{mn})$ is an exponential family distribution with natural parameter ϕ_{mn}, $q(\pi_1|\gamma_1)$ and $q(\pi_2|\gamma_2)$ are K and L dimensional Dirichlet distributions with parameters γ_1 and γ_2 respectively while the cluster assignments z_{1m} and z_{2n} follow discrete distributions over K and L clusters with parameters r_{1m} and r_{2n} respectively. The variational parameters $(\gamma_1, \gamma_2, \phi_{mn}, r_{1m}, r_{2n})$ are then given by (see Appendix for derivation, and [47] for further detail on the variational derivation):

$$\phi_{mn} = \sum_{k=1}^{K} \sum_{l=1}^{L} r_{1mk} r_{2nl} \left(\beta_{kl}^{\dagger} x_{mn} \right) \tag{2}$$

$$\gamma_{1k} = \sum_{m=1}^{M} r_{1mk} + \alpha_{1k} \tag{3}$$

$$\gamma_{2l} = \sum_{n=1}^{N} r_{2nl} + \alpha_{2l} \tag{4}$$

$$\log r_{1mk} \propto \log p_{\psi_1}(x_{1m}|\theta_{1k}) + \Psi(\gamma_{1k}) +$$

$$\sum_{n=1}^{N} \sum_{l=1}^{L} r_{2nl} \left(w_{mn} \log p_{\psi_y}(y_{mn}|\beta_{kl}^{\dagger} x_{mn}) + (1 - w_{mn}) \mathbb{E}_q \left[\log p_{\psi_y}(y_{mn}|\beta_{kl}^{\dagger} x_{mn}) \right] \right) \tag{5}$$

$$\log r_{2nl} \propto \log p_{\psi_2}(x_{2n}|\theta_{2l}) + \Psi(\gamma_{2l}) +$$

$$\sum_{m=1}^{M} \sum_{k=1}^{K} r_{1mk} \left(w_{mn} \log p_{\psi_y}(y_{mn}|\beta_{kl}^{\dagger} x_{mn}) + (1 - w_{mn}) \mathbb{E}_q \left[\log p_{\psi_y}(y_{mn}|\beta_{kl}^{\dagger} x_{mn}) \right] \right) \tag{6}$$

The optimal lower bound on the observed log-likelihood with respect to the variational distribution in (1) is then given by:

$$\log p(\mathcal{Y}_{obs}, \mathcal{X}_1, \mathcal{X}_2 | \alpha_1, \alpha_2, \Theta_1, \Theta_2, \beta)$$
$$\geq H[q^*] + \mathbb{E}_{q^*}[\log p(\mathcal{Y}_{obs}, \mathcal{X}_1, \mathcal{X}_2, \mathcal{Z}_1, \mathcal{Z}_2, \pi_1, \pi_2 | \alpha_1, \alpha_2, \Theta_1, \Theta_2, \beta)]$$

This bound can be maximized with respect to the free model parameters to get their improved estimates. Taking partial derivatives of the bound with respect to the model parameters and setting them to zero, we obtain the following updates (see Appendix for details):

$$\theta_{1k} = \nabla \psi_1^{-1} \left(\frac{\sum_{m=1}^{M} r_{1mk} x_{1m}}{\sum_{m=1}^{M} r_{1mk}} \right) \tag{7}$$

$$\theta_{2l} = \nabla \psi_2^{-1} \left(\frac{\sum_{n=1}^{N} r_{2nl} x_{2n}}{\sum_{n=1}^{N} r_{2nl}} \right) \tag{8}$$

$$\boldsymbol{\beta}_{kl} = \arg\max_{\boldsymbol{\beta}\in\mathbb{R}^D} \sum_{m=1}^{M} \sum_{n=1}^{N} r_{1mk} r_{2nl} \left[\left\langle (w_{mn}y_{mn} + (1-w_{mn})\nabla\psi_y(\phi_{mn})), \boldsymbol{\beta}^\dagger x_{mn} \right\rangle - \psi_y\left(\boldsymbol{\beta}^\dagger x_{mn}\right) \right]$$

(9)

$$\alpha_1 = \arg\max_{\alpha_1\in\mathbb{R}_{++}^K} \left(\log \frac{\Gamma(\sum_{k=1}^{K} \alpha_{1k})}{\prod_{k=1}^{K} \Gamma(\alpha_{1k})} + \sum_{k=1}^{K} \left(\alpha_{1k} + \sum_{m=1}^{M} r_{1mk} - 1 \right) \left(\Psi(\gamma_{1k}) - \Psi\left(\sum_{k'=1}^{K} \gamma_{1k'}\right) \right) \right)$$

(10)

$$\alpha_2 = \arg\max_{\alpha_2\in\mathbb{R}_{++}^L} \left(\log \frac{\Gamma(\sum_{l=1}^{L} \alpha_{2l})}{\prod_{l=1}^{L} \Gamma(\alpha_{2l})} + \sum_{l=1}^{L} \left(\alpha_{2l} + \sum_{n=1}^{N} r_{2nl} - 1 \right) \left(\Psi(\gamma_{2l}) - \Psi\left(\sum_{l'=1}^{L} \gamma_{2l'}\right) \right) \right)$$ (11)

The updates for the parameters of the Dirichlet distributions α_1 and α_2, can be efficiently performed using the Newton-Raphson's method. An EM algorithm for learning the model parameters of LD-AA-BAE is given in algorithm 1.

Algorithm 1. Learn LD-BAE

Input: $\mathcal{Y}_{\text{obs}}, \mathcal{X}_1, \mathcal{X}_2, K, L$
Output: $\alpha_1, \alpha_2, \boldsymbol{\Theta}_1, \boldsymbol{\Theta}_2, \boldsymbol{\beta}$
    $[m]_1^M, [n]_1^N, [k]_1^K, [l]_1^L$

Step 0: Initialize $\alpha_1, \alpha_2, \boldsymbol{\Theta}_1, \boldsymbol{\Theta}_2, \boldsymbol{\beta}$
Until Convergence
    Step 1: E-Step
        Step 1a: Initialize r_{1mk}, r_{2nl}
        Until Convergence
        Step 1b: Update ϕ_{mn} using equation (2)
        Step 1c: Update γ_{1k} using equation (3)
        Step 1d: Update γ_{2l} using equation (4)
        Step 1e: Update r_{1mk} using equation (5)
        Step 1f: Update r_{2nl} using equation (6)
    Step 2: M-Step
        Step 2a: Update θ_{1k} using equation (7)
        Step 2b: Update θ_{2l} using equation (8)
        Step 2c: Update $\boldsymbol{\beta}_{kl}$ using equation (9)
        Step 2d: Update α_1 using equation (10)
        Step 2e: Update α_2 using equation (11)

---

5 Concluding Remarks and Future Work

The side information aware Bayesian affinity estimation approach introduced in this article is a promising framework that efficiently incorporates attribute information within dyadic data. The approach can be readily generalized to where there are more than two modes, or sets of interacting items. Moreover, the graphical model can be further elaborated on to accommodate other types of side information including temporal information, and/or neighborhood structures. The use of exponential family distributions for modeling entity attributes as well as the affinity relationships renders great flexibility for modeling diverse data types in numerous domains. The approach can also be extended to non-parametric models by replacing the Dirichlet distribution priors with the corresponding process prior, which is useful when the desired number of clusters is not known. But clearly there is much work to be done on this line of models, in terms of both algorithmic development and applications.

A common feature of affinity datasets is sparsity - often only a very small percentage of the affinities are known. In the derivation provided in this paper, one carries around the unobserved affinities as well, which adds to computational demands and does not benefit from sparsity. In several settings however, one can simply ignore these values and just build a model based on the observed values, since conditioning on the unobserved affinity values does not effect any of the posterior distributions. This observation can be exploited to develop more efficient versions of LD-BAE.

Acknowledgements. This research was supported by NSF grants IIS-0713142 and IIS-1016614, and by NHARP. We thank Meghana Deodhar for her collaboration on SCOAL.

References

1. Abernethy, J., Bach, F., Evgeniou, T., Vert, J.P.: A new approach to collaborative filtering: Operator estimation with spectral regularization. The Journal of Machine Learning Research 10, 803–826 (2009)
2. Agarwal, D., Chen, B.: Regression-based latent factor models. In: KDD 2009, pp. 19–28 (2009)
3. Agarwal, D., Chen, B., Elango, P.: Spatio-temporal models for estimating click-through rate. In: WWW 2009: Proceedings of the 18th International Conference on World Wide Web, pp. 21–30 (2009)
4. Agarwal, D., Chen, B.: flda: matrix factorization through latent dirichlet allocation. In: Proc. ACM International Conference on Web Search and Data Mining 2010, pp. 91–100 (2010)
5. Agarwal, D., Merugu, S.: Predictive discrete latent factor models for large scale dyadic data. In: KDD 2007, pp. 26–35 (2007)
6. Dempster, A.P., Laird, N., Rubin, D.: Maximum likelihood from incomplete data via the em algorithm. J. Royal Statistical Society, Series B(Methodological) 39(1), 1–38 (1977)
7. Banerjee, A., Merugu, S., Dhillon, I., Ghosh, J.: Clustering with Bregman divergences. Jl. Machine Learning Research (JMLR) 6, 1705–1749 (2005)
8. Banerjee, A., Basu, S., Merugu, S.: Multi-way clustering on relation graphs. In: SDM (2007)
9. Basilico, J., Hofmann, T.: Unifying collaborative and content-based filtering. In: ICML (2004)
10. Bertsekas, D.: Nonlinear Programming. Athena Scientific (1999)
11. Chamberlain, D.E., Gough, S., Vickery, J.A., Firbank, L.G., Petit, S., Pywell, R., Bradbury, R.B.: Rule-based predictive models are not cost-effective alternatives to bird monitoring on farmland. Agriculture, Ecosystems & Environment 101(1), 1–8 (2004)
12. Deodhar, M., Ghosh, J.: A framework for simultaneous co-clustering and learning from complex data. In: KDD 2007, pp. 250–259 (2007)

13. Deodhar, M., Ghosh, J.: Simultaneous co-clustering and modeling of market data. In: Workshop for Data Mining in Marketing, Industrial Conf. on Data Mining 2007, pp. 73–82 (2007)
14. Deodhar, M., Ghosh, J.: Simultaneous co-segmentation and predictive modeling for large, temporal marketing data. In: Data Mining for Marketing Workshop, ICDM 2008 (2008)
15. Deodhar, M., Ghosh, J.: Mining for most certain predictions from dyadic data. In: Proc. 15th ACM SIGKDD Conf. on Knowledge Discovery and Data Mining, KDD 2009 (2009)
16. Deodhar, M., Ghosh, J., Tsar-Tsansky, M.: Active learning for recommender systems with multiple localized models. In: Proc. Fifth Symposium on Statistical Challenges in Electronic Commerce Research, SCECR 2009 (2009)
17. Dietterich, T.G., Domingos, P., Getoor, L., Muggleton, S., Tadepalli, P.: Structured machine learning: the next ten years. Machine Learning 73(1), 3–23 (2008)
18. Blei, D.M., Ng, A.Y., Jordan, M.I.: Latent dirichlet allocation. JMLR 3, 993–1022 (2003)
19. Dzeroski, S.: Multi-relational data mining: an introduction. SIGKDD Explorations 5(1), 1–16 (2003)
20. Airoldi, E., Blei, D.M., Fienberg, S.E., Xing, E.P.: Mixed membership stochastic blockmodels. JMLR 9, 1981–2014 (2008)
21. Gelman, A., Hill, J.: Data Analysis Using Regression and Multilevel/Hierarchical Models. Cambridge University Press (2007)
22. George, T., Merugu, S.: A scalable collaborative filtering framework based on co-clustering. In: Proceedings of the Fifth IEEE International Conference on Data Mining, pp. 625–628 (2005)
23. Getoor, L., Friedman, N., Koller, D., Taskar, B.: Learning probabilistic models of relational structure. In: Proc. 18th International Conf. on Machine Learning, pp. 170–177. Morgan Kaufmann, San Francisco (2001), citeseer.ist.psu.edu/article/getoor01learning.html
24. Grover, R., Srinivasan, V.: A simultaneous approach to market segmentation and market structuring. Journal of Marketing Research, 139–153 (1987)
25. Hastie, T., Tibshirani, R., Friedman, J.: The Elements of Statistical Learning, 2nd edn. Springer, Heidelberg (2009)
26. Herlocker, J., Konstan, J., Borchers, A., Riedl, J.: An algorithmic framework for performing collaborative filtering. In: SIGIR 1999: Proceedings of the 22nd Annual International ACM SIGIR Conference on Research and Development in Information Retrieval, pp. 230–237. ACM, Berkeley (1999)
27. Kim, B., Rossi, P.: Purchase frequency, sample selection, and price sensitivity: The heavy-user bias. Marketing Letters, 57–67 (1994)
28. Kim, B., Sullivan, M.: The effect of parent brand experience on line extension trial and repeat purchase. Marketing Letters, 181–193 (1998)
29. Kolda, T.: Tensor decompositions and data mining. In: Tutorial at ICDM (2007)
30. Kolda, T.G., Sun, J.: Scalable tensor decompositions for multi-aspect data mining. In: ICDM, pp. 363–372 (2008)
31. Lim, Y., Teh, Y.: Variational bayesian approach to movie rating prediction. In: Proc. KDD Cup and Workshop (2007)
32. Lokmic, L., Smith, K.A.: Cash flow forecasting using supervised and unsupervised neural networks. IJCNN 06, 6343 (2000)
33. Lu, Z., Agarwal, D., Dhillon, I.: A spatio-temporal approach to collaborative filtering. In: RecSys 2009 (2009)
34. Madeira, S.C., Oliveira, A.L.: Biclustering algorithms for biological data analysis: A survey. IEEE/ACM Trans. Comput. Biology Bioinform. 1(1), 24–45 (2004)
35. Moe, W., Fader, P.: Modeling hedonic portfolio products: A joint segmentation analysis of music compact disc sales. Journal of Marketing Research, 376–385 (2001)

36. Munson, M.A., et al.: The ebird reference dataset. Tech. Report, Cornell Lab of Ornithology and National Audubon Society (June 2009)
37. Murray-Smith, R., Johansen, T.A.: Multiple Model Approaches to Modelling and Control. Taylor and Francis, UK (1997)
38. Nowicki, K., Snijders, T.A.B.: Estimation and prediction for stochastic blockstructures. Journal of the American Statistical Association 96(455), 1077–1087 (2001), http://www.ingentaconnect.com/content/asa/jasa/2001/00000096/00000455/art00025
39. Oh, K., Han, I.: An intelligent clustering forecasting system based on change-point detection and artificial neural networks: Application to financial economics. In: HICSS-34, vol. 3, p. 3011 (2001)
40. Reutterer, T.: Competitive market structure and segmentation analysis with self-organizing feature maps. In: Proceedings of the 27th EMAC Conference, pp. 85–115 (1998)
41. Salakhutdinov, R., Mnih, A.: Probabilistic matrix factorization. In: NIPS 2007 (2007)
42. Salakhutdinov, R., Mnih, A.: Bayesian probabilistic matrix factorization using markov chain monte carlo. In: Proc. ICML 2008, pp. 880–887 (2008)
43. Sanderson, F.J., Kloch, A., Sachanowicz, K., Donald, P.F.: Predicting the effects of agricultural change on farmland bird populations in poland. Agriculture, Ecosystems & Environment 129(1-3), 37–42 (2009)
44. Seetharaman, P., Ainslie, A., Chintagunta, P.: Investigating household state dependence effects across categories. Journal of Marketing Research, 488–500 (1999)
45. Shan, H., Banerjee, A.: Residual bayesian co-clustering and matrix approximation. In: Proc. SDM 2010, pp. 223–234 (2010)
46. Shan, H., Banerjee, A.: Bayesian co-clustering. In: ICDM, pp. 530–539 (2008)
47. Sharma, A., Ghosh, J.: Side information aware bayesian affinity estimation. Technical Report TR-11, Department of ECE, UT Austin (2010)
48. Takcs, G., Pilszy, I., NÈmeth, B., Tikk, D.: Investigation of various matrix factorization methods for large recommender systems. In: 2nd KDD-Netflix Workshop (2008)
49. Vasilescu, M.A.O., Terzopoulos, D.: Multilinear Analysis of Image Ensembles: TensorFaces. In: Heyden, A., Sparr, G., Nielsen, M., Johansen, P. (eds.) ECCV 2002, Part I. LNCS, vol. 2350, pp. 447–460. Springer, Heidelberg (2002)
50. Wainwright, M.J., Jordan, M.I.: Graphical models, exponential families, and variational inference. Foundations and Trends in Machine Learning 1(1-2), 1–305 (2008)

Appendix A: Variational Inference Using Mean Field Approximation (MFA)

A maximum likelihood approach to parameter estimation generally involves maximization of the observed log-likelihood $\log p(X|\Theta)$ with respect to the free model parameters, i.e.,

$$\Theta^*_{ML} = \arg\max_{\Theta} \log p(X|\Theta) \tag{A1}$$

$$= \arg\max_{\Theta} \log \int_Z p(X, Z|\Theta) dZ \tag{A2}$$

where X and Z are sets of observed and hidden variables respectively. In the presence of hidden variables, the maximum likelihood estimate is often done using the Expectation-Maximization (EM) algorithm [6]. The following lemma forms the basis of the EM algorithm [50].

Lemma 1. *Let* $\mathcal{X}$ *denote a set of all the observed variables and* $\mathcal{Z}$ *a set of the hidden variables in a Bayesian network. Then, the observed log-likelihood can be lower bounded as follows*

$$\log p(\mathcal{X}, \mathcal{Z}|\boldsymbol{\Theta}) \geq \mathcal{F}(Q, \boldsymbol{\Theta})$$

where

$$\mathcal{F}(Q, \boldsymbol{\Theta}) = -\int_{\mathcal{Z}} Q(\mathcal{Z}) \log Q(\mathcal{Z}) d\mathcal{Z} + \int_{\mathcal{Z}} Q(\mathcal{Z}) \log p(\mathcal{X}, \mathcal{Z}|\boldsymbol{\Theta}) d\mathcal{Z} \qquad \text{(A3)}$$

for some distribution Q and the free model parameters $\boldsymbol{\Theta}$.

Proof. The proof follows from the Jensen's inequality and the concavity of the log function.

$$\begin{aligned}
\log p(\mathcal{X}|\boldsymbol{\Theta}) &= \log \int_{\mathcal{Z}} \frac{Q(\mathcal{Z})}{Q(\mathcal{Z})} p(\mathcal{X}, \mathcal{Z}|\boldsymbol{\Theta}) d\mathcal{Z} \\
&\geq \int_{\mathcal{Z}} Q(\mathcal{Z}) \log \frac{p(\mathcal{X}, \mathcal{Z}|\boldsymbol{\Theta})}{Q(\mathcal{Z})} d\mathcal{Z} \\
&= -\int_{\mathcal{Z}} Q(\mathcal{Z}) \log Q(\mathcal{Z}) d\mathcal{Z} + \int_{\mathcal{Z}} Q(\mathcal{Z}) \log p(\mathcal{X}, \mathcal{Z}|\boldsymbol{\Theta}) d\mathcal{Z} \\
&= \mathcal{F}(Q, \boldsymbol{\Theta})
\end{aligned}$$

Starting from an initial estimate of the parameters, $\boldsymbol{\Theta}_0$, the EM algorithm alternates between maximizing the lower bound $\mathcal{F}$ with respect to Q (E-step) and $\boldsymbol{\Theta}$ (M-step), respectively, holding the other fixed. The following lemma shows that maximization the lower bound with respect to the distribution Q in the E-step makes the bound exact, so that the M-step is guranteed to increase the observed log-likelihood with respect to the parameters.

Lemma 2. *Let* $\mathcal{F}(Q, \boldsymbol{\Theta})$ *denote a lower bound on the observed log-likelihood of the form in* (A3), *then*

$$Q^* = p(\mathcal{Z}|\mathcal{X}, \boldsymbol{\Theta}) = \arg\max_{Q} \mathcal{F}(Q, \boldsymbol{\Theta})$$

and $\mathcal{F}(Q^*, \boldsymbol{\Theta}) = \log p(\mathcal{X}|\boldsymbol{\Theta})$.

Proof. The lower bound on the observed log-likelihood is

$$\begin{aligned}
\mathcal{F}(Q, \boldsymbol{\Theta}) &= -\int_{\mathcal{Z}} Q(\mathcal{Z}) \log Q(\mathcal{Z}) d\mathcal{Z} + \int_{\mathcal{Z}} Q(\mathcal{Z}) \log p(\mathcal{X}, \mathcal{Z}|\boldsymbol{\Theta}) d\mathcal{Z} \\
&= -\int_{\mathcal{Z}} Q(\mathcal{Z}) \log \frac{Q(\mathcal{Z})}{p(\mathcal{Z}|\mathcal{X}, \boldsymbol{\Theta})} d\mathcal{Z} + \int_{\mathcal{Z}} Q(\mathcal{Z}) \log \frac{p(\mathcal{X}, \mathcal{Z}|\boldsymbol{\Theta})}{p(\mathcal{Z}|\mathcal{X}, \boldsymbol{\Theta})} d\mathcal{Z} \\
&= \log p(\mathcal{X}|\boldsymbol{\Theta}) - \mathrm{KL}(Q \parallel p(\mathcal{Z}|\mathcal{X}, \boldsymbol{\Theta}))
\end{aligned}$$

Maximum is attained when the KL-divergence $\mathrm{KL}(Q \parallel p(\mathcal{Z}|\mathcal{X}, \boldsymbol{\Theta}))$ is zero, which is uniquely achieved for $Q^* = p(\mathcal{Z}|\mathcal{X}, \boldsymbol{\Theta})$ at which point the bound becomes an equality for $\log p(\mathcal{X}|\boldsymbol{\Theta})$.

However, in many cases, computation of the true posterior distribution, $p(\mathcal{Z}|X, \boldsymbol{\Theta})$ is intractable. To overcome this problem, the distribution Q is restricted to a certain family of distributions. The optimal distribution within this restricted class is then obtained by minimizing the KL-divergence to the true posterior distribution. The approximating distribution is known as a *variational distribution* [50].

There are a number of ways in which the family of possible distributions can be restricted. One way of restricting the approximating distributions is to use a parameteric distribution $Q(\mathcal{Z}|\boldsymbol{\Phi})$ determined by a set of parameters $\boldsymbol{\Phi}$, known as *variational parameters*. In the E-step, the lower bound then becomes a function of variational parameters, and standard non-linear optimization methods can be employed to obtain the optimal values of these parameters. Yet another way to restrict the family of approximationg distributions is to assume a certain conditional independence structure over the hidden variables $\mathcal{Z}$. For example, one can assume a family of fully factorized distributions of the following form

$$Q = \prod_i q_i(z_i) \tag{A4}$$

This fully factorized assumption is often known as a *mean field approximation* in statistical mechanics. The following lemma derieves the expression for optimal variational distribution subject to a full factorization assumption.

Lemma 3. *Let $\mathcal{Q} = \{Q\}$ be a family of factorized distributions of the form in (A4). Then the optimal factorized distribution corresponding to the tightest lower bound is given by,*

$$Q^* = \prod_i q_i^*(z_i) = \underset{Q \in \mathcal{Q}}{\arg\max} \mathcal{F}(Q, \boldsymbol{\Theta}) \quad such\ that \quad q_i^*(z_i) \propto \exp\left(\mathbb{E}_{-i}[\log p(X, \mathcal{Z}|\boldsymbol{\Theta})]\right)$$

where $\mathbb{E}_{-i}[\log p(X, \mathcal{Z}|\boldsymbol{\Theta})]$ denotes a conditional expectation conditioned on z_i.

Proof. Using lemma 2, the optimal distribution $Q \in \mathcal{Q}$ is given by

$$Q^* = \underset{Q \in \mathcal{Q}}{\arg\min} \quad KL(Q \| p(\mathcal{Z}|X, \boldsymbol{\Theta}))$$

where the KL-divergence can be expressed as

$$KL(Q \| p(\mathcal{Z}|X, \boldsymbol{\Theta})) = \sum_i \int_{z_i} q_i(z_i) \log q_i(z_i) dz_i - \int_{z_i} q_i(z_i) \left\{ \int_{\mathcal{Z}_{-i}} \log p(\mathcal{Z}|X, \boldsymbol{\Theta}) \prod_{j \neq i} q_j(z_j) d\mathcal{Z}_{-i} \right\} dz_i$$

$$= \sum_{j \neq i} \int_{z_j} q_j(z_j) \log q_j(z_j) dz_j + \int_{z_i} q_i(z_i) \log \frac{q_i(z_i)}{\exp\left(\mathbb{E}_{-i}[\log p(X, \mathcal{Z}|\boldsymbol{\Theta})]\right)} dz_i$$

The second term in the above expression is a KL-divergence. Keeping $\{q_{j \neq i}(z_j)\}$ fixed, the optimum with respect to $q_i(z_i)$ is attained when KL-divergence is zero, i.e. $q_i^*(z_i) \propto \exp\left(\mathbb{E}_{-i}[\log p(X, \mathcal{Z}|\boldsymbol{\Theta})]\right)$.

The above lemma shows that the optimal variational distribution subject to the factorization constraint is given by a set of consistency conditions over different factors of the hidden variables. These coupled equations are known as *mean field equations* and can be satisfied iteratively. Convergence is guaranteed because the bound $\mathcal{F}$ is convex with respect to each of the factors [10].

Appendix B: Mean Field Approximation (MFA) for Bayesian Affinity Estimation

This appendix illustrates the derivation of a MFA based expectation maximization algorithm for parameter estimation of a Latent Dirichlet Attribute Aware Bayesian Affinity Estimation framework (LD-AA-BAE). The techniques introduced in this appendix are also used for derieving updates for rest of the models in the paper and the same analysis can be easily extended. For the purpose of exposition, we however, concentrate only on the LD-AA-BAE model.

The joint distribution over all observable and latent variables for the LD-AA-BAE model is given by:

$$p(\mathcal{Y}, \mathcal{X}_1, \mathcal{X}_2, \mathcal{Z}_1, \mathcal{Z}_2, \pi_1, \pi_2 | \alpha_1, \alpha_2, \Theta_1, \Theta_2, \beta) =$$

$$p(\pi_1|\alpha_1)p(\pi_2|\alpha_2)\left(\prod_m p(z_{1m}|\pi_1)p_{\psi_1}(x_{1m}|\theta_{1z_{1m}})\right)\left(\prod_n p(z_{2n}|\pi_2)p_{\psi_2}(x_{2n}|\theta_{2z_{2n}})\right)\left(\prod_{m,n} p_{\psi_y}(y_{mn}|\beta_{z_{1m}z_{2n}}^\dagger x_{mn})\right)$$

(B1)

The approximate variational distribution Q over the hidden variables is

$$Q(\mathcal{Y}_{unobs}, \mathcal{Z}_1, \mathcal{Z}_2, \pi_1, \pi_2) = q(\pi_1|\gamma_1)q(\pi_2|\gamma_2)\left(\prod_{\substack{m,n \\ y_{mn} \in \mathcal{Y}_{unobs}}} q(y_{mn}|\phi_{mn})\right)\left(\prod_m q(z_{1m}|r_{1m})\right)\left(\prod_n q(z_{2n}|r_{2n})\right)$$

(B2)

The updates for factors corresponding to the optimal variational distribution is obtained using lemma 3.

E-step Update for $q^*(y_{mn}|\phi_{mn})$: Collecting terms containing the affinities y_{mn} in the conditional expectation of the complete log-likelihood, we obtain

$$q^*(y_{mn}) \propto p_0(y_{mn}) \exp\left(\sum_{K,L=1}^{K,L} r_{1mk}r_{2nl}\langle y_{mn}, \beta_{kl}^\dagger x_{mn}\rangle\right)$$

which shows that variational distribution for the missing affinities is an exponential family distribution having the same form as the one assumed for the affinities with the natural parameter given by:

$$\phi_{mn} = \sum_{k,l=1}^{K,L} r_{1mk}r_{2nl}\left(\beta_{kl}^\dagger x_{mn}\right)$$

(B3)

E-step Updates for $q^*(\pi_1|\gamma_1)$ **and** $q^*(\pi_2|\gamma_2)$: Conditional expectation with respect to the mixing coefficients π_1 yields,

$$q^*(\pi_1) \propto \exp\left(\sum_{k=1}^K \left(\alpha_{1k} + \sum_{m=1}^M r_{1mk}\right)\log \pi_{1k}\right)$$

$$= \prod_{k=1}^K (\pi_{1k})^{\left(\alpha_{1k} + \sum_{m=1}^M r_{1mk}\right)}$$

Easy to see that, the optimal variational distribution $q^*(\pi_1|\gamma_1)$ is a Dirichlet distribution over a K-simplex with parameters given by:

$$\gamma_{1k} = \alpha_{1k} + \sum_{m=1}^{M} r_{1mk} \tag{B4}$$

Similarly, $q^*(\pi_2|\gamma_2)$ is a Dirichlet distribution over a L-simplex with parameters:

$$\gamma_{2l} = \alpha_{2l} + \sum_{n=1}^{N} r_{2nl} \tag{B5}$$

E-step Updates for $q(z_{1m}|r_{1m})$ and $q(z_{2n}|r_{2n})$: Conditional expectation with respect to discrete cluster assignment variable z_{1mk} for the cluster k results in the following update:

$$q^*(z_{1mk} = 1) = r_{1mk} \propto \exp\left[\log p_{\psi_1}(x_{1m}|\theta_{1k}) + \Psi(\gamma_{1k}) - \Psi\left(\sum_{k'=1}^{K} \gamma_{1k'}\right) + \right.$$

$$\left. \sum_{n=1}^{N}\sum_{l=1}^{L} r_{2nl}\left(w_{mn}\log p_{\psi_y}(y_{mn}|\beta_{kl}^{\dagger}x_{mn}) + (1 - w_{mn})\mathbb{E}_q\left[\log p_{\psi_y}(y_{mn}|\beta_{kl}^{\dagger}x_{mn})\right]\right)\right] \tag{B6}$$

The first term is the log-likelihood of the entity attributes, the second term is the expectation of $\log \pi_{1k}$ with respect to the variational Dirichlet distribution while the last term involves the log-likelihood of all the affinities associated with the entity e_{1m}. The known log-likelihood is used if the affinity is observed ($w_{mn} = 0$), while the log-likelihood for the missing affinities is replaced by the corresponding expecations under the variational distribution $q^*(y_{mn}|\phi_{mn})$. Analogously, the update equation for the cluster assignment variable $q^*(z_{2nl} = 1)$ is given by:

$$q^*(z_{2nl} = 1) = r_{2nl} \propto \exp\left[\log p_{\psi_2}(x_{2n}|\theta_{2l}) + \Psi(\gamma_{2l}) - \Psi\left(\sum_{l'=1}^{L} \gamma_{2l'}\right) + \right.$$

$$\left. \sum_{m=1}^{M}\sum_{k=1}^{K} r_{1mk}\left(w_{mn}\log p_{\psi_y}(y_{mn}|\beta_{kl}^{\dagger}x_{mn}) + (1 - w_{mn})\mathbb{E}_q\left[\log p_{\psi_y}(y_{mn}|\beta_{kl}^{\dagger}x_{mn})\right]\right)\right] \tag{B7}$$

M-step Updates for θ_{1k} and θ_{2l}: Taking expectation of the complete log-likelihood with respect to the variational distribution, we obtain the following expression for the lower bound $\mathcal{F}$ as a function of the entity attributes parameters:

$$\mathcal{F}(\Theta_1, \Theta_2) = \sum_{m=1}^{M}\sum_{k=1}^{K} r_{1mk} \log p_{\psi_1}(x_{1m}|\theta_{1k}) + \sum_{n=1}^{N}\sum_{l=1}^{L} r_{2nl} \log p_{\psi_2}(x_{2n}|\theta_{2l})$$

Taking partial derivatives with respect to θ_{1k} and θ_{2l}, we obtain the following updates:

$$\theta_{1k} = \nabla\psi_1^{-1}\left(\frac{\sum_{m=1}^{M} r_{1mk}x_{1m}}{\sum_{m=1}^{M} r_{1mk}}\right) \tag{B8}$$

$$\theta_{2l} = \nabla\psi_2^{-1}\left(\frac{\sum_{n=1}^{N} r_{2nl}x_{2n}}{\sum_{n=1}^{N} r_{2nl}}\right) \tag{B9}$$

M-step Updates for β_{kl}: Collecting terms containing the GLM coefficients in the lower bound, we obtain:

$$\mathcal{F}(\beta_{kl}) = \sum_{m=1}^{M} \sum_{n=1}^{N} r_{1mk} r_{2nl} \left[\left\langle (w_{mn} y_{mn} + (1 - w_{mn}) \nabla \psi_y(\phi_{mn})), \beta^\dagger x_{mn} \right\rangle - \psi_y \left(\beta^\dagger x_{mn} \right) \right]$$

As earlier, the missing affinities are replaced by corresponding expected values under the variational exponential family distribution. The lower bound can be maximized using a gradient ascent method. The expressions for the gradient and the gradient-ascent updates are obtained as follows:

$$\nabla \mathcal{F}(\beta_{kl}) = \sum_{m=1}^{M} \sum_{n=1}^{N} r_{1mk} r_{2nl} \left[(w_{mn} y_{mn} + (1 - w_{mn}) \nabla \psi_y(\phi_{mn})) - \nabla \psi_y \left(\beta^\dagger x_{mn} \right) \right] x_{mn}$$

(B10)

$$\beta_{kl}^{t+1} = \beta_{kl}^{t} + \eta \nabla \mathcal{F}(\beta_{kl})$$

(B11)

where η is the step-size for the update.

M-step Updates for α_1 and α_2: The expression for the lower bound as a function of the Dirichlet parameters α_1 is:

$$\mathcal{F}(\alpha_1) = \log \frac{\Gamma(\sum_{k=1}^{K} \alpha_{1k})}{\prod_{k=1}^{K} \Gamma(\alpha_{1k})} + \sum_{k=1}^{K} \left(\alpha_{1k} + \sum_{m=1}^{M} r_{1mk} - 1 \right) \left(\Psi(\gamma_{1k}) - \Psi \left(\sum_{k'=1}^{K} \gamma_{1k'} \right) \right)$$

Taking derivative with respect to α_{1k} yield:

$$\frac{\partial \mathcal{F}}{\partial \alpha_{1k}} = \Psi \left(\sum_{k'=1}^{K} \alpha_{1k} \right) - \Psi(\alpha_{1k}) + \Psi \left(\sum_{k'=1}^{K} \gamma_{1k} \right) - \Psi(\gamma_{1k})$$

Note that the update for α_{1k} depends on $\{\alpha_{1k'}, [k']_1^K, k' \neq k\}$, so a closed form solution cannot be obtained. Following [18], Newton-Raphson's method can then be used to update the parameters. The Hessian H is given by

$$H(k, k) = \frac{\partial^2 \mathcal{F}}{\partial \alpha_{1k}^2} = \Psi' \left(\sum_{k'=1}^{K} \alpha_{1k} \right) - \Psi'(\alpha_{1k})$$

$$H(k, k') = \frac{\partial^2 \mathcal{F}}{\partial \alpha_{1k} \partial \alpha_{1k'}} = \Psi' \left(\sum_{k'=1}^{K} \alpha_{1k} \right) \quad (k' \neq k)$$

The update can then be obtained as follows:

$$\alpha_1^{t+1} = \alpha_1^{t} + \eta H^{-1} \nabla(\alpha_1)$$

(B12)

The step-size η can be adapted to satisfy the positivity constraint for the Dirichlet parameters. A Similar method is followed for update of α_2.

Improving the Semantics of a Conceptual Schema of the Human Genome by Incorporating the Modeling of SNPs

Óscar Pastor, Matthijs van der Kroon, Ana M. Levin, Matilde Celma, and Juan Carlos Casamayor

Centro de Investigación en Métodos de Producción de Software -PROS-
Universidad Politécnica de Valencia, Camino de Vera s/n, 46022 Valencia, Valencia, Spain

Abstract. In genetic research, the concept known as SNP, or single nucleotide polymorphism, plays an important role in detection of genes associated with complex ailments and detection of hereditary susceptibility of an individual to a specific trait. Discussing the issue, as it surfaced in the development of a conceptual schema for the human genome, it became clear a high degree of conceptual ambiguity surrounds the term. Solving this ambiguity has lead to the main research question: *What makes a genetic variation, classified as a SNP different from genetic variations, not classified as SNP?*. For optimal biological research to take place, an unambiguous conceptualization is required. Our main contribution is to show how conceptual modeling techniques applied to human genome concepts can help to disambiguate and correctly represent the relevant concepts in a conceptual schema, thereby achieving a deeper and more adequate understanding of the domain.

1 Introduction

Todays genomic domain evolves around insecurity. Genomics is often not an exact science due to the immense complexity of nature and its processes. Basic concepts like genes, alleles and mutations are frequently variable in their definition. Their exact denotation often depends on both context and position in time. A gene for instance can be defined as a locatable region of genomic sequence, corresponding to a unit of inheritance, which is associated with regulatory regions, transcribed regions, and or other functional sequence regions. However, the existence of splicing, in which gene expression changes at runtime, complicates this view as is very well described by [1]: *"in eukaryotes, the gene is, in most cases, not yet present at DNA-level. Rather, it is assembled by RNA processing."*. [2] and [3] provide recent insights on the evolution of the gene concept. For efficient research to take place, it is obvious that clear definitions of concepts are crucial. This is especially the case in research where various separate research groups are collaborating and exchanging information on a subject. Formally describing concepts, ruling out ambiguity and relating concepts to each other and their context is the main objective of model driven software development. A conceptual model is a simplified representation of reality, devised for a certain purpose and seen from a certain point of view.

This paper converges on a single element of ambiguity existing in the genomic domain, applying a model driven approach to it. The single nucleotide polymorphism, or

A. Fred et al. (Eds.): IC3K 2010, CCIS 272, pp. 23–37, 2013.
© Springer-Verlag Berlin Heidelberg 2013

SNP concept, widely used in the genomic domain is crucial in detection and localization of hereditary deviations [4], understanding the molecular mechanisms of mutation [5] [6] and deducing the origins of modern human populations [7] [8] [9]. Like many other concepts in the domain, it is poorly defined in terms of rigor. Present day solutions exist in natural text definitions, or in terms of ontologies [10]. These solutions have yet to reach general consensus and the ambiguity continues to exist. The weakness associated to these types of solutions to the problem is twofold: on one hand they fail to describe the concept to its fullest and in a manner that allows for computational processing. This is the case, for instance, in the natural language definition commonly accepted among biologists. On the other hand, these methods often fail to relate the concept to its genomic context. Defining a concept intrinsically in terms of its characteristics often proves to be insufficient. The way it behaves in its respective context is often equally as important, especially in complex, interrelated domains like genomics. This paper is an represents and effort to solve the ambiguity of the SNP concept by capturing it in a conceptual model, a method proven to be very effective at capturing domain knowledge in Information Systems (IS) design [11]. By providing the SNP concept as an extension to the already existing conceptual model of the human genome [12] [13] both criteria at defining the concept are satisfied: rigorously describing what a SNP is, and relating it to its context.

Explaining the various uses of the SNP concept, separating them into their appropriate conceptual context and presenting a conceptual modeling approach will allow solving this ambiguity. This is achieved by answering the main research question; *What makes a genetic variation, classified as a SNP different from those genetic variations not classified as SNP?*. The main contribution is to show how a model driven approach to the genomic domain can aid in the understanding and adequate representation of relevant concepts. Previous solutions applied to the problem space, including ontologies [10] and literal descriptions in natural language, often fail to comprehensively model the concept. As a result of the ambiguity intrinsically associated to natural language, defining the concept exhaustively this way has proven to be a daunting task. In addition to this, relating the concept to its context is equally as important as defining it intrinsically, and conventional methods often fail exactly at satisfying this requirement. The contribution of this work is twofold however, on the one hand fixing the semantics of the concept in a conceptual model while on the second hand enforcing the use of conceptual models in the bioinformatics domain. Bioinformatics has seen a rapid evolution in the past decade, starting with the sequencing of the human genome in 2000 by Venter and Collins [14], and the amount of data being generated is constantly growing. Managing these data has been done by ad-hoc solutions mostly until now. These solutions often do not cope well with changing environments –one that bioinformatics is without a doubt– scale poorly and do not manage reusability well. Conceptual modeling, as the cornerstone of the Model Driven Architecture, has been proven to show significant improvements in engineering Information Systems [15]. Conventional solutions require a large amount of manual coding effort each time the domain changes in order to keep implementations in line. The conceptual modeling based solution allows for a process much closer to the problem space by allowing the engineer to create models that can be understood by the domain expert. From these models partial, or in some cases full, implementations can be generated following sets of rules and formal transformation processes, greatly reducing the effort required to maintain implementations consistent with a changing domain.

The paper starts with a section covering the context of the domain; *The genomic domain*. In this section effort is put into clarifying relevant concepts, and describing the problem space in detail. The various existing definitions of the SNP concept are discussed, along with an explanation on how the concept is used in the domain, in order to achieve a deep understanding of the biological concept. The lessons learned from this research are incorporated in a conceptual schema in the *CSHG: Incorporation of the SNP concept* section. The *Related work* section provides context by discussing other efforts in the domain to resolving the same problem. Finally in the *Conclusion* section the research question is answered and the contribution of this paper is summarized. Results of the research are discussed and further research suggested.

2   The Genomic Domain

Differences observed between individuals, like hair color, eye color but also development of hereditary diseases are a result of a complex interaction between genes and the environment. Purely looking at genetic differences, we see that only approximately 1% of the base pairs in the genetic sequence is different among individuals. These differences, or polymorphisms, between individuals can be classified as being negative, positive or neutral. Following Darwin's theory of natural selection, a negative effect, or deleterious effect, usually results in elimination of the individual from the population before breeding, and along with it the genetic polymorphism that caused the disadvantage. [16] provides an exhaustive introduction to cell biology.

What naturally flows from this predicate is that a positive effect, by allowing the carrier to produce offspring, will usually result in conservation of the hereditary trait in the population. Examples of positive polymorphisms include the ones associated to immunity for disease. A neutral polymorphism has neither a positive, nor negative effect and is thus not filtered out of the population. Rather, neutral polymorphisms are considered evolutionarily stable not changing much from generation to generation. Examples include the natural differences that exist between individuals. Since they provoke neither significant disadvantages, nor advantages to the individual they are sustained over various generations, and therefore accumulate. For this reason, polymorphisms with neutral phenotypic effect are relatively common and happen approximately once every 1200 base pairs [17], and in at least 1% of the observed population [18].

Due to these characteristics, polymorphisms with neutral phenotypic effect have been used as markers in detection of hereditary susceptibility to a trait. These traits include hereditary diseases such as cancer, diabetes, vascular disease, and some forms of mental illness. In order to identify whether an individual possesses genetic predisposition to hereditary disease, the first step is sequencing the individuals genome. At present day, due to high cost associated to genetic sequencing, instead of the entire DNA sequence, so-called markers are used. A marker is a segment of DNA with an identifiable location on a chromosome and whose inheritance can be followed. A marker can be a gene, or it can be some section of DNA with no known function. As technology advances and costs drop, it is expected the full sequencing of an individuals genome will be economically feasible. Either way, the obtained sequence is compared against a reference sequence, usually NCBI refSeq. The sequence comparison allows for detection of deviations in the individuals DNA sequence, so called polymorphisms. As a result of phenotypic differences amongst individuals, relatively much variation is expected. Not every polymorphism has a negative effect on phenotype, but in order to find the few pathogenic mutations among the vast amount of neutral polymorphisms a way of separating these, both technically and conceptually, is required.

In literature, these polymorphisms with neutral effect of phenotype are often referred to as SNPs, or single nucleotide polymorphism. The conventional definition of SNP indicates a single nucleotide polymorphisms, happening in a significant portion of the population. However, other definitions, much broader then the conventional are used. dbSNP for instance refers to a wide variety of genetic variations when using the SNP label. Also, what is meant exactly by a significant portion of the population is not entirely clear and commonly accepted percentages range from 0.5% and 1%. Clearly ambiguity exists and considering the importance of the term in the domain, this is undesirable.

2.1 Existing Definitions of the SNP Concept

The dbSNP data repository [19] is an NCBI [20] driven effort to collect and integrate all available data on genetic polymorphisms. The name suggests emphasis on single nucleotide polymorphisms, but reality is slightly different. As is mentioned on the dbSNP website: *"This collection of polymorphisms includes single-base nucleotide substitutions (also known as single nucleotide polymorphisms or SNPs), small-scale multibase deletions or insertions (also called deletion insertion polymorphisms or DIPs), and retroposable element insertions and microsatellite repeat variations (also called short tandem repeats or STRs)"*. Clearly NCBI is stretching the definition of SNP to fit any type of polymorphism, thereby blurring the concept. Confusing as it may be, we believe dbSNP should actually be called dbVariation, as stated by NCBI itself in the Frequently Asked Questions section of their website. [21] uses a very similar definition of SNP, stating that a SNP can be either neutral or deleterious. This resource appears to be using the literal meaning of the label, simply indicating every polymorphism that influences a single nucleotide as being a SNP, regardless of its effect on phenotype.

A different definition originates from the Department of Biological Sciences, Oakland University, Rochester, MI, USA [22]: *"Single nucleotide polymorphism (SNP) is the simplest form of DNA variation among individuals. These simple changes can be of transition or transversion type and they occur throughout the genome at a frequency of about one in 1,000 bp. They may be responsible for the diversity among individuals, genome evolution, the most common familial traits such as curly hair, interindividual differences in drug response, and complex and common diseases such as diabetes, obesity, hypertension, and psychiatric disorders. SNPs may change the encoded amino acids (nonsynonymous) or can be silent (synonymous) or simply occur in the noncoding regions. They may influence promoter activity (gene expression), messenger RNA (mRNA) conformation (stability), and subcellular localization of mRNAs and/or proteins and hence may produce disease"*. It resembles the conventional definition loosely, as it mentions a frequency of occurrence in an individual (*about one in 1,000 bp*) and pathogenicity (*and complex and common diseases such as*). Depending on interpretation, one could state that also the size dimension is mentioned implicitly, considering the definition handles single nucleotide polymorphism.

2.2 SNP Characteristics

During the investigation of various definitions of the SNP concept existing in todays genomic domain, the determinant characteristics, or dimensions, have been identified. They form the conceptual spectrum in which the SNP concept is positioned. Each concrete definition represents an instance of these abstract dimensions, thereby fixing a position within the gamut. It is expected that by incorporating these dimensions into a database, instead of fixing the concept on one definition, flexibility is acquired and data can be queried dynamically depending on the desired SNP interpretation. The commonly accepted definition for SNP is explicitly based on 2 dimensions: occurrence in the population and genotypic size. According to [18] a SNP is a single base change in a DNA sequence. It is considered that the least frequent allele should have a frequency of 1% or greater. The first dimension, occurrence in population, hides a third and fourth characteristic. As a direct result of natural selection, for a polymorphism to happen more than a given percentage in the population, it needs to have either a neutral or a positive phenotypic effect. A negative, or deleterious effect is impossible to happen in a substantial portion of the population as it eliminates itself. Therefore, SNPs considered according to the conventional definition have no direct relation to disease. For this very same reason, SNPs as defined in the conventional sense are relatively common among each individual. On average, 8.33 SNPs happen every 10kb [17].

Various literature sources use various definitions of the SNP concept. dbSNP forms one extremity of this spectrum by defining the concept very loosely. The conventional definition then describes the other side of the spectrum by being more rigorous. Following the various definitions leads to the identification of four dimensions on which SNPs are commonly defined: occurrence in population, occurrence in sequence, pathogenicity and size. Every dimension will now be clarified briefly.

Occurrence in Population. This dimension entails the aspect of occurrence of a SNP in a given population. Examples of different populations include Asian, African or European. A higher population occurrence than a fixed low percentage, suggests a polymorphism with neutral effect on phenotype. This barrier is not defined rigorously, but commonly accepted percentages vary from 0.5% to 1%.

Occurrence in Sequence. Due to natural selection, polymorphisms with deleterious or negative effects are usually filtered out of a population. This process conserves neutral polymorphisms without observable effect and polymorphisms with positive effect on phenotype. As a result of this mechanism, these type of polymorphisms stack up in the genome of an individual. According to conventional definition, in the human genome every 1200 base pairs a SNP is expected as stated by [17]. This dimension however, describes a characteristic of the SNP concept in general. Indeed, a specific SNP does not happen once every 1200 base pairs, rather a different SNP is expected to be occurring with this frequency. It is therefore a characteristic of a genome; stating the amount of SNPs happening in it, rather than being a defining criterium whether a specific polymorphism is considered to be a SNP or not.

Pathogenicity. The relation of a SNP to hereditary disease is captured in this dimension. The general belief is that SNPs are not causatively associated to disease. Rather, due to a phenomenon called Linkage Disequilibrium [23] and by using SNPs as markers, susceptibility to hereditary disease can be detected without having to sequence an entire individuals genome. In short, the SNP itself has no relation to a pathogenic effect. SNPs used as marker, however, predict a different mutation, happening at another position. This second, predicted, mutation usually does have a direct pathogenic effect. This process will be discussed in more detail in the *SNPs uses* section.

Size. Clearly, a single nucleotide polymorphism should involve only one nucleotide, hence the name. However, considering the definition in which the SNP label is used to identify non deleterious polymorphisms this description might be stretched. Considering that DNA is read in triplets (combinations of three nucleotides), the polymorphisms that are not a multiple of three often result in frame-shift, changing the transcribed mRNA and thus the resulting protein. Frame-shift often leads to an entirely different protein, and is thus very likely to result in negative phenotype. A non-deleterious polymorphism, in theory, can thus be any variation in which the affected nucleotides are a multiple of three. However, every changed triplet, if non-synonymous, also changes an amino-acid eventually leading to an increased chance of significantly changing the resulting protein. Thus, although in theory a non-malicious polymorphisms might include more then one nucleotide, it is most likely to involve a single nucleotide change.

2.3   Uses of the SNP Concept

For a proper understanding of the genomic concept known as SNP, and to define it properly in terms of a conceptual model, it is necessary to know what purpose the term serves. Not only needs to be investigated what the domain exactly considers to be a SNP, but also how this concept is used in the domain. This exact understanding of the concept, will allow for an adequate application of the conceptual modeling approach. After applying the conceptual modeling perspective to the problem of properly specifying SNPs, the concepts behind them will less ambiguous and clearer. By more clear is meant that specific uses in particular contexts are labeled with an adequate term, while the SNP notion emerges clearly from the conceptual schema. Three distinct uses of the SNP concept have been identified. (i) dbSNP, (ii) distinguishing harmless polymorphisms from harmful variations and (iii) as a genetic marker.

dbSNP uses the term to cover a wide variety of polymorphisms known to happen in the human genome. It thereby stretches the conventional definition of the term, and should thus actually be called dbVariation, as it mentions itself. Another application of the term is identified when assuming SNP to be defined according to measures on the dimensions shown in [table 1]. When sequencing a human genome, various polymorphisms are usually encountered. Every 1200 base pairs, a mismatch happens between the sequence used as reference and the sample at hand. Most of these polymorphisms happen without negative effect on the phenotype, as has been explained earlier. Considering that DNA is read in triplets, or combinations of three nucleotides, the polymorphisms that are not a multiple of three often result in frame-shift, rendering the transcribed mRNA sequence senseless. This taken into account, the most probable size in which a polymorphism is expected to not alter the genetic sequence is one nucleotide, or a multiple

Table 1. SNP concept defined to discriminate polymorphisms with neutral effect on phenotype from polymorphisms with negative effect on phenotype

property	value
Occurrence in population	>0.5%
Pathogenicity	None
Size	1 nucleotide

of three. In the absence of negative effect, they are likely to happen in more than 0.5% of the population. Hidden among these relatively innocent polymorphism, lurks the one that is associated to the development of disease. More than one polymorphism with negative effect on phenotype, happening in any individual is very unlikely. Discriminating between polymorphisms with negative effects, and polymorphisms without is thus very important. The SNP concept, shaped in the way as is done in [table 1] serves this purpose elegantly.

SNPs are also commonly used as genetic markers. Due to relative high pricing of DNA sequencing, ways of using the available sequencing capacity efficiently had to be come up with. One approach to this cost-saving strategy, is making use of linkage disequilibrium. LD describes the nonrandom correlation that exists between a marker and the presence of polymorphisms at a locus [23]. Following LD it is possible to asses the probability of a polymorphism happening in a subjects genetic code, by knowing the state of a marker position. [18] describes the use of SNPs as molecular markers in animal genetics. The use of SNPs as genetic markers can thus be considered an imperfect solution to high costs associated to DNA sequencing.

3   CSHG: Incorporation of the SNP Concept

This section handles the incorporation of the SNP concept in the conceptual model of the human genome (CSHG), as devised by [13]. First, the existing model of the human genome is presented and clarified, after which the suggested incorporation of the SNP concept is presented. Devising a conceptual schema representing the human genome holds great potential. Not only will it provide a standard structure in which genomic data can be stored and exchanged, it might also cause a paradigm shift. In software design, conceptual modeling has long proven to be an essential step in creating reliable, high quality products. Conceptual models are used to direct the process of software creation, by functioning as the equivalent of a blue print used in traditional construction. It also functions as a tool to gain a deeper understanding of the domain at hand, in order to drive effective design of these very same blue prints. Essentially by decompiling the biological processes and structures underlying the most complex software possibly ever to have existed, an attempt is made at uncovering the source code of life itself. At the same time, this is also an strategy to show how the application of conceptual modeling principles can improve the way in which information is managed in the Bioinformatics domain, where too often data repositories are exploited without having the required clear definition of the conceptual schema they are supposed to be a representation of.

Initially, an ideal model [figure 1] was created, essentially describing how the concepts should be. However, as data was matched from various external sources to this ideal model, it soon became clear a dichotomy existed between the ideal model and the way data was represented in the real world. A second model was created, logically named the real model. This real model serves as a practical tool of resolving the encountered limitations, it therefore compromises on the aspect of understanding the domain. The intention of the real model is to adapt the modeling elements included in the ideal model to the way in which we found that data are stored and managed in practical settings. Even if there is always a conceptual mapping between concepts in the ideal model and how they appear in real models, for the purpose of incorporating the SNP concept we focus in this work in the ideal model. The ideal model exists of three sub views; (i) the gene-mutation view, (ii) the genome view and (iii) the transcription view. Considering the SNP concept forms part of the gene-mutation view, we will constrain ourselves to this part of the model. [figure 1] provides an overview of the present day model relevant concepts. The *Gene* class captures the general concept of a gene: *ID_symbol* represents an alphanumeric code for the gene according to HGNC [24], it also functions as the primary key; *ID_HUGO*, a numeric code assigned to the gene by HGNC; *official_name*, the full name of the gene; *summary*, a short description; *chromosome*, the *chromosome* on which the gene is located and *locus*, representing the location of the gene within the chromosome.

The *Allele* class then represents the various encountered instances of a gene. It stores information about the start and the end of the allele; the *start_position* and *end_position* attributes respectively, and which strand (+ or -) it is located: the *strand* attribute. The *ord_num* attribute functions as an internal identifier. An allele can either be an *Allelic variant* or an *Allelic reference* type. The two entities contain the same *Sequence* attribute which contains a nucleotide string. Each gene has at least one reference allele associated to it, which is obtained from external sources, usually NCBI. The allelic variant tuples then, are derived from this allele reference in combination with information contained in the *Variation* entity. The *Variation* entity stores information about changes in a certain allele in respect to the reference, the allelic reference type. It has an *id_variation* attribute for internal identifying purposes. *id_variation_db* refers to the identification used in the external source. It further holds a description, meant to store a small description about the variation.

The entities specifying the *Variation* concept through generalization can then be classified into three categories; (i) effect, (ii) location and (iii) description, each representing a specific type of polymorphism. The location classes store information about whether the variation affects one or more genes. In the case of a *genic* location, only one gene is affected, while in the case of *chromosomal*, multiple genes might be influenced. The effect entities specify the variations effect on phenotype. This can either be *mutant*, a *neutral polymorphism* or the effect might be *unknown*. The *splicing, regulatory, missense* and *others* concepts are considered to be mutations since they have a negative effect, hence they are a specialization of the mutant concept. Ultimately, the description classes include descriptive information about the variation. Depending on the degree to which the data on the variation is precise, it falls into either the *precise* or *imprecise* class. When imprecise, the entity only stores a general description. In the case of

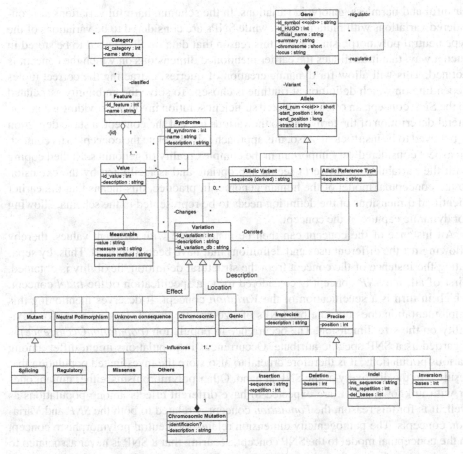

Fig. 1. Conceptual model of the human genome

precise data, it stores the position of the variation and further specifies the nature of the variation into four classes: *insertion*, *deletion*, *indel* and *inversion*. Each of these concepts store information about this specific type of variation and the exact attributes vary from type to type.

The *Category*, *Feature*, *Value*, *Measurable* and *Syndrome* concepts associate a *Variation* to phenotype. A variation is thus associated to a *Value*, which in turn is associated to a *Syndrome*. A *Syndrome* is considered to be a negative phenotypic effect, or disease. Every *Value* is composed of *Features*. These *Features* in turn, are classified by *Categories*, which have a recursive property indicated by the self-referencing relation.

3.1   SNPs Incorporation

Incorporating SNPs into the human genome conceptual model is essential. [figure 2] demonstrates a representation of the extension. It provides means of discriminating various types of polymorphisms. The most important being the separation between harmful and harmless genotypic variations. In the schema, harmful variations are considered variations with mutant effect, while SNPs are considered to be variations of the type neutral polymorphism. It is for this reason that data on SNPs need to be stored in such a way, that it facilitates the earlier mentioned dimensions on which the concept is defined. This will allow for dynamic creation of queries, extracting the correct tuples depending on which definition at runtime is chosen. To solve the ambiguity associated to the SNP concept, an often considered sufficient solution involves providing a concise, literal description of the term. Due to the various uses of the concept, a static definition is believed to be insufficient. Also, this approach fails to relate the concept to its context, an aspect considered very important in the complex reality of genomics. Rather, coping with the variability of the term seems appropriate and is facilitated by the extension to the conceptual model of the human genome. In practice, this means that the earlier identified dimensions of the definition needs to be represented in the schema, allowing for dynamic capture of the concept.

An instance of the concept can then be instantiated with different values, thereby allowing for the different uses and definitions that have been identified. Thus, by separating the instance of the concept from the structural definition, flexibility is obtained. First of all, the *SNP* concept is considered to be a specification of the *Indel* concept, which in turn is a specification of the *Variation* concept. It deserves mentioning this implementation does not rule out SNPs of size greater then 1, leaving room for flexibility on the size dimension. The occurrence in population (*populationOccurrence*) is regarded as a SNP specific attribute. Occurrences in populations might differ among various populations, it is therefore crucial to also store the investigated population that resulted in the discovery of the SNP at hand. Other polymorphisms, either mutant ones or with unknown effect, are expected to have different effects among populations as well. It is for this reason, the *Population* concept is related to both the *SNP* and Variation concepts. The pathogenicity dimension relates the neutral polymorphism concept in the conceptual model to the SNP concept, securing that a SNP is never associated to disease. According to the schema, a polymorphism associated to negative phenotypic effects is considered to be mutant and is thus captured already in the form of a *Variation*, the *SNP* super-type.

As one might notice, the dimension occurrence in sequence is not incorporated in the schema. The occurrence in sequence represents not a characteristic of a single SNP, rather a characteristic of a specific genome. Indeed, a specific SNP does not happen every 1200 base pairs, there is simply a different SNP happening with this frequency. It is therefore not associated to the SNP concept and thus left out of the schema. Also, it is considered unnecessary to be stored, for the reason that it can be derived for every genome from the database itself, simply combining the data on the total amount of nucleotides in an individuals genome, and the amount of found SNPs.

To incorporate the marker functionality of the SNP concept, a new relation would need to be made between *SNP* and *Variation*. This relation would consist of a probability score, indicating the probability to which the SNP at hand predicts the variation. This use of the concept is considered to be a result of the immatureness of the domain. Using SNPs as markers merely satisfies a cost-efficiency strategy, and is expected to be rendered useless in the near future by advances in DNA sequencing technology.

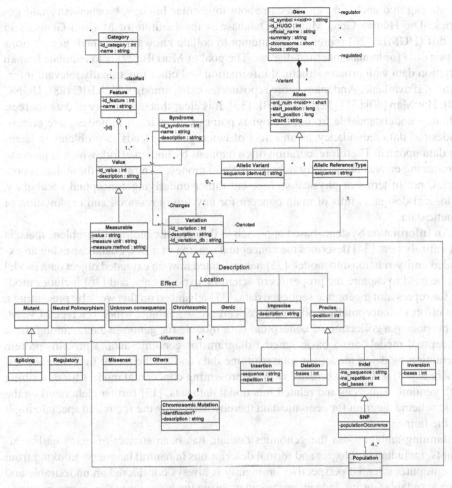

Fig. 2. Conceptual model of the human genome including the *SNP* concept

Although a correlation exists between neutral SNP occurrence and malicious polymorphisms, there appears to be no causal relation. It is for this reason, that has been decided not to incorporate this use of the SNP concept in the conceptual model of the human genome.

4   Related Work

Todays geneticists need to negotiate a wide variety of genomic data repositories. Present day genomics is closely tied to diversity. As the subject under investigation often relates to genetic diversity, at the same time storage of current genetic knowledge is highly dispersed over a variety of resources. The National Center for Biotechnology Information (NCBI) [25] is a United States government run initiative to create automated systems for storing and analyzing knowledge about molecular biology, biochemistry and genetics. The Human Gene Mutation Database at the Institute of Medical Genetics in Cardiff (HGMD) [26] represents an attempt to collate known (published) gene lesions responsible for human inherited disease. The goal of MutDB [27] is to annotate human variation data with protein structural information and other functionally relevant information, if available. And many more repositories exist, among which BIC [28], HGNC [29], HapMap [30] [31] [32] and EMBL [33]. It is clear that this variety of data sources, although understandable from a biologists point of view, leads to undesirable effects. Undesired data redundancy, as an effect of heterogeneity, leads to problems in keeping data updated. Therefore, curation often happens by human effort, which is prone to introducing errors and is costly in both time and money. Centralizing these data repositories, not in terms of physical storage but rather underlying conceptual vocabulary, or logical design, is thus of main concern for any future research and exploitation of genetic data.

An Information Systems based approach to this specific biological problem space is not entirely new. [34] describes the conceptual schema of a DNA database using an extended entity-relationship model. [35] has indicated how an extended object data model can be used to capture the properties of scientific experiments, and [36] includes models for representing genomic sequence data. [37] advanced on this work by presenting a first effort in conceptually modeling the *S. cerevisiae* genome, which is a type of yeast, by proposing a collection of conceptual data models for genomic data. Among these conceptual models are a basic schema diagram for genomic data, a protein- protein interaction model, a model for transcriptome data and a schema for modeling alleles. Whereas [38] provides a broader view by presenting conceptual models for describing both genome sequences and related functional data sets, [13] further elaborated on the basic schema diagram for genomic data thereby narrowing the focus and specializing it for the human genome.

Banning ambiguity in the genomics domain has been subject of many earlier attempts, including ontologies and formal descriptions in natural language. Looking from the computer scientist perspective, ambiguity is always considered an undesirable and often avoidable feature. Indeed, in computer design the behavior of the system is always intended to be known. In biology, and especially genomics, this is simply not the case. Complexity derived from the randomness which created the conditions allowing life to emerge is today obscuring the processes driving this very same system. Earlier attempts at solving this undesired ambiguity include the application of ontologies. [39] provides an overview of Gene Ontology (GO) and how it is used to solve ambiguity in the genomics domain in general. [40] describes the application of the Gene Ontology to the SNP concept. [41] then provides an ontology-based converter that allows for solving the notational problems associated to heterogeneous SNP descriptions. But when one wants to enter in details, all these approaches are far from providing what we could call a precise conceptual definition. Generally, the application of an ontology is considered necessary, however only forms part of the solution.

It is interesting to remark that, although many databases and ontologies exist and they include a lot of diverse genomic information, it is hard to see their corresponding, precise conceptual models that would enable characterization, comparison and evaluation of the quality of their data. It is the main goal of the next section to introduce such a Conceptual Model as the main contribution of this chapter, to show that it is possible to provide the required formal structure for all the data that are considered relevant to understand and interpret this huge amount of genomic information.

The main difference between ontologies and conceptual models thus seems to exist in the intention with which they are created. Ontologies are considered descriptive by nature and usually serve to ensure concepts are indicated by the same terms, essentially disambiguating domain jargon. Models are prescriptive, commonly defined as part of a design process. Their function is to disambiguate, and at the same time simulate reality. [42] states that two ontologies can be different in the vocabulary used (using English or Italian words, for instance), while sharing the same conceptualization. The vocabulary used in this work to capture the ontological concepts of the genomic domain comes in the form of a conceptual model. The conceptual modeling approach taken in this work, thus serves to specify an ontological description. It shows that representing concepts in a descriptive manner, as is common in creation of ontologies, is very well facilitated by the application of a conceptual model.

5 Conclusions

This paper has contributed a conceptual modeling approach applied to the ambiguity associated to the SNP concept. The existence of ambiguity surrounding the term has been explained, and various existing solutions to this have been discussed. It has been clarified how and why a conceptual modeling approach is structurally different from existing solutions, like ontologies, and that only conceptual modeling techniques allow for providing a precise definition. By researching the uses and existing definitions of the term, a conceptualization of the domain was achieved. A clear understanding of the concept has thus been established, and modeled accordingly as an extension to the already existing conceptual model of the human genome. The dimensions on which the SNP concept is commonly defined have been identified and formalized in the conceptual model, allowing for dynamically querying the resulting data repository according for each type of definition. Coming back to the main research question: *What makes a genetic variation, classified as a SNP different from genetic variations, not classified as SNP?*, the answer is clear. A genetic variation identified as a SNP is a single nucleotide polymorphism, not causatively associated to disease and (therefore) happening in more then 0.5% of a population. At the same time, SNPs are common due to their characteristics and for this reason happen about once every 1200 base pairs. Therefore, the dbSNP data repository should not be considered a collection of SNPs, but rather one of genetic polymorphisms. The authors of this paper stress the importance of a name change to dbVariation, in order to avoid future confusion. This paper further demonstrates the capabilities of a conceptual modeling approach to solving ambiguity in a domain. Especially the Bioinformatics domain, where large amounts of data have to be properly managed, and where, unfortunately, too often sound and well-known Information Systems concepts required to perform an effective and efficient data exploitation, are not correctly applied.

References

1. Scherren, K., Jost, J.: Gene and genon concept: coding versus regulation. Theory in Biosciences 126(2-3), 65–113 (2007)
2. Gerstein, M.B., Bruce, C., Rozowosky, J., Zheng, D., Du, J., Korbel, J., Emanuelson, O., Zhang, Z., Weissman, S., Snyder, M.: What is a gene, post-ENCODE? Genome Research 17(6), 669–681 (2007)
3. Pearson, H.: Genetics: What is a gene? Nature 441(7092), 398–402 (2006)
4. Risch, N., Merikangas, K.: The future of genetic studies of complex human diseases. Science 273, 1516–1517 (1996)
5. Li, W.-H., Wu, C.-I., Luo, C.-C.: Nonrandomness of point mutation as reflected in nucleotide substitutions in pseudogenes and its evolutionary implications. Journal of Molecular Evolution 21, 58–71 (1984)
6. Zhao, Z., Boerwinkle, E.: Neighboring-nucleotide effects on single nucleotide polymorphisms: a study of 2.6 million polymorphisms across the human genome. Genome Research 12, 1679–1686 (2002)
7. Kaessmann, H., Heißig, F., von Haeseler, A., Pääbo, S.: DNA sequence variation in a noncoding region of low recombination on the human X chromosome. Natural Genetics 22, 78–81 (1999)
8. Zhao, Z., Li, J., Fu, Y.-X., et al.: Worldwide DNA sequence variation in a 10-kilobase noncoding region on human chromosome 22. Proceedings of the National Academy of Sciences USA 97, 11354–11358 (2000)
9. Jorde, L.B., Watkins, W.S., Bamshad, M.J.: Population genomics: a bridge from evolutionary history to genetic medicine. Human Molecular Genetics 10, 2199–2207 (2001)
10. Schwarz, D.F., Hädicke, O., Erdmann, J., Ziegler, A., Bayer, D., Möller, S.: SNPtoGO: characterizing SNPs by enriched GO terms. Bioinformatics 24(1), 146 (2008)
11. Selic, B.: The Pragmatics of Model-Driven Development. IEEE Software 20(5), 19–26 (2003)
12. Pastor, O.: Conceptual Modeling Meets the Human Genome. In: Li, Q., Spaccapietra, S., Yu, E., Olivé, A. (eds.) ER 2008. LNCS, vol. 5231, pp. 1–11. Springer, Heidelberg (2008)
13. Pastor, O., Levin, A.M., Celma, M., Casamayor, J.C., Eraso Schattka, L.E., Villanueva, M.J., Perez-Alonso, M.: Enforcing Conceptual Modeling to Improve the Understanding of the Human Genome. In: Procs. of the IVth Int. Conference on Research Challenges in Information Science, RCIS 2010, Nice, France. IEEE Press (2010) ISBN #978-1-4244-4840-1
14. Venter, C., Adams, M.D., Myers, E.W., et al.: The Sequence of the Human Genome. Science 291(5507), 1304–1351 (2000)
15. Pastor, O., Molina, J.C.: Model-driven architecture in practice: a software production environment based on conceptual modeling. Springer, Heidelberg (2007)
16. Alberts, B., Bray, D., Hopkin, K., Johnson, A., Lewis, J., Raff, M., Roberts, K., Walter, P.: Essential Cell Biology. Zayatz, E., Lawrence, E. (eds.), 2nd edn., Garland Science USA (2003)
17. Zhao, Z., Fu, Y.-X., Hewett-Emmett, D., Boerwinkle, E.: Investigating single nucleotide polymorphism (SNP) density in the human genome and its implications for molecular evolution. Gene 312, 207–213 (2003)
18. Vignal, A., Milan, D., SanCristobal, M., Eggen, A.: A review on SNP and other types of molecular markers and their use in animal genetics. Genetics, Selection, Evolution 34(3), 275 (2002)
19. dbSNP, http://www.ncbi.nlm.nih.gov/projects/SNP/
20. National Center for Biotechnology Information, http://www.ncbi.nlm.nih.gov/

21. Yue, P., Moult, J.: Identification and analysis of deleterious human SNPs. Journal of Molecular Biology 356(5), 1263–1274 (2006)
22. Shastry, B.S.: SNPs: Impact on gene function and phenotype. Methods in Molecular Biology 578, 3–22 (2009)
23. Devlin, B., Risch, N.: A comparison of Linkage Disequilibrium measures for fine-scale mapping. Genomics 29(2), 311–322 (1995)
24. HUGO Gene Nomenclature Committee, http://www.genenames.org/
25. Maglott, D., Ostell, J., Pruitt, K.D., Tatusova, T.: Entrez Gene: gene-centered information at NCBI. Nucleic Acids Research 35, 26–32 (2006)
26. Stenson, P.D., Mort, M., Ball, E.V., Howells, K., Phillips, A.D., Thomas, N.S.T., Cooper, D.N.: The Human Gene Mutation Database: 2008 update. Genome Medicine 1, 13 (2009)
27. Mooney, S.D., Altman, R.B.: MutDB: annotating human variation with functionally relevant data. Bioinformatics 19, 1858–1860 (2003)
28. Szabo, C., Masiello, A., Ryan, J.F., Brody, L.C.: The Breast Cancer Information Core: Database design, structure, and scope. Human Mutation 16, 123–131 (2000)
29. Povey, S., Lovering, R., Bruford, E., Wright, M., Lush, M., Wain, H.: The HUGO Gene Nomenclature Committee (HGNC). Human Genetics 109, 678–680 (2001)
30. The HapMap project, http://www.hapmap.org
31. International HapMap Consortium. A second generation human haplotype map of over 3.1 million SNPs. Nature 449, 851–862 (2007)
32. Gibbs, R.A., Belmont, J.W., Hardenbol, P., Willis, T.D., Yu, F., et al.: The International HapMap project. Nature 426, 789–796 (2003)
33. Stoesser, G., Tuli, M.A., Lopez, R., Sterk, P.: The EMBL Nucleotide Sequence Database. Nucleic Acids Research 27, 18–24 (1999)
34. Okayama, T., Tamura, T., Gojobori, T., Tateno, Y., Ikeo, K., Miyazaki, S., Fukami-Kobayashi, K., Sugawara, H.: Formal design and implementation of an improved DDBJ DNA database with a new schema and object-oriented library. Bioinformatics 14(6), 472 (1998)
35. Chen, I.M.A., Markowitz, V.: Modeling scientific experiments with an object data model. In: Proceedings of the SSDBM, pp. 391–400. IEEE Press (1995)
36. Medigue, C., Rechenmann, F., Danchin, A., Viari, A.: Imagene, an integrated computer environment for sequence annotation and analysis. Bioinformatics 15(1), 2 (1999)
37. Paton, N.W., Khan, S.A., Hayes, A., Moussouni, F., Brass, A., Eilbeck, K., Goble, C.A., Hubbard, S.J., Oliver, S.G.: Conceptual modeling of genomic information. Bioinformatics 16(6), 548–557 (2000)
38. Pastor, M.A., Burriel, V., Pastor, O.: Conceptual Modeling of Human Genome Mutations: A Dichotomy Between What we Have and What we Should Have. BIOSTEC Bioinformatics, 160–166 (2010) ISBN: 978-989-674-019-1
39. Ashburner, M., Ball, C.A., Blake, J.A.: Gene Ontology: tool for the unification of biology. Nature Genetics 25(1), 25–30 (2000)
40. Schwarz, D.F., Hdicke, O., Erdmann, J., Ziegler, A., Bayer, D., Mller, S.: SNPtoGO: characterizing SNPs by enriched GO terms. Bioinformatics 24(1), 146 (2008)
41. Coulet, A., Smaïl-Tabbone, M., Benlian, P., Napoli, A., Devignes, M.-D.: SNP-Converter: An Ontology-Based Solution to Reconcile Heterogeneous SNP Descriptions for Pharmacogenomic Studies. In: Leser, U., Naumann, F., Eckman, B. (eds.) DILS 2006. LNCS (LNBI), vol. 4075, pp. 82–93. Springer, Heidelberg (2006)
42. Guarino, N.: Formal Ontology in Information Systems. In: Bennett, B., Fellbaum, C. (eds.) Proceedings of the Fourth International Conference (FOIS 2006), vol. 150. IOS Press (1998/2006)

Part I

Knowledge Discovery and Information Retrieval

A Spatio-anatomical Medical Ontology and Automatic Plausibility Checks

Manuel Möller, Daniel Sonntag, and Patrick Ernst

German Research Center for AI (DFKI)
Stuhlsatzenhausweg 3, 66123 Saarbrücken, Germany
manuelm@manuelm.org, {sonntag,patrick.ernst}@dfki.de

Abstract. In this paper, we explain the peculiarities of medical knowledge management and propose a way to augment medical domain ontologies by spatial relations in order to perform automatic plausibility checks. Our approach uses medical expert knowledge represented in formal ontologies to check the results of automatic medical object recognition algorithms for spatial plausibility. It is based on the comprehensive Foundation Model of Anatomy ontology which we extend with spatial relations between a number of anatomical entities. These relations are learned inductively from an annotated corpus of 3D volume data sets. The induction process is split into two parts. First, we generate a quantitative anatomical atlas using fuzzy sets to represent inherent imprecision. From this atlas we then abstract the information further onto a purely symbolic level to generate a generic qualitative model of the spatial relations in human anatomy. In our evaluation we describe how this model can be used to check the results of a state-of-the-art medical object recognition system for 3D CT volume data sets for spatial plausibility. Our results show that the combination of medical domain knowledge in formal ontologies and sub-symbolic object recognition yields improved overall recognition precision.

Keywords: Medical imaging, Semantic technologies, Spatial reasoning, Formal ontologies, Plausibility.

1 Introduction

During the last decades a great deal of effort went into the development of automatic object recognition techniques for medical images. Today a huge variety of available algorithms solve this task very well. The precision and sophistication of the different image parsing techniques have improved immensely to cope with the increasing complexity of medical imaging data. There are numerous advanced object recognition algorithms for the detection of particular objects on medical images. However, the results of the different algorithms are neither stored in a common format nor extensively integrated with patient and image metadata.

At the same time the biomedical informatics community managed to represent enormous parts of medical domain knowledge in formal ontologies [11]. Today, comprehensive ontologies cover large parts of the available taxonomical as well as mereological (part-of) knowledge of human anatomy [3,23].

A. Fred et al. (Eds.): IC3K 2010, CCIS 272, pp. 41–55, 2013.

Our approach is to augment medical domain ontologies and allow for an automatic detection of anatomically implausible constellations in the results of a state-of-the-art system for automatic object recognition in 3D CT scans. The output of our system provides feedback which anatomical entities are most likely to have been located incorrectly. The necessary spatio-anatomical knowledge is learned from a large corpus of annotated medical image volume data sets. The spatial knowledge is condensed into a digital anatomical atlas using fuzzy sets to represent the inherent variability of human anatomy.

Our main contributions are (i) the inductive learning of a spatial atlas of human anatomy, (ii) its representation as an extension of an existing biomedical ontology, and (iii) an application of this knowledge in an automatic semantic image annotation framework to check the spatio-anatomical plausibility of the results of medical object recognition algorithms. Our approach fuses a statistical object recognition and reasoning based on a formal ontology into a generic system. In our evaluation we show that the combined system is able to rule out incorrect detector results with a precision of 85.6% and a recall of 65.5% and can help to improve the overall performance of the object recognition system.

2 Peculiarities of Medical Knowledge Management and Related Work

The technological progress of the last fifty years poses new challenges for clinical knowledge management. Immense amounts of medical images are generated every day resulting in considerable problems for knowledge identification and preservation.

Computer-aided medical imaging has been a very active area over the last 30 years. The invention of CT and its first application for medical diagnosis was a major breakthrough. It allows 3D navigation through stacks of images exhibiting the patient's anatomy. Also in the 1970ies, PET added a technique which allows visualizing functional processes in the human body by detecting pairs of gamma rays emitted indirectly by a positron-emitting radionuclide which is introduced into the body on a biologically active molecule. These advances in medical imaging have enormously increased the volume of images produced in clinical facilities. They can provide incredible detail and wealth of information about the human anatomy and associated diseases.

An informal study at the Department of Radiology at the Mayo Clinic in Jacksonville, FL, revealed an increase of radiological images made per day from 1,500 in 1994 to 16,000 images in 2002, an almost 11-fold increase over 9 years [1]. Another clinic, the University Hospital of Erlangen, Germany, has a total of about 50 TB of medical images in digital format. Currently they have approximately 150,000 medical examinations producing 13 TB per year [19]. Meanwhile, the image resolution has increased the data amount even more. Today "image overload may be the single largest challenge to effective, state-of-the-art practice in the delivery of consistent and well-planned radiological services in health care" [1]. The amount of available medical data during a clinical assessment and especially the image data present an information overload for clinicians. First, there are images from the current patient which can easily exceed a gigabyte. Second, there are immense amounts of data from previous patients which had potentially similar symptoms and whose clinical record could thus be relevant for the current examination.

This increase in the volume of data has brought about significant advances in techniques for analyzing such data. During the last decades a lot of effort went into the development of automatic object recognition techniques and today there is a huge variety of algorithms available solving this task very well. The precision and sophistication of different image understanding techniques, such as object recognition and image segmentation, have also improved to cope with the increasing amount of data. There are numerous algorithms for the detection of particular objects on medical images. However, "spatial relations between objects or object parts have been neglected." [5]

These improvements in analysis have not yet resulted in more flexible or generic image understanding techniques. Instead, the analysis techniques are object specific and not generic enough to be applied for multiple applications. We address this fact as *lack of scalability*. Consequently, current image search techniques, both for Web sources and for medical PACS, still depend on the subjective association of keywords to images for retrieval. This severely hampers the knowledge utilization step of a knowledge management cycle.

One of the important reasons behind the lack of scalability in image understanding techniques has been the absence of *generic* information representation structures capable of overcoming the feature-space complexity of image data. Indeed, most current content-based image search applications are focused on indexing syntactic image features that do not generalize well across domains. As a result, current image search does not operate at the *semantic* level and, hence, is not scalable. As other authors already have argued (cf. [10] as mentioned above), the complicated access to knowledge leads to severe problems in clinical health care.

To cope with the data increase (semi-)automatic image segmentation and understanding techniques from computer vision are applied to ease the work of radiological personnel during image assessment and annotation. However, these systems are usually based on statistical algorithms. Thus, the detection and localization of anatomical structures can only be performed with limited precision or recall. The outcome is a certain number of incorrect results.

Our primary source of medical domain knowledge is the Foundational Model of Anatomy (FMA) [21], the most comprehensive formal ontology of human anatomy available. However, the number of spatial relations in the FMA is very limited and covers only selected body systems [18]. Thus, our approach is to infer additional spatial relations between the concepts defined in the FMA by learning from annotated medical volume data sets.

In [15] the authors describe a hybrid approach which also uses metadata extracted from the medical image headers in combination with low-level image features. However, their aim is to speed up content-based image retrieval by restricting the search space by leveraging metadata information.

The approach in [12] is complementary to our work in so far as the authors also propose to add spatial relations to an existing anatomical ontology. Their use case is the automatic recognition of brain structures in 3D MRI scans. However, they generate the spatial relations manually, while a major aspect of our approach is the automatic learning from a large corpus.

Quantitative spatial models are the foundation of digital anatomical atlases. Fuzzy logic has been proven as an appropriate formalism which allows for quantitative representations of spatial models [8]. In [14] the authors expressed spatial features and relations of object regions using fuzzy logic. In [9] and [7] the authors describe generalizations of this approach and compare different options to express relative positions and distances between 3D objects with fuzzy logic.

3   System Architecture

Figure 1 shows an abstraction of the distributed system architecture. It is roughly organized in the order of the data processing horizontally from left to right. All parsing results are stored in a custom-tailored spatial database.

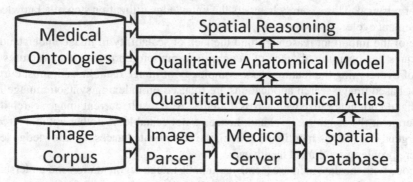

Fig. 1. System architecture overview

3.1   Image Parser

To represent the results of the automatic object recognition algorithms in the format of our ontology we had to integrate rather disparate techniques into a hybrid system. The automatic object recognition performs an abstraction process from simple low-level features to concepts represented in the FMA.

For automatic object recognition we use a state-of-the-art anatomical landmark detection system described in [22]. It uses a network of 1D and 3D landmarks and is trained to quickly parse 3D CT volume data sets and estimate which organs and landmarks are present as well as their most probable locations and boundaries. Using this approach, the segmentation of seven organs and detection of 19 body landmarks can be obtained in about 20 seconds with state-of-the-art accuracy below 3 millimeters mean mesh error and has been validated on 80 CT full or partial body scans [22].

The image parsing algorithm generates two fundamentally different output formats: *Point3D* for landmarks and *Mesh* for organs. Apart from their geometric features, they always point to a certain anatomical concept which is hard-wired to the model that the detection/segmentation algorithm has used to generate them. A landmark is a point in 3D without spatial extension. Usually it represents an extremal point of an anatomical entity with a spatial extension. Sometimes these extremal points are not part of the official FMA. In these cases we modeled the respective concepts as described in [18]. In total we were able to detect 22 different landmarks from the trunk of the human body. Examples are the bottom tip of the sternum, the tip of the coccyx, or the top point of the liver.

Organs, on the contrary, are approximated by polyhedral surfaces. Such a surface, called *mesh*, is a collection of vertices, edges, and faces defining the shape of the object in 3D. For the case of the urinary bladder, the organ segmentation algorithm uses the prototype of a mesh with 506 vertices which are then fitted to the organ surface of the current patient. Usually, vertices are used for more than one triangle. Here, these 506 vertices form 3,024 triangles. In contrast to the Point3D data, meshes are used to segment organs. For our test, the following organs were available: left/right kidney, left/right lung, bladder, and prostate.

3.2 Medico Server

One of the main challenges was to combine the C++ code for volume parsing with the Java-based libraries and applications for handling data in Semantic Web formats. We developed a distributed architecture with the *MedicoServer* acting as a middleware between the C++ and Java components using CORBA [20].

3.3 Spatial Database

As we have seen in the section about the image parsing algorithms, the automatic object recognition algorithms generate several thousand points per volume data set. Storage and efficient retrieval of this data for further processing made a spatial database management system necessary. Our review of available open-source databases with support for spatial data types revealed that most of them now also have support for 3D coordinates. However, the built-in operations ignore the third dimension and thus yield incorrect results, e. g., for distance calculations between two points in 3D. Eventually we decided to implement a light-weight spatial database supporting the design rationales of simplicity and scalability for large numbers of spatial entities.

4 Corpus

The volume data sets of our image corpus were selected primarily by the first use case of MEDICO which is support for *lymphoma* diagnosis. The selected data sets were picked randomly from a list of all available studies in the medical image repositories of the University Hospital in Erlangen, Germany. The selection process was performed by radiologists at the clinic. All images were available in the Digital Imaging and Communications in Medicine (DICOM) format, a world wide established format for storage and exchange of medical images [16].

Table 1 summarizes major quantitative features of the available corpus. Out of 6,611 volume data sets in total only 5,180 belonged to the modality CT which is the only one currently processible by our volume parser. Out of these, the number of volumes in

Table 1. Summary of corpus features

volume data available in total	777 GB
number of distinct patients	377
volumes (total)	6,611
volumes (modality CT)	5,180
volumes (parseable)	3,604
volumes (w/o duplicates)	2,924
landmarks	37,180
organs	7,031

which at least one anatomical entity was detected by the parser was 3,604. This results from the rationale of the parser which was in favor of precision and against recall. In our subsequent analysis we found that our corpus contained several DICOM volume data sets with identical Series ID. The most likely reason for this is that an error occurred during the data export from the clinical image archive to the directory structure we used to store the image corpus. To guarantee for consistent spatial entity locations, we decided to delete all detector results for duplicate identifiers. This further reduced the number of available volume data sets to 2,924.

4.1 Controlled Corpus

Due to the statistical nature of the object detection algorithm used for annotating the volume data sets, we have to assume that we have to deal with partially incorrect results. Hence, we decided to conduct manual corpus inspections using a 3D detect result visualization. The goal was to identify a reasonable set of controlled training examples suitable for generation and evaluation of a quantitative anatomical atlas and a qualitative model. These manual inspections turned out to be very time consuming. For each volume in the corpus a 3D visualization had to be generated and manually navigated to verify the correct location of landmarks and organ meshes. After some training we were able to process approximately 100 volume data sets per hour. For higher accuracy, all manual inspection results were double checked by a second person resulting in a bisection of the per-head processing rate to about 50 per hour.

During our inspection we found that the quality of the detector results exhibits a high variability. Subsequently, we distinguish three quality classes: clearly incorrect, sufficiently correct, and perfectly correct. The visualizations in Figure 2 show one example for each class.

To have a solid basis for the generation of the spatio-anatomical model we decided to label a reasonable subset of the available volume data sets manually. We ended up with more than 1,000 manually labeled volume data sets. Table 2 summarizes the results quantitatively. All quantitative evaluations of the performance of the spatial consistency check are based on this corpus.

We consider a detector result *incorrect* if a spatial entity configuration has been detected that is clearly contradictory to human anatomy. Figure 2 (a) shows such an example with arbitrarily deformed lungs. Normally, the lungs should be located vertically at about the same level. Here, this is not the case. Additionally, the prostate has been

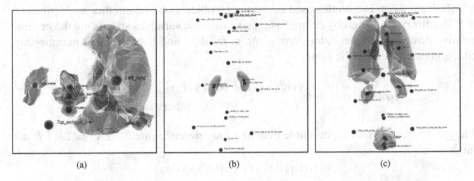

(a)   (b)   (c)

Fig. 2. Visualizations of detector results: (a) incorrect; (b) sufficient; (c) perfect

Table 2. Summary of the manual corpus inspection

detector results inspected in total	1,119
apparently incorrect volume data sets	482 (43%)
sufficiently correct detector results	388 (34%)
perfect detector results	147 (13%)
volumes containing meshes	946 (85%)
volumes containing landmarks	662 (59%)

located on the top right side of the right lung although it belongs to a completely different body region.

A detector result is considered as *sufficiently correct* if it contains a reasonable number of landmarks and/or meshes. The following flaws distinguish them from perfect detector results (at least one condition is met): (i) It contains either only landmarks or only meshes. (ii) A minor number of anatomical entities has been detected at slightly incorrect positions. (iii) The overall number of detected anatomical entities in the detector result is rather low.

A *perfectly correct* detector result has to contain both landmarks and meshes. In addition, none of the landmarks or meshes is allowed to be located incorrectly. The anatomical atlas is learned only from detector results labeled as either sufficiently or perfectly correct. Incorrect detector results are discarded during model generation.

5   Quantitative Anatomical Atlas

Based on the spatial entities in the corpus we distinguish between two different types of relations to build up a quantitative atlas, namely: (i) *elementary relations* directly extracted from 3D data and represented as fuzzy sets, and (ii) *derived relations* which are defined using fuzzy logic and based on one or more elementary relations.

5.1   Elementary Relations

Orientation. The orientation or relative position of objects to each other is important to describe spatial coherencies. We use a typical fuzzy representation of the orientation which depends on two angles used to rotate two objects on one another as decribed in [6]. The fuzzy set is thereby defined using six linguistic variables specifying the general relative positions: *above, below, left, right, in front of*, and *behind*. Their membership functions are basically the same.

$$\mu_{rel}(\alpha_1, \alpha_2) = \begin{cases} \cos^2(\alpha_1)\cos^2(\alpha_2) & \text{if } \alpha_{1,2} \in \left[-\frac{\pi}{2}, \frac{\pi}{2}\right] \\ 0 & \text{otherwise} \end{cases}$$

They only vary in a direction angle denoting the reference direction, e. g., for *left* the angle is π.

$$\mu_{left}(\alpha_1, \alpha_2) = \mu_{rel}(\alpha_1 - \pi, \alpha_2)$$

More details about this approach can be found in [6]. The definition of complex objects' relative positions is not straightforward. One option is to use centroids. Mirtich et al. describe a fast procedure for the computation of centroids [17]. However, complex objects are reduced to single points and therefore information is lost. As the authors of [4] state: "This still limits the capability to distinguish perceptually dissimilar configurations." For this reason we decided to use 3D angle histograms providing a richer quantitative representation for the orientation (cf. [13] for a good overview). A histogram H_A^R stores the relative number of all angles between a reference object R and a target A. The degree of membership is then obtained by computing the fuzzy compatibility between H_A^R and a particular directional relation μ_{rel}. Thus, we achieve a compatibility fuzzy set describing the membership degree.

$$\mu_{CP(\mu_{rel},H)}(u) = \begin{cases} 0 & \text{if } H_A^{R-1}(\alpha_1, \alpha_2) = 0 \\ \sup\limits_{(\alpha_1,\alpha_2), u = H_A^R(\alpha_1,\alpha_2)} \mu_{rel}(\alpha_1, \alpha_2) & \bot \end{cases}$$

To compute a single value the center of gravity of $\mu_{CP(\mu_{rel},H)}$ is determined by

$$\mu_{rel}^R(A) = \frac{\int\limits_0^1 u \mu_{CP(\mu_{rel},H)}(u)du}{\int\limits_0^1 \mu_{CP(\mu_{rel},H)}(u)du}$$

Using this approach the orientation relations are now depending on the entire shape of the spatial entities. In addition, these histograms capture distance information. For example, if one object is moved closer or further away from another, the angles will also change according to the distance. Unfortunately, the membership degree computation is more complex compared to using centroids. However, since we are relying exclusively on the surface points of meshes, the computation time is acceptable with an average of 33 seconds for an entire volume.

Intersection. The detection of organ borders is a very difficult task in medical image understanding because it is mainly based on the tissue density [2]. However, adjacent organs can have very similar densities. Thus, detection is sometimes error-prone and objects may intersect. To check for such inconsistencies we are determining the degree of intersection between two spatial entities A and B. On that account, a new mesh or point is generated describing the intersection Int, so that the degree of intersection is determined by dividing the volume of Int with the minimum volume of A and B.

$$\mu_{int}(A, B) = \frac{V_{Int}}{\min\{V_A, V_B\}}$$

Inclusion. The inclusion of two spatial entities is similarly defined as the intersection. We say that a spatial entity B is included in an entity A, if

$$\mu_{inc}(A, B) = \frac{V_{Int}}{V_B}$$

Compared to intersection inclusion only considers the volume of the entity being included. For that reason this relation is not symmetrical contrary to all other relations described in this work.

5.2 Derived Relations

Adjacency. Many anatomical entities in the human body exist which share a common border or adjoin to each other, e. g., the border of the prostate and urinary bladder. These adjacent coherencies are represented using a trapezoid neighborhood measure depicted in Figure 3. Two spatial entities are fully neighbored if the distance between them is less than 2 millimeters. After that border the neighborhood decreases to a distance of 4.5 millimeters at which spatial entities are not considered as neighbored anymore. However, for an appropriate representation of adjacency the intersection between two objects has to be incorporated. This is important since if two spatial entities intersect, they are not adjacent anymore. To formulate those circumstances using fuzzy sets, we comprised the degree of non-intersection and the neighborhood measure using a fuzzy t-norm [13]:

$$\mu_{adj}(A, B) = t[\mu_{\neg Int}(A, B), \sup_{x \in A} \sup_{y \in B} n_{xy}]$$

where the non-intersection is computed using the fuzzy logical *not*. Currently, three different t-norm based logic definitions are implemented, namely Lukasiewicz Logic, Gdel Logic, and Product Logic. The details of their definitions can be derived from [13]. Table 3 compares the average and standard deviations between the different logics. We decided to use the Lukasiewicz logic because it provides the highest average of actual adjacent concepts determined during a manual data examination. Additionally, the logic also yields the lowest standard deviations in comparison to the average value.

6   Qualitative Anatomical Model

Figure 4 illustrates our modeling of instantiated fuzzy spatial relations. It is loosely oriented on the formalism in the FMA for storing spatial relations. However, the value for each spatial relation is stored separately. Another difference is the representation

Table 3. Mean values and standard deviations for fuzzy membership values for the adjacency relation depending on the choice of the t-norm

	$\top_{Goedel}$		$\top_{prod}$		$\top_{Lukas}$	
Relation	avg	stddev	avg	stddev	avg	stddev
Bronchial bifurcation, Right lung	0.0485	0.2001	0.0485	0.2001	0.0485	0.2001
Hilum of left kidney, Left kidney	0.0043	0.0594	0.0043	0.0594	0.0043	0.0594
Hilum of right kidney, Right kidney	0.0032	0.0444	0.0032	0.0444	0.0032	0.0444
Left kidney, Left lung	0.0427	0.1753	0.0427	0.1753	0.0427	0.1753
Left lung, Right lung	0.1556	0.3319	0.1556	0.3319	0.2617	0.3967
Left lung, Top of left lung	0.2322	0.3526	0.2322	0.3526	0.2322	0.3526
Prostate, Top of pubic symphysis	0.0116	0.0922	0.0116	0.0922	0.0116	0.0922
Prostate, Urinary bladder	0.2647	0.4035	0.2647	0.4035	0.7442	0.3408
Right kidney, Right lung	0.0376	0.1788	0.0376	0.1788	0.0383	0.1796
Right lung, Top of right lung	0.2900	0.3985	0.2900	0.3985	0.2900	0.3985
Right lung, Top point of liver	0.2288	0.3522	0.2288	0.3522	0.2288	0.3522
Top of pubic symphysis, Urinary bladder	0.0114	0.0918	0.0114	0.0918	0.0114	0.0918

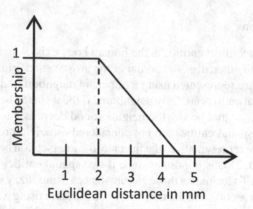

Fig. 3. Graph of the fuzzy membership function for the linguistic variable *adjacent*

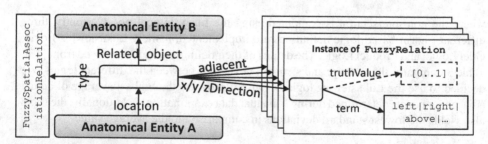

Fig. 4. Illustration of the extended structure for storing the six linguistic variables separately and represent truth values in the ontology

with a term further qualifying the relation together with a truth value in a separate instance. Currently, we integrate orientation and adjacency in a qualitative model.

In order to create a qualitative anatomical model, we extracted instances containing the spatial relations described in Sect. 5. An instance describes the relation between two spatial entities occurring in a volume data set. To transform a relation into the model, a truth value is computed representing the mean of all extracted values of this relation. Thereby, the orientation is stated using a directional term, i. e., *left*, *right*, *in front*, etc. determined by the linguistic variables. On the other hand, the adjacency only gets a simple boolean qualifier. We determined a threshold of 0.2 (see Table 3) to distinguish between adjacent and not adjacent.

7 Evaluation and Results

When an actual detector result is to be checked against the generic qualitative anatomical model, we first represent all its inherent spatial relations using the same formalism that we use for the generic anatomical model. This yields a set of OWL instances. Next, we iterate over all instances of the detector result and compare their directions and truth values to the generic model. We consider a spatial relation instance to be *not conform with the model* if the truth values differ by at least 50%. We then count the occurrences of the anatomical concepts among the non-conform instances. The higher this number is for a given anatomical concept, the more likely the respective organ has been located incorrectly.

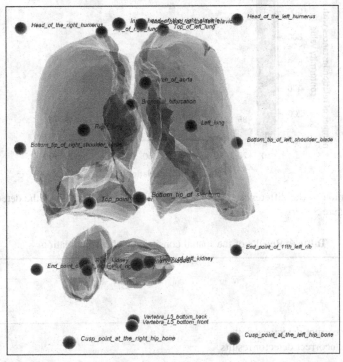

Fig. 5. Visualization of the organ and landmark locations for an incorrect detector results (cf. the location of the urinary bladder)

Figure 5 shows the visualization of an incorrect detector result. In the upper part you can see the two lungs and a number of landmarks. In the lower half you see one kidney and, to the right of the kidney, the urinary bladder has been located. This is clearly incorrect; in fact the urinary bladder should lie much further below. The other kidney has not been detected at all. Figure 6 shows a histogram of the differences in percent between the model and the spatial relation instances of this volume data set. Apparently, most of the relation instances have a comparably low difference to the model. Among all relation instances with a difference to the model of more than 50%, those with relation to urinary bladder account for 11 out of 16. This information gives evidence that the location of the urinary bladder is very likely to be incorrect.

Validation on Controlled Corpus. We performed a systematic evaluation of the spatial consistency check on our manually labeled corpus using four-fold cross evaluation. Our results show that the average difference in percent between the spatial relation instances in the learned model and the instances generated for an element from the evaluation set is an appropriate measure for the spatial consistency. The average difference to the truth value in the model for correct detector results was 2.77% whereas the average difference

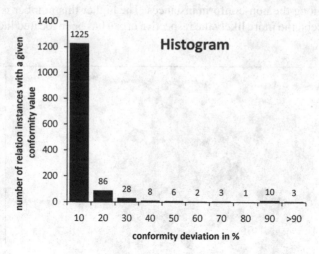

Fig. 6. Distribution of the differences of the truth values between model and the detector result presented in Figure 5

Table 4. Results of the spatial consistency check evaluation

true positives	407
true negatives	431
false positives	67
false negatives	213
avg. difference correct detector results	2.7%
avg. difference incorrect detector results	9.0%
precision	85.7%
recall	65.5%

to the truth value in the model for incorrect detector results was 9%. Using 5% as a threshold to distinguish spatially consistent ($< 5\%$) from inconsistent ($>= 5\%$) yields a precision of 85.7% with a recall of 65.5% for the detection of spatially inconsistent detector results.

8   Conclusions and Future Work

We explained the peculiarities of medical knowledge management in detail and presented an approach fusing state-of-the-art object recognition algorithms for 3D medical volume data sets with technologies from the Semantic Web. In a two-stage process we augmented the FMA as the most comprehensive reference ontology for human anatomy with spatial relations. These relations were acquired inductively from a corpus of semantically annotated CT volume data sets. The first stage of this process abstracted relational information using a fuzzy set representation formalism. In the second stage we further abstracted from the fuzzy anatomical atlas to a symbolic level using an extension of the spatial relation model of the FMA.

In our evaluation we were able to show that this spatio-anatomical model can be applied successfully to check the results of automatic object detection algorithms. The detection of incorrect object recognition constellations can be performed with a high precision of 85.6% and a recall of 65.5%. The presented method can thus improve existing statistical object recognition algorithms by contributing a method to sort out incorrect results and increase the overall performance by reducing the number of incorrect results. Currently our anatomical model only covers directional information for pairs of spatial entities in our corpus.

Among our next steps is also a user evaluation of clinical applications making use of the reasoning, e. g., to support radiologists by suggesting anatomical concepts and relations during manual image annotation. Furthermore, our approach could be used to generate warnings for manually generated image annotations in case they do not conform to the spatial anatomical model. A clinical evaluation of these features is planned in the near future.

Acknowledgements. This research has been supported in part by the research program THESEUS in the MEDICO project, which is funded by the German Federal Ministry of Economics and Technology under the grant number 01MQ07016. The responsibility for this publication lies with the authors.

References

1. Andriole, K.P., Morin, R.L., Arenson, R.L., Carrino, J.A., Erickson, B.J., Horii, S.C., Piraino, D.W., Reiner, B.I., Seibert, J.A., Siegel, E.L.: Addressing the coming radiology crisis - the society for computer applications in radiology transforming the radiological interpretation process trip^{tm} initiative. J. Digital Imaging 17(4), 235–243 (2004)
2. Bankman, I. (ed.): Handbook of medical imaging: processing and analysis. Elsevier (January 2000), http://books.google.com/books?id=UHkkPBnhT-MC&printsec=frontcover

3. Bateman, J., Farrar, S.: Towards a generic foundation for spatial ontology. In: Proc. of the International Conference on Formal Ontology in Information Systems (FOIS 2004), Trento, Italy, pp. 237–248 (2004)

4. Berretti, S., Bimbo, A.D.: Modeling spatial relationships between 3d objects. In: 18th International Conference on Pattern Recognition, ICPR 2006, vol. 1, pp. 119–122 (2006), `http://ieeexplore.ieee.org/search/srchabstract.jsp?tp=& arnumber=1698847&queryText%253D%2528%2528Document+Title %253AModeling+Spatial+Relationships+between+3D+Objects%2529 %2529%2526openedRefinements%253D%2526sortType%253Ddesc_ Publication+Year%2526matchBoolean%253Dtrue%2526rowsPerPage %253D50%2526searchField%253DSearch+All`

5. Berretti, S., Bimbo, A.D.: Modeling Spatial Relationships between 3D Objects. In: ICPR, August 20-24, pp. 119–122. IEEE Computer Society, Hong Kong (2006), `http://dblp.uni-trier.de/db/conf/icpr/icpr2006-1.html #BerrettiB06`

6. Bloch, I.: Fuzzy relative position between objects in image processing: new definition and properties based on a morphological approach. International Journal Of Uncertainty, Fuzziness And Knowledge Based Systems 7, 99–134 (1999)

7. Bloch, I.: On fuzzy distances and their use in image processing under imprecision. Pattern Recognition 32(11), 1873–1895 (1999)

8. Bloch, I.: Fuzzy spatial relationships for image processing and interpretation: a review. Image and Vision Computing 23(2), 89–110 (2005)

9. Bloch, I., Ralescu, A.: Directional relative position between objects in image processing: a comparison between fuzzy approaches. Pattern Recognition 36(7), 1563–1582 (2003)

10. Chen, H., Fuller, S.S., Friedman, C.P.: Medical Informatics: Knowledge Management and Data Mining in Biomedicine. Integrated Series in Information Systems. Springer, Heidelberg (2005), `http://www.amazon.com/exec/obidos/redirect?tag= citeulike07-20&path=ASIN/038724381X`

11. Davies, J.: Semantic Web Technologies: Trends and Research in Ontology-based Systems. John Wiley & Sons (July 2006), `http://www.amazon.ca/exec/obidos/ redirect?tag=citeulike09-20&amp;path=ASIN/0470025964`

12. Hudelot, C., Atif, J., Bloch, I.: Fuzzy spatial relation ontology for image interpretation. Fuzzy Sets Syst. 159(15), 1929–1951 (2008)

13. Klir, G.J., Yuan, B.: Fuzzy sets and fuzzy logic: theory and applications. Prentice-Hall, Inc., Upper Saddle River (1994)

14. Krishnapuram, R., Keller, J.M., Ma, Y.: Quantitative analysis of properties and spatial relations of fuzzy image regions. IEEE Transactions on Fuzzy Systems 1(3), 222–233 (1993)

15. da Luz, A., Abdala, D.D., Wangenheim, A.V., Comunello, E.: Analyzing dicom and non-dicom features in content-based medical image retrieval: A multi-layer approach. In: 19th IEEE International Symposium on Computer-Based Medical Systems, CBMS 2006, pp. 93–98 (2006), `http://ieeexplore.ieee.org/xpls/abs_all.jsp?arnumber=1647552`

16. Mildenberger, P., Eichelberg, M., Martin, E.: Introduction to the DICOM standard. European Radiology 12(4), 920–927 (2002), `http://www.springerlink.com/content/yj705ftj58u738gh/`

17. Mirtich, B.: Fast and accurate computation of polyhedral mass properties. Graphics tools: The jgt editors' choice (January 2005), `http://books.google.com/books?hl=en&lr=&id=87NzFbSROUYC&oi= fnd&pg=PA109&dq=%2522Fast+and+accurate+computation+polyhedral +mass+properties%2522&ots=I961J6HGK2& sig=M-JTC9q08yIX5pkgnxCKqqVANxw`

18. Möller, M., Folz, C., Sintek, M., Seifert, S., Wennerberg, P.: Extending the foundational model of anatomy with automatically acquired spatial relations. In: Proc. of the International Conference on Biomedical Ontologies, ICBO (July 2009), http://precedings.nature.com/documents/3471/version/1

19. Möller, M., Sintek, M.: A generic framework for semantic medical image retrieval. In: Proc. of the Knowledge Acquisition from Multimedia Content (KAMC) Workshop, 2nd International Conference on Semantics And Digital Media Technologies, SAMT (November 2007)

20. Object Management Group, I.: Common object request broker architecture: Core specification, version 3.0.3 (March 2004), http://www.omg.org/docs/formal/04-03-12.pdf

21. Rosse, C., Mejino, J.L.V.: The Foundational Model of Anatomy Ontology. In: Anatomy Ontologies for Bioinformatics: Principles and Practice, vol. 6, pp. 59–117. Springer, Heidelberg (2007), http://sigpubs.biostr.washington.edu/archive/00000204/01/FMA_Chapter_final.pdf

22. Seifert, S., Kelm, M., Möller, M., Mukherjee, S., Cavallaro, A., Huber, M., Comaniciu, D.: Semantic annotation of medical images. In: Proc. of SPIE Medical Imaging, San Diego, CA, USA (2010)

23. Sonntag, D., Wennerberg, P., Buitelaar, P., Zillner, S.: Pillars of ontology treatment in the medical domain. Journal of Cases on Information Technology (JCIT) 11(4), 47–73 (2009)

Experimentally Studying Progressive Filtering in Presence of Input Imbalance

Andrea Addis, Giuliano Armano, and Eloisa Vargiu

Dept. of Electrical and Electronic Engineering, University of Cagliari, Cagliari, Italy
{addis,armano,vargiu}@diee.unica.it
http://iasc.diee.unica.it

Abstract. Progressively Filtering (PF) is a simple categorization technique framed within the local classifier per node approach. In PF, each classifier is entrusted with deciding whether the input in hand can be forwarded or not to its children. A simple way to implement PF consists of unfolding the given taxonomy into pipelines of classifiers. In so doing, each node of the pipeline is a binary classifier able to recognize whether or not an input belongs to the corresponding class. In this chapter, we illustrate and discuss the results obtained by assessing the PF technique, used to perform text categorization. Experiments, on the Reuters Corpus (RCV1- v2) dataset, are focused on the ability of PF to deal with input imbalance. In particular, the baseline is: (i) comparing the results to those calculated resorting to the corresponding flat approach; (ii) calculating the improvement of performance while augmenting the pipeline depth; and (iii) measuring the performance in terms of generalization- / specialization- / misclassification-error and unknown-ratio. Experimental results show that, for the adopted dataset, PF is able to counteract great imbalances between negative and positive examples. We also present and discuss further experiments aimed at assessing TSA, the greedy threshold selection algorithm adopted to perform PF, against a relaxed brute-force algorithm and the most relevant state-of-the-art algorithms.

Keywords: Progressive filtering, Threshold selection, Hierarchical text categorization, Input imbalance.

1 Introduction

According to the "divide et impera" philosophy, the main advantage of the hierarchical perspective is that the problem is partitioned into smaller subproblems, each being effectively and efficiently managed. Therefore, it is not surprising that in the Web 2.0 age people organize large collections of web pages, articles or emails in hierarchies of topics or arrange a large body of knowledge in ontologies. Such organization allows to focus on a specific level of details ignoring specialization at lower levels and generalization at upper levels. In this scenario, the main goal of automatic categorization systems is to deal with reference taxonomies in an effective and efficient way.

In this chapter, we are aimed at assessing the effectiveness of the "Progressive Filtering" (PF) technique, applied to text categorization in presence of input imbalance. In its simplest setting, PF decomposes a given rooted taxonomy into pipelines, one for

A. Fred et al. (Eds.): IC3K 2010, CCIS 272, pp. 56–71, 2013.

each path that exists between the root and any given node of the taxonomy, so that each pipeline can be tuned in isolation. To this end, a Threshold Selection Algorithm (TSA) has been devised, aimed at finding a sub-optimal combination of thresholds for each pipeline. The chapter extends and revises our previous work [2]. The main extensions consist of: (i) a more detailed presentation of TSA; and (ii) further experiments aimed at assessing the effectiveness of TSA with respect to three state-of-the-art threshold selection algorithms.

The chapter is organized as follows. In Section 2, we give a brief survey of relevant work on: (i) hierarchical text categorization; (ii) threshold selection strategies; and (iii) input imbalance. Section 3 concentrates on PF and TSA. In Section 4, we present and discuss experimental results. Section 5 ends the chapter with conclusions.

2 Background

The most relevant issues that help to clarify the contextual setting of the chapter are: (i) the work done on HTC, (ii) the work done on thresholding selection, and (iii) the work done on the input imbalance.

2.1 Hierarchical Text Categorization

In recent years several researchers have investigated the use of hierarchies for text categorization.

Until the mid-1990s researchers mostly ignored the hierarchical structure of categories that occur in several domains. In 1997, Koller and Sahami [17] carry out the first proper study about HTC on the Reuters-22173 collection. Documents are classified according to the given hierarchy by filtering them through the best-matching first-level class and then sending them to the appropriate second level. This approach shows that hierarchical models perform well when a small number of features per class is used, as no advantages were found using the hierarchical model for large numbers of features. Mc Callum et al. [23] propose a method based on naïve Bayes. The authors compare two techniques: (i) exploring all possible paths in the given hierarchy and (ii) greedily selecting at most two branches according to their probability, as done in [17]. Results show that the latter is more error prone while computationally more efficient. Mladenic and Grobelink [24] use the hierarchical structure to decompose a problem into a set of subproblems, corresponding to categories (i.e., the nodes of the hierarchy). For each subproblem, a naïve Bayes classifier is generated, considering examples belonging to the given category, including all examples classified in its subtrees. The classification applies to all nodes in parallel; a document is passed down to a category only if the posterior probability for that category is higher than a user defined threshold. D'Alessio et al. [12] propose a system in which, for a given category, the classification is based on a weighted sum of feature occurrences that should be greater than the category threshold. Both single and multiple classifications are possible for each document to be tested. The classification of a document proceeds top-down possibly through multiple paths. An innovative contribution of this work is the possibility of restructuring a given hierarchy or building a new one from scratch. Dumas and Chen [13] use the hierarchical structure for two purposes: (i) training several SVMs, one for each intermediate node and (ii) classifying documents by combining scores from SVMs at different levels. The sets of positive and negative examples are built considering documents that belong to categories at the same level, and different feature sets are built, one for each category. Several combination rules have also been assessed. In the work of Ruiz and Srinivasan [26], a variant of the Hierarchical Mixture of Experts model is used. A hierarchical classifier combining several neural networks is also proposed in [30]. Gaussier et al. [15] propose a hierarchical generative model for textual data, i.e., a model for hierarchical clustering and categorization of co-occurrence data, focused on documents organization. In [25], a kernel-based approach for hierarchical text classification in a multi-label context is presented. The work demonstrates that the use of the dependency structure of microlabels (i.e., unions of partial paths in the tree) in a Markovian Network framework leads to improved prediction accuracy on deep hierarchies. Optimization is made feasible by utilizing decomposition of the original problem and making incremental conditional gradient search in the subproblems. Ceci and Malerba [9] present a comprehensive study on hierarchical classification of Web documents. They extend a previous work [8] considering: (i) hierarchical feature selection mechanisms; (ii) a naïve Bayes algorithm aimed at avoiding problems related to different document lengths; (iii) the validation of their framework for a probabilistic SVM-based classifier; and (iv) an automated threshold selection algorithm. More recently, in [14], the authors propose a multi-label hierarchical text categorization algorithm consisting of a hierarchical variant of ADABOOST.MH, a well-known member of the family of "boosting" learning algorithms. Bennet and Nguyen [6] study the problem of the error propagation under the assumption that the "higher" the node is in the hierarchy, the worse the mistake is. The authors also study the problem of dealing with increasingly complex decision surfaces. Brank et al. [7] deal with the problem of classifying textual documents into a topical hierarchy of categories. They construct a coding matrix gradually, one column at a time, each new column being defined in a way that the corresponding binary classifier attempts to correct the most common mistakes of the current ensemble of binary classifiers. The goal is to achieve good performance while keeping reasonably low the number of binary classifiers.

2.2   Threshold Selection Strategies

In TC, the three most commonly used thresholding strategies are RCut, PCut, and SCut [35]. For each document, RCut sorts categories by score and selects the t top-ranking categories, with $t \geq 1$ (however, as noted in [28], RCut is not a strict thresholding policy). For each category C_j, PCut sorts the documents by score and sets the threshold of C_j so that the number of documents accepted by C_j corresponds to the number of documents assigned to C_j. For each category C_j, SCut scores a validation set of documents and tunes the threshold over the local pool of scores, until a suboptimal, though satisfactory, performance of the classifier is obtained for C_j.

Few threshold selection algorithms have been proposed for HTC [12] [27] [9]. The algorithm proposed by D'Alessio et al. [12] tunes the thresholds by considering categories in a top-down fashion. For each category C_j, the search space is visited by incrementing the corresponding threshold with steps of 0.1. For each threshold value, the number of True Positives (TP) and False Positives (FP), i.e., the number of documents that would be correctly and incorrectly placed in C_j, is calculated. The utility function, i.e., the goodness measure that must be maximized for each threshold, is $TP - FP$ (in the event that the same value of the utility function occurs multiple times, the lowest threshold with that value is selected). Ruiz [27] selects thresholds that optimize the F_1 values for the categories, using the whole training set to identify the (sub)optimal thresholds. His expert-based system takes a binary decision at each expert gate and then optimizes the thresholds using only examples that reach leaf nodes. This task is performed by grouping experts into levels and finding the thresholds that maximize F_1. The best results are selected upon trials performed with each combination of thresholds from the vector $[0.005, 0.01, 0.05, 0.10]$ for level 1 and $[0.005, 0.01, 0.05, 0.10, 0.15, \ldots, 0.95]$ for levels 2, 3, 4. The algorithm proposed by Ceci and Malerba [9] is based on a recursive bottom-up threshold determination. The algorithm takes as input a category C and the set of thresholds already computed for some siblings of C and their descendants. It returns the union of the input set with the set of thresholds computed for all descendants of C. In particular, if C' is a direct subcategory of C, the threshold associated to C' is determined by examining the sorted list of classification scores and by selecting the middle point between two values in the list, to minimize the expected error. The error function is estimated on the basis of the distance between two nodes in a tree structure (TD), the distance being computed as the sum of the weights of all edges of the unique path connecting the two categories in the hierarchy (a unit weight is associated to each edge).

2.3 Input Imbalance

High imbalance occurs in real-world domains where the decision system is aimed at detecting rare but important cases [18]. Imbalanced datasets exist in many real-world domains, such as spotting unreliable telecommunication customers, detection of oil spills in satellite radar images, learning word pronunciations, text classification, detection of fraudulent telephone calls, information retrieval and filtering tasks, and so on [32] [34]. Japkowicz [16] contributed to study the class imbalance problem in the context of binary classification, the author studied the problem related to domains in which one class is represented by a large number of examples whereas the other is represented by only a few.

A number of solutions to the class imbalance problem have been proposed both at the data- and algorithmic-level [20] [10] [19]. Data-level solutions include many different forms of resampling such as random oversampling with replacement, random undersampling, directed oversampling, directed undersampling, oversampling with informed generation of new samples, and combinations of the above techniques. To counteract the class imbalance, algorithmic-level solutions include adjusting the costs of the various classes, adjusting the decision threshold, and adopting recognition-based, rather than discrimination-based, learning. Hybrid approaches have also been used to deal with the class imbalance problem.

3   Progressive Filtering

Progressive Filtering (PF) is a simple categorization technique framed within the local classifier per node approach, which admits only binary decisions. In PF, each classifier is entrusted with deciding whether the input in hand can be forwarded or not to its children. The first proposals in which sequential boolean decisions are applied in combination with local classifiers per node can be found in [12], [13], and [29]. In [31], the idea of mirroring the taxonomy structure through binary classifiers is clearly highlighted; the authors call this technique "binarized structured label learning".

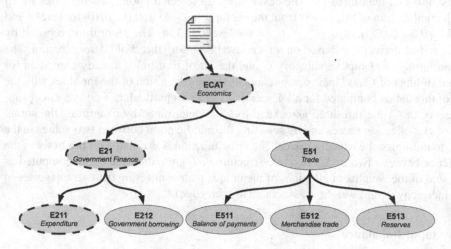

Fig. 1. An example of PF (highlighted with bold-dashed lines)

In PF, given a taxonomy, where each node represents a classifier entrusted with recognizing all corresponding positive inputs (i.e., interesting documents), each input traverses the taxonomy as a "token", starting from the root. If the current classifier recognizes the token as positive, it is passed on to all its children (if any), and so on. A typical result consists of activating one or more branches within the taxonomy, in which the corresponding classifiers have accepted the token. Figure 1 gives an example of how PF works. A theoretical study of the approach is beyond the scope of this chapter, the interested reader could refer to [4] for further details.

3.1   The Approach

A simple way to implement PF consists of unfolding the given taxonomy into pipelines of classifiers, as depicted in Figure 2. Each node of the pipeline is a binary classifier able to recognize whether or not an input belongs to the corresponding class (i.e., to the corresponding node of the taxonomy). Partitioning the taxonomy in pipelines gives rise to a set of new classifiers, each represented by a pipeline.

Finally, let us note that the implementation of PF described in this chapter performs a sort of "flattening" though *preserving* the information about the hierarchical relationships embedded in a pipeline. For instance, the pipeline $\langle ECAT, E21, E211 \rangle$ actually

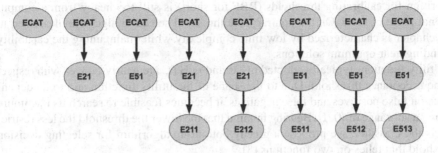

Fig. 2. The pipelines corresponding to the taxonomy in Figure 1

represents the classifier $E211$, although the information about the existing subsumption relationships are preserved (i.e., $E211 \prec E21 \prec ECAT$, where "$\prec$" denotes the usual covering relation).

3.2 The Adopted Threshold Selection Algorithm

According to classical text categorization, given a set of documents D and a set of labels C, a function $CSV_i : D \rightarrow [0, 1]$ exists for each $c_i \in C$. The behavior of c_i is controlled by a threshold θ_i, responsible for relaxing or restricting the acceptance rate of the corresponding classifier. Let us recall that, with $d \in D$, $CSV_i(d) \geq \theta_i$ is interpreted as a decision to categorize d under c_i, whereas $CSV_i(d) < \theta_i$ is interpreted as a decision not to categorize d under c_i.

In PF, we assume that CSV_i exists, with the same semantics adopted by the classical setting. Considering a pipeline π, composed by n classifiers, the acceptance policy strictly depends on the vector of thresholds $\theta = \langle \theta_1, \theta_2, \ldots, \theta_n \rangle$ that embodies the thresholds of all classifiers in π. In order to categorize d under π, the following constraint must be satisfied: for $k = 1 \ldots n$, $CSV_i(d) \geq \theta_k$. On the contrary, d is not categorized under c_i in the event that a classifier in π rejects it. Let us point out that we allow different behaviors for a classifier, depending on which pipeline it is embedded in. As a consequence, each pipeline can be considered in isolation from the others. For instance, given $\pi_1 = \langle ECAT, E21, E211 \rangle$ and $\pi_2 = \langle ECAT, E21, E212 \rangle$, the classifier $ECAT$ is not compelled to have the same threshold in π_1 and in π_2 (the same holds for $E21$).

In PF, given a utility function[1], we are interested in finding an effective and computationally "light" way to reach a sub-optimum in the task of determining the best vector of thresholds. Unfortunately, finding the best acceptance thresholds is a difficult task. In fact, exhaustively trying each possible combination of thresholds (brute-force approach) is unfeasible, the number of thresholds being virtually infinite. However, the brute-force approach can be approximated by defining a granularity step that requires to check only a finite number of points in the range $[0, 1]$, in which the thresholds are permitted to vary with step δ. Although potentially useful, this "relaxed" brute force

---

[1] Different utility functions (e.g., precision, recall, F_β, user-defined) can be adopted, depending on the constraint imposed by the underlying scenario.

algorithm for calibrating thresholds (RBF for short) is still too heavy from a computational point of view. On the contrary, the threshold selection algorithm described in this chapter is characterized by low time complexity while maintaining the capability of finding near-optimum solutions.

Utility functions typically adopted in TC and in HTC are nearly-convex with respect to the acceptance threshold. Due to the shape of the utility function and to its dependence on false positives and false negatives, it becomes feasible to search its maximum around a subrange of $[0, 1]$. Bearing in mind that the lower the threshold the less restrictive is the classifier, we propose a greedy bottom-up algorithm for selecting decision threshold that relies on two functions [3]:

– *Repair* ($\mathcal{R}$), which operates on a classifier C by increasing or decreasing its threshold –i.e., $\mathcal{R}(up, C)$ and $\mathcal{R}(down, C)$, respectively– until the selected utility function reaches and maintains a local maximum.
– *Calibrate* ($\mathcal{C}$), which operates going downwards from the given classifier to its offspring by repeatedly calling $\mathcal{R}$. It is intrinsically recursive and, at each step, calls $\mathcal{R}$ to calibrate the current classifier.

Given a pipeline $\pi = \langle C_1, C_2, \ldots, C_L \rangle$, TSA is defined as follows (all thresholds are initially set to 0):

$$TSA(\pi) := for\ k = L\ downto\ 1\ do\ \mathcal{C}(up, C_k) \tag{1}$$

which asserts that $\mathcal{C}$ is applied to each node of the pipeline, starting from the leaf ($k = L$).

Under the assumption that p is a structure that contains all information about a pipeline, including the corresponding vector of thresholds and the utility function to be optimized, the pseudo-code of TSA is:

```
function TSA(p:pipeline):
    for k := 1 to p.length do p.thresholds[i] = 0
    for k := p.length downto 1 do Calibrate(up,p,k)
    return p.thresholds
end TSA
```

The *Calibrate* function is defined as follows:

$$\begin{aligned}
\mathcal{C}(up, C_k) &:= \mathcal{R}(up, C_k), \quad k = L \\
\mathcal{C}(up, C_k) &:= \mathcal{R}(up, C_k); \mathcal{C}(down, C_{k+1}), \quad k < L
\end{aligned} \tag{2}$$

and

$$\begin{aligned}
\mathcal{C}(down, C_k) &:= \mathcal{R}(down, C_k), \quad k = L \\
\mathcal{C}(down, C_k) &:= \mathcal{R}(down, C_k); \mathcal{C}(up, C_{k+1}), \quad k < L
\end{aligned} \tag{3}$$

where ";" denotes a sequence operator, meaning that in "$a;b$" action a is performed *before* action b.

The pseudo-code of *Calibrate* is:

```
function Calibrate(dir:{up,down}, p:pipeline, level:integer):
   Repair(dir,p,level)
   if level < p.length
     then Calibrate(toggle(dir),p,level+1)
 end Calibrate
```

where *toggle* is a function that reverses the current direction (from *up* to *down* and vice versa). The reason why the direction of threshold optimization changes at each call of *Calibrate* (and hence of *Repair*) lies in the fact that increasing the threshold θ_{k-1} is expected to forward less FP to C_k, which allows to decrease θ_k. Conversely, decreasing the threshold θ_{k-1} is expected to forward more FP to C_k, which must react by increasing θ_k.

The pseudo-code of *Repair* is:

```
function Repair(dir:{up,down}, p:pipeline, level:integer):
   delta := (dir = up) ? p.delta : -p.delta
   best_threshold := p.thresholds[level]
   max_uf := p.utility_function()
   uf := max_uf
   while uf >= max_uf * p.sf and p.thresholds[level] in [0,1]
      do p.thresholds[level] += delta
         uf := p.utility_function()
         if uf < max_uf then continue
         max_uf := uf
         best_threshold := p.thresholds[level]
   p.thresholds[level] := best_threshold
 end Repair
```

The scale factor ($p.sf$) is used to limit the impact of local minima during the search, depending on the adopted utility function (e.g., a typical value of $p.sf$ for F_1 is 0.8).

It is worth pointing out that, as also noted in [21], the sub-optimal combination of thresholds depends on the adopted dataset, hence the sub-optimal combination of thresholds need to be recalculated for each dataset.

Searching for a sub-optimal combination of thresholds in a pipeline π is characterized by high time complexity. In particular, two sources of intractability hold: (i) the optimization problem that involves the thresholds and (ii) the need of retraining classifiers after modifying thresholds. In this work, we concentrate on the former issue while deciding not to retrain the classifiers. In any case, it is clear that the task of optimizing thresholds requires a solution that is computationally light. To calculate the computational complexity of TSA, let us define a granularity step that requires to visit only a finite number of points in a range $[\rho_{min}, \rho_{max}]$, $0 \leq \rho_{min} < \rho_{max} \leq 1$, in which the thresholds vary with step δ. As a consequence, $p = \lfloor \delta^{-1} \cdot (\rho_{max} - \rho_{min}) \rfloor$ is the maximum number of points to be checked for each classifier in a pipeline. For a pipeline π of length L, the expected running time for TSA, say $T_{TSA}(\pi)$, is proportional to $(L + L^2) \cdot p \cdot (\rho_{max} - \rho_{min})$. This implies that TSA has complexity $O(L^2)$,

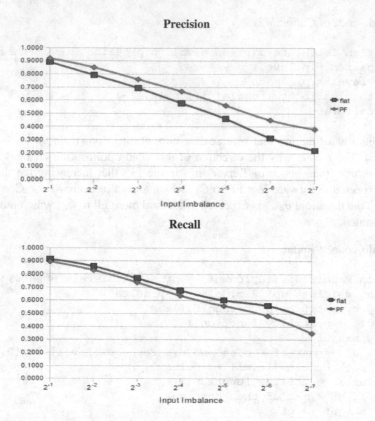

Fig. 3. Comparison of precision and recall between PF and flat classification

i.e., quadratic with the number of classifiers embedded by a pipeline. A comparison between TSA and the brute-force approach is unfeasible, as the elements of the threshold vector are real numbers. However, a comparison between TSA and RBF is feasible, although RBF is still computationally heavy. Assuming that p points are checked for each classifier in a pipeline, the expected running time for RBF, $T_{RBF}(\pi)$, is proportional to p^L, which implies that its computational complexity is $O(p^L)$.

4   Experiments and Results

The Reuters Corpus Volume I (RCV1-v2) [22] has been chosen as benchmark dataset. In this corpus, stories are coded into four hierarchical groups (a fragment of the taxonomy is reported in Figure 1: Corporate/Industrial (CCAT), Economics (ECAT), Government/Social (GCAT), and Markets (MCAT). Although the complete list consists of 126 categories, only some of them have been used to test our hierarchical approach. The total number of codes actually assigned to the data is 93, whereas the overall number of documents is about 803,000. Each document belongs to at least one category and, on average, to 3.8 categories.

Reuters dataset has been chosen because it allows us to perform experiments with pipelines up to level 4 while maintaining a substantial number of documents along the pipeline. To perform experiments, we used, about 2,000 documents per each leaf category. We considered the 24 pipelines of depth 4 yielding a total of about 48,000 documents.

Experiments have been performed on a SUN Workstation with two Opteron 280, 2Ghz+ and 8Gb Ram. The system used to perform benchmarks has been implemented using X.MAS [1], a generic multiagent architecture built upon JADE [5] and devised to make it easier the implementation of information retrieval/filtering applications.

Experiments have been carried out by using classifiers based on the *wk-NN* technology [11], which do not require specific training and are very robust with respect to noisy data. As for document representation, we adopted the bag of words approach, a typical method for representing texts in which each word from a vocabulary corresponds to a feature and a document to a feature vector. After determining the overall sets of features, their values are computed for each document resorting to the well-known TFIDF method. To reduce the high dimensionality of the feature space, we select the features that represent a node by adopting the information gain method.

Experiments have been performed to validate the proposed approach with respect to the impact of PF in the input imbalance. To this end, three series of experiments have been performed: first, performances calculated resorting to PF have been compared with those calculated by resorting to the corresponding flat approach. Then, PF has been tested to assess the improvement of performances while augmenting the pipeline depth. Finally, performances have been calculated in terms of generalization- / specialization- / misclassification-error and unknown-ratio.

Furthermore, to show that the overall performances of PF are not worsened by the adoption of TSA, we performed further experiments, on a balanced dataset of 2000 documents for each class, focused on (i) comparing the running time and F_1 of TSA vs. RBF, and on (ii) comparing TSA with the selected state-of-the-art algorithms, i.e., those proposed by D'Alessio et al. [12], by Ruiz [27], and by Ceci and Malerba [9].

4.1 Experimenting Progressive Filtering in Presence of Input Imbalance

The main issue being investigated is the effectiveness of PF with respect to flat classification. In order to make a fair comparison, the same classification system has been adopted, i.e., a classifier based on the *wk*-NN technology [11]. The motivation for the adoption of this particular technique stems from the fact that it does not require specific training and is very robust with respect to noisy data. In fact, as demonstrate in [33] *wk*-NN-based approaches can reduce the error rate due to robustness against outliers.

During the training activity, first, each classifier is trained with a balanced data set of 1000 documents by using 200 (TFIDF) features selected in accordance with their information gain. For any given node, the training set contains documents taken from the corresponding subtree and documents of the sibling subtrees –as positive and negative examples, respectively. Then, the best thresholds are selected. Both the thresholds of the pipelines and of the flat classifiers have been chosen by adopting F_1 as utility function[2]. As for pipelines, we used a step δ of 10^{-4} for TSA.

---

[2] The utility function can be adopted depending on the constraint imposed by the given scenario. For instance, F_1 is suitable if one wants to give equal importance to precision and recall.

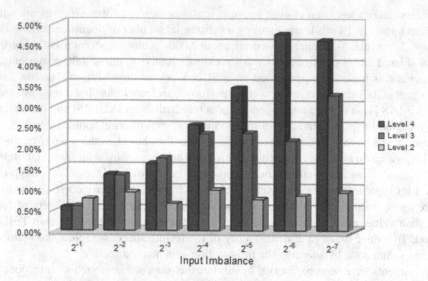

Fig. 4. Performance improvement

Experiments have been performed by assessing the behavior of the proposed hierarchical approach in presence of different ratios of positive examples versus negative examples, i.e., from 2^{-1} to 2^{-7}. We considered only pipelines that end with a leaf node of the taxonomy. Accordingly, for the flat approach, we considered only classifiers that correspond to a leaf.

PF vs Flat Classification. Figure 3 shows macro-averaging of precision and recall. Precision and recall have been calculated for both the flat classifiers and the pipelines by varying the input imbalance. As pointed out by experimental results (for precision), the distributed solution based on pipelines has reported better results than those obtained with the flat model. On the contrary, results on recall are worse than those obtained with the flat model.

Improving Performance along the Pipeline. Figure 4 shows the performance improvements in terms of F_1 of the proposed approach with respect to the flat one. The improvement has been calculated in percentage with the formula $(F_1(pipeline) - F_1(flat)) \times 100$. Experimental results –having the adopted taxonomies a maximum depth of five– show that PF performs always better than the flat approach.

Hierarchical Metrics. Figure 5 depicts the results obtained varying the imbalance. Analyzing the results, it is easy to note that the generalization-error and the misclassification-error grow with the imbalance, whereas the specialization-error and the unknown-ratio decrease. As for the generalization-error, it depends on the overall number of false negatives (FNs), the greater the imbalance the greater the amount of FNs. Hence, the generalization-error increases with the imbalance. In presence of input imbalance, the trend of the generalization-error is similar to the trend of the recall measure. As for the specialization-error, it depends on the overall number of false positives (FPs), the greater the imbalance, the lower the amount of FPs. Hence, the specialization-error

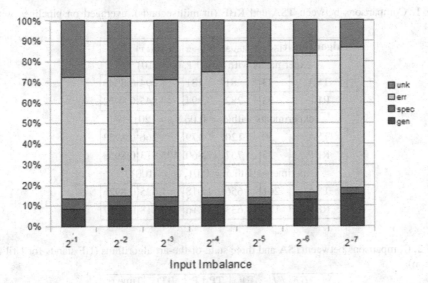

Fig. 5. Hierarchical measures

decreases with the imbalance. In presence of input imbalance, the trend of the specialization-error is similar to the trend of the precision. As final remark, let us note that different utility functions can be adopted, depending on which aspect or unwanted effect one wants to improve or mitigate. For instance, recall could be adopted as utility function while trying to reduce the number of FNs. This is due to the fact that recall optimization is biased against FNs (with the typical unwanted effect of increasing FPs). In this way, the unknown-ratio (which depends on FNs) decreases, while the misclassification-error (which depends on FPs) increases.

4.2 Comparative Experiments of Threshold Selection Strategies

TSA vs. RBF. Experiments performed to compare TSA with RBF have been carried out calculating the time (in milliseconds) required to set the optimal vector of thresholds for both algorithms, i.e., the one that reaches the optimal value in term of F_1, the selected utility function. Different values of δ (i.e., $0.1, 0.05, 0.01$) have been adopted to increment thresholds during the search. Each pair of rows in Table 1 reports the comparison in terms of the time spent to complete each calibrate step ($t_{lev4} \dots t_{lev1}$), together with the corresponding F_1. Results clearly show that the cumulative running time for RBF tends to rapidly become intractable[3], whereas the values of F_1 are comparable.

TSA vs. State-of-the-art Algorithms. As already pointed out, to compare TSA we considered the algorithms proposed by D'Alessio et al. [12], by Ruiz [27], and by Ceci and Malerba [9]. We used $\delta = 10^{-3}$ for TSA. Let us note that Ruiz uses the same threshold value for level 3 and level 4, whereas we let its algorithm to search on the entire space of thresholds. In so doing, the results in terms of utility functions cannot be

---

[3] Note that 1.9E+8 millisecond are about 54.6 hours.

Table 1. Comparisons between TSA and RBF (in milliseconds), averaged on pipelines with $L = 4$

Algorithm	t_{lev4}	t_{lev3}	t_{lev2}	t_{lev1}	F_1
experiments with $\delta = 0.1, (p = 10)$					
TSA	33	81	131	194	0.8943
RBF	23	282	3,394	43,845	0.8952
experiments with $\delta = 0.05, (p = 20)$					
TSA	50	120	179	266	0.8953
RBF	35	737	17,860	405,913	0.8958
experiments with $\delta = 0.01, (p = 100)$					
TSA	261	656	1,018	1,625	0.8926
RBF	198	17,633	3.1E+6	1.96E+8	0.9077

Table 2. Comparisons between TSA and three state-of-the-art algorithms (UF stands for Utility Function)

UF: F1	F1	TP-FP	TD	Time (s)
TSA	**0.9080**	814.80	532.24	1.74
Ceci & Malerba	**0.0927**	801.36	545.44	0.65
Ruiz	**0.8809**	766.72	695.32	29.39
D'Alessio et al.	**0.9075**	812.88	546.16	14.42
UF: TP-FP	F1	TP-FP	TD	Time (s)
TSA	0.9050	**813.36**	521.48	1.2
Ceci & Malerba	0.9015	**802.48**	500.88	1.14
Ruiz	0.8770	**764.08**	675.20	24.4
D'Alessio at al.	0.9065	**812.88**	537.48	11.77
UF: TD	F1	TP-FP	TD	Time (s)
TSA	0.8270	704.40	**403.76**	1.48
Ceci & Malerba	0.8202	694.96	**404.96**	0.62
Ruiz	0.7807	654.72	**597.32**	26.31
D'Alessio et al.	0.8107	684.78	**415.60**	13.06

worse than those calculated by means of the original algorithm. However, the running time is an order of magnitude greater than the original algorithm.

For each algorithm, we performed three sets of experiments in which a different utility function has been adopted. The baseline of our experiments is a comparison among the four algorithms. In particular, we used: $F1$, according to the metric adopted in a previous work on PF [2] and in [27]; $TP - FP$, according to the metric adopted in [12]; and TD, according to the metric adopted in [9].

As reported in Table 2, for each experimental setting, we calculated the performance and the time spent for each selected metric. Table 2 summarizes the results. For each experimental setting, the most relevant results (highlighted in bold in the table) correspond to the metric used as utility function. As shown, TSA always performs better in terms of $F1$, $TP - FP$, and TD. As for the running time, the algorithm proposed by Ceci and Malerba shows the best performance.

5 Conclusions

In this chapter, we made experiments on PF to investigate how the ratio between positive and negative examples affects the performances of a classifier system and how these performances can be improved by adopting PF instead of a classical flat approach. Results show that the proposed approach is able to deal with high imbalance between negative and positive examples.

Acknowledgements. This research was partly sponsored by the Autonomous Region of Sardinia (RAS), through a grant financed with the "Sardinia POR FSE 2007-2013" funds and provided according to the L.R. 7/2007 "Promotion of the Scientific Research and of the Technological Innovation in Sardinia".

References

1. Addis, A., Armano, G., Vargiu, E.: From a generic multiagent architecture to multiagent information retrieval systems. In: AT2AI-6, Sixth International Workshop, From Agent Theory to Agent Implementation, pp. 3–9 (2008)
2. Addis, A., Armano, G., Vargiu, E.: Assessing progressive filtering to perform hierarchical text categorization in presence of input imbalance. In: Proceedings of International Conference on Knowledge Discovery and Information Retrieval, KDIR 2010 (2010)
3. Addis, A., Armano, G., Vargiu, E.: A Comparative Experimental Assessment of a Threshold Selection Algorithm in Hierarchical Text Categorization. In: Clough, P., Foley, C., Gurrin, C., Jones, G.J.F., Kraaij, W., Lee, H., Mudoch, V. (eds.) ECIR 2011. LNCS, vol. 6611, pp. 32–42. Springer, Heidelberg (2011)
4. Armano, G.: On the progressive filtering approach to hierarchical text categorization. Tech. rep., DIEE - University of Cagliari (2009)
5. Bellifemine, F., Caire, G., Greenwood, D. (eds.): Developing Multi-Agent Systems with JADE (Wiley Series in Agent Technology). John Wiley and Sons (2007)
6. Bennett, P.N., Nguyen, N.: Refined experts: improving classification in large taxonomies. In: SIGIR 2009: Proceedings of the 32nd international ACM SIGIR conference on Research and development in information retrieval, pp. 11–18. ACM, New York (2009)
7. Brank, J., Mladenic, D., Grobelnik, M.: Large-scale hierarchical text classification using svm and coding matrices. In: Large-Scale Hierarchical Classification Workshop (2010)
8. Ceci, M., Malerba, D.: Hierarchical Classification of HTML Documents with WebClassII. In: Sebastiani, F. (ed.) ECIR 2003. LNCS, vol. 2633, pp. 57–72. Springer, Heidelberg (2003)
9. Ceci, M., Malerba, D.: Classifying web documents in a hierarchy of categories: a comprehensive study. Journal of Intelligent Information Systems 28(1), 37–78 (2007)
10. Chawla, N.V., Bowyer, K.W., Hall, L.O., Kegelmeyer, W.P.: SMOTE: Synthetic minority over-sampling technique. Journal of Artificial Intelligence Research 16, 321–357 (2002)
11. Cost, R.S., Salzberg, S.: A weighted nearest neighbor algorithm for learning with symbolic features. Machine Learning 10, 57–78 (1993)
12. D'Alessio, S., Murray, K., Schiaffino, R.: The effect of using hierarchical classifiers in text categorization. In: Proceedings of of the 6th International Conference on Recherche dInformation Assiste par Ordinateur (RIAO), pp. 302–313 (2000)
13. Dumais, S.T., Chen, H.: Hierarchical classification of Web content. In: Belkin, N.J., Ingwersen, P., Leong, M.K. (eds.) Proceedings of 23rd ACM International Conference on Research and Development in Information Retrieval, SIGIR 2000, pp. 256–263. ACM Press, New York (2000)

14. Esuli, A., Fagni, T., Sebastiani, F.: Boosting multi-label hierarchical text categorization. Inf. Retr. 11(4), 287–313 (2008)
15. Gaussier, É., Goutte, C., Popat, K., Chen, F.: A Hierarchical Model for Clustering and Categorising Documents. In: Crestani, F., Girolami, M., van Rijsbergen, C.J.K. (eds.) ECIR 2002. LNCS, vol. 2291, pp. 229–247. Springer, Heidelberg (2002)
16. Japkowicz, N.: Learning from imbalanced data sets: a comparison of various strategies. In: AAAI Workshop on Learning from Imbalanced Data Sets (2000)
17. Koller, D., Sahami, M.: Hierarchically classifying documents using very few words. In: Fisher, D.H. (ed.) Proceedings of 14th International Conference on Machine Learning, ICML 1997, pp. 170–178. Morgan Kaufmann Publishers, San Francisco (1997)
18. Kotsiantis, S., Kanellopoulos, D., Pintelas, P.: Handling imbalanced datasets: a review. GESTS International Transactions on Computer Science and Engineering 30, 25–36 (2006)
19. Kotsiantis, S., Pintelas, P.: Mixture of expert agents for handling imbalanced data sets. Ann Math. Comput. Teleinformatics 1, 46–55 (2003)
20. Kubat, M., Matwin, S.: Addressing the curse of imbalanced training sets: One-sided selection. In: Proceedings of the Fourteenth International Conference on Machine Learning, pp. 179–186. Morgan Kaufmann (1997)
21. Lewis, D.D.: Evaluating and optimizing autonomous text classification systems. In: SIGIR 1995: Proceedings of the 18th Annual International ACM SIGIR Conference on Research and Development in Information Retrieval, pp. 246–254. ACM, New York (1995)
22. Lewis, D.D., Yang, Y., Rose, T., Li, F.: RCV1: A new benchmark collection for text categorization research. Journal of Machine Learning Research 5, 361–397 (2004)
23. McCallum, A.K., Rosenfeld, R., Mitchell, T.M., Ng, A.Y.: Improving text classification by shrinkage in a hierarchy of classes. In: Shavlik, J.W. (ed.) Proceedings of 15th International Conference on Machine Learning, ICML 1998, pp. 359–367. Morgan Kaufmann Publishers, San Francisco (1998)
24. Mladenic, D., Grobelnik, M.: Feature selection for classification based on text hierarchy. In: Text and the Web, Conference on Automated Learning and Discovery CONALD 1998 (1998)
25. Rousu, J., Saunders, C., Szedmak, S., Shawe-Taylor, J.: Learning hierarchical multi-category text classification models. In: ICML 2005: Proceedings of the 22nd International Conference on Machine Learning, pp. 744–751. ACM, New York (2005)
26. Ruiz, M.E., Srinivasan, P.: Hierarchical text categorization using neural networks. Information Retrieval 5(1), 87–118 (2002)
27. Ruiz, M.E.: Combining machine learning and hierarchical structures for text categorization. Ph.D. thesis, supervisor-Srinivasan, Padmini (2001)
28. Sebastiani, F.: Machine learning in automated text categorization. ACM Computing Surveys (CSUR) 34(1), 1–55 (2002)
29. Sun, A., Lim, E.: Hierarchical text classification and evaluation. In: ICDM 2001: Proceedings of the 2001 IEEE International Conference on Data Mining, pp. 521–528. IEEE Computer Society, Washington, DC, USA (2001)
30. Weigend, A.S., Wiener, E.D., Pedersen, J.O.: Exploiting hierarchy in text categorization. Information Retrieval 1(3), 193–216 (1999)
31. Wu, F., Zhang, J., Honavar, V.: Learning Classifiers using Hierarchically Structured Class Taxonomies. In: Zucker, J.-D., Saitta, L. (eds.) SARA 2005. LNCS (LNAI), vol. 3607, pp. 313–320. Springer, Heidelberg (2005)
32. Wu, G., Chang, E.Y.: Class-boundary alignment for imbalanced dataset learning. In: ICML 2003 Workshop on Learning from Imbalanced Data Sets, pp. 49–56 (2003)

33. Takigawa, Y., Hotta, S., Kiyasu, S., Miyahara, S.: Pattern classification using weighted average patterns of categorical k-nearest neighbors. In: Proceedings of the 1th International Workshop on Camera-Based Document Analysis and Recognition, pp. 111–118 (2005)
34. Yan, R., Liu, Y., Jin, R., Hauptmann, A.: On predicting rare classes with svm ensembles in scene classification. In: Proceedings of 2003 IEEE International Conference on Acoustics, Speech, and Signal Processing, (ICASSP 2003), vol. 3, pp. III-21–III-4 (April 2003)
35. Yang, Y.: An evaluation of statistical approaches to text categorization. Information Retrieval 1(1/2), 69–90 (1999)

Semantic Web Search System Founded on Case-Based Reasoning and Ontology Learning

Hajer Baazaoui-Zghal[1], Nesrine Ben Mustapha[1], Manel Elloumi-Chaabene[1],
Antonio Moreno[2], and David Sanchez[2]

[1] Riadi Lab., ENSI Campus Universitaire de la Manouba, 2010 Tunis, Tunisie
[2] ITAKA Research Group, Departament d'Enginyeria Informatica i Matematiques
Universitat Rovira i Virgili, Av. Paisos Catalans, 26. 43007, Tarragona, Spain
{hajer.baazaouizghal,nesrine.benmustapha}@riadi.rnu.tn
{antonio.moreno,david.sanchez}@urv.cat
manel.elloumichaabene@gmail.com

Abstract. With the continuous growth of data volume on the Web, the search for information has become a challenging task. Ontologies are used to improve the accuracy of information retrieval from the web by incorporating a degree of semantic analysis during the search. However, manual ontology building is time consuming. An automatic approach may aid to solve this problem by analyzing implicitly available knowledge such as the users' search feedback. In this context, we propose a semantic web search system founded on Case-Based-Reasoning (CBR) and ontology learning that aims to enrich automatically the ontologies by using previous search queries performed by the user. Some experiments and results obtained with the proposed system are also presented, which show an improvement on the precision of the Web search and ontology enrichment.

Keywords: Semantic web search, Ontologies, Ontology learning, Case-Based-Reasoning.

1 Introduction

With the definition of the semantic web, ontologies have been used to improve web information retrieval by exploiting modelled knowledge during the search process. The need of domain ontologies in Information Retrieval (IR) has been explored by some approaches to better answer users' queries. The specific representation of knowledge is made through ontologies that are an explicit specification and a formal shared conceptualization. Ontologies have contributed to the emergence of semantic search engines. They are used to enhance query formulation, document indexing and conceptual classification of results.

However, the problem of query formulation is becoming a challenging point. A good query formulation must include the necessary features to retrieve the relevant information. For this reason, adaptation and learning strategies have long been viewed as crucial parts of IR systems. In this work, we rely on the *Case-Based reasoning* (CBR) model to improve Web IR. In fact, Case-based Reasoning (CBR) is a methodology for problem solving reusing previous experience [1] and also for collecting new experience

A. Fred et al. (Eds.): IC3K 2010, CCIS 272, pp. 72–86, 2013.

since every new problem case, once solved, becomes a new case that may be stored and reused. CBR presents the advantage of its easy exploitation with respect to other learning methods.

Given that the main problems of ontology-based IR are the reformulation of the users' queries and the difficulties to build and manage ontologies, the idea of embedding the CBR mechanism in an ontology-based IR system can be a promising approach to improve query reformulation, attaining good precision in the search results. On the other hand, ontologies are available on the Web, but it is still difficult to find one for every domain associated to a user's query. However, it is obvious that building ontologies for all domains, so that they can be exploited by semantic search engines, is difficult. So, using previous experiences of users (CBR) can help to perform ontology learning.

Our objective is to develop a semantic search system based on the use of many domain ontologies in order to meet any user's need. In fact, our motivation is to use past and current queries to improve the precision of retrieved results for the following users and to index Web documents using submitted queries and discovered ontology elements from these documents.

The rest of the paper is organized as follows. Section 2 presents an overview of related works in the area of semantic search and ontology learning. The third section describes the proposed Semantic Web Search System founded on CBR and Ontology Learning. The last section presents some evaluations and tests. Finally, we conclude and discuss directions for future research.

2 Related Works

In this section we present an overview of related works in the areas of ontology learning and CBR for semantic search systems.

2.1 Ontology Learning for Semantic Search

Ontologies [2] are used to represent a shared conceptualization of a knowledge domain, adding a semantic layer to computer systems. In other words, an ontology is a formal representation of the main concepts in a domain and their interrelationships. Ontologies are often specified in a declarative form by using semantic markup languages [3] such as RDF [4], OIL [5], XOL [6] and OWL [7]. Several types of ontologies exist, like generic ontology, application ontology, domain ontology, modular ontology, etc. Ontology building tools provide automatic or semi-automatic support for the construction of ontologies.

Ontology Learning (OL) aims at building ontologies from knowledge sources using a set of machine learning techniques and knowledge acquisition methods. OL from texts has been widely used in the knowledge engineering community since texts are semantically richer than the other data sources. The corpus should be representative of the domain used to build the ontology. By applying a set of text mining techniques, a granular ontology is enriched with concepts and relationships discovered from textual data. Using a set of techniques, we try to project knowledge contained in texts to the ontology by extracting concepts and relations. We distinguish mainly four categories of approaches:

– Pattern-based approaches [8][9]: a relation between a pair of concepts is recognized when a sequence of words in the text matches a pattern implicitly representing a semantic relationship (e.g. typically taxonomical).
– Conceptual clustering [10]: concepts are grouped according to the semantic distance between each other to make up hierarchies. Right now, there are still several problems in using this method which restrict its usability [11] as its inefficiency in high dimensional spaces.[12].
– Association rule-based techniques [13].
– Hybrid approaches [14], combining several of the above techniques.

In these approaches, human intervention is commonly required to validate the relevance of learned concepts and relationships.

In the last decade, with the enormous growth of Web information, the Web has become an important source of information for knowledge acquisition. Its main advantages are its huge size and its large degree of heterogeneity.

OL from Web documents requires the same techniques as those used for ontology extraction from texts. Several approaches are based on eliminating tags from documents to obtain plain texts to which traditional text mining texts could be applied. These approaches are dedicated to ontology building from the Web and are based on the generation of taxonomies without using a priori knowledge or natural language processing techniques.

In [15], an incremental approach for ontology learning from Web is proposed. A study of several types of available Web search engines and how they can be used to assist the learning process (searching Web resources and computing IR measures) is described. The proposed learning process is based on four steps. The first one is a Taxonomic learning step where the user starts by specifying keywords used as a seed for the learning process using a Web search engine. The output of this step is one-level taxonomy and a set of verbs appearing in the same context as the extracted concepts. Secondly, non-taxonomic relations learning are carried out. The verbs and keywords lists are used as bootstrap for building domain-related patterns as well as for defining queries addressed to a search engine. The third step is the recursive learning task where the two previous learning stages are recursively executed for each discovered concept. Finally, the post-processing step consists in refining and evaluating the obtained ontology. This approach is domain-independent and incremental.

In the same context, our previous work [16] has proposed an incremental approach of ontology learning from Web. We combined several text mining techniques and used an ontology-based IR System to classify Web documents. Our experiments have shown that ontology enrichment from documents resulting from an ontology based-search system is more accurate and that the relevance of search results is improved. So, our objective is to integrate ontology learning within semantic search systems.

2.2   CBR for Search Systems

Query formulation for information retrieval accuracy improvement is a challenging task. A good query formulation method must include all the necessary features to retrieve the relevant information, which is not an obvious task [17]. Thus, the use of an iterative process of trial and error is necessary to improve query formulation. For this reason, adaptation and learning have long been viewed as crucial parts of IR systems [18]. Therefore, using a CBR model in an IR process is a promising idea.

CBR is a problem-solving method [19] based on the concept of "case", which is the description of a problem and its solution. The main idea under CBR consists in storing experiences as cases and problem-solving processes as instances of cases. When a new problem is found, the system uses the relevant past stored cases to interpret or to solve it [17]. The system performance increases with the growth of stored cases. In [20], many approaches have investigated the use of the CBR model to overcome classic IR problems. Given that CBR is based on retrieving similar cases by using similarity measures between terms, it is obvious that semantic knowledge can improve the results of this process. In the same context, the idea of combining ontologies (domain knowledge) with CBR-based systems for knowledge management has been proposed by many authors [21][22][23].

Using a CBR mechanism in an ontology-based IR system may improve semantic indexing and query reformulation in a question-answer service [24], improve recommendations results [25] and obtain good precision of search results [26]. Therefore, our underlying hypothesis is that CBR supported by ontology technology is a promising approach for achieving semantic-aware search and ontology learning. In fact, our motivation is to use past and current queries to improve the precision of provided results to the next users and to index Web documents using submitted queries and discovered ontology elements from these documents.

3 Semantic Web Search System Based on CBR and Ontology Learning

To overcome the problems introduced in section 2, we have developed a semantic search system based on CBR and ontology learning. In this section, we present the generic approach [27], which allows any search engine to develop its semantic layer by building ontologies and indexing a document base. Our contribution here consists on integrating CBR with ontologies to improve the semantic search and add an automatic ontology learning component.

3.1 Repository of Ontology Fragments

We have proposed in previous works the extensive use of domain ontologies [16]. Our choice is motivated by the fact that domain ontologies are restricted to the concepts related to a specific knowledge field. This has the advantage of limiting the ambiguity of terms defined in the ontology, facilitating their detection in documents. We can represent a relationship between domain ontologies and ontology fragments. In fact, a domain ontology consists on a set of ontology fragments. Two domain ontologies can share the same modular ontology. For example, we can define two domains such as "Tourism" and "Computer Science", which can contain a shared ontology module like "Conference".

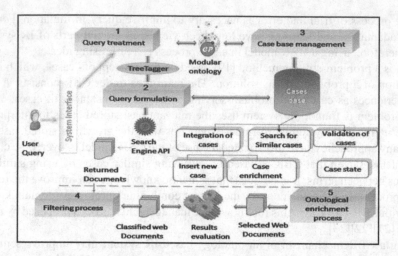

Fig. 1. Semantic Web Search System founded on CBR and Ontology Learning Modules

3.2 Proposed System Description

The proposed system is composed of four main modules (see Figure 1). The first module is responsible of the treatment and reformulation of the users' queries, and the second one manages the case base, using ontologies. After that, the system filters and classifies the documents selected by the users. Finally, the system performs an ontology enrichment process based on the analysis of the text of the selected Web documents. The first step consists on processing the user's search query in order to match it with a topic covered in the ontology (module). The choice of the module from the ontology can help to retrieve the exact meaning of the search using WordNet[1][28]. Then, the user query is processed by the query treatment module. In the case that the search topic is new, the system provides an initial list of Web documents. If a similar case (search) is found in the case base, the system displays a list of already retrieved documents containing relevant results and adds the list from the search engine. Then, these documents are classified by means of the filtering process. Finally, the documents that are more similar to the query are displayed to the user. After this, the reformulation query module uses the case base (more precisely, the semantic signature) to provide a new query. The reformulated query allows launching a new search instance to refine the results. The final task of the system is the enrichment of the ontology from the Web documents assessed as relevant, using text-based Ontology Learning techniques.

3.3 User Query Reformulation

The *Query Reformulation* step aims to extend the original query with additional information to obtain more accurate results, in order to solve the difficulties faced by the users when constructing their queries. Our system is based on the enrichment of the

---

[1] *WordNet* is an on-line lexical reference system where English nouns, verbs, and adjectives are organized into synonym sets, each representing one underlying lexical concept.

query using the information available from past cases. The submitted query is extended on the basis of the concepts and relations of the domain ontology. These concepts and relations represent the semantic signature of the pivotal concept, which is a list of terms that appear frequently with the chosen concept. More precisely, two tasks are performed to reformulate the query:

- *Morphological analysis*: It allows the recognition of different forms of words from a lexicon (dictionary or thesaurus). Stemming allows the transformation of conjugate or flexed forms to their canonical form or lemma. In our approach the stemming of the query is performed by tree tagger.
- *Adding terms from semantic signature*: Each query is referenced by the semantic signature of a case, whose terms can be used to enrich the query. They represent the concepts used more often for the search topic and the given query. Each term is described by a weight calculated from the relevant documents of the case.

3.4 Ontology-Based CBR

The combination of ontologies (as semantic background) and the CBR mechanism (to enrich the ontology from search feedback) can improve the performance of semantic Web search. However, it faces the difficulties of knowledge modeling from textual data and the representation of search cases.

Case Modeling. Typically, a case contains at least two parts: a description of a situation representing a "problem", and a "solution" used to remedy this situation. We have modeled a case with the structure shown in Figure 2.

- **Problem Modeling.** A problem is composed of five parts [16]. The first one represents the *search goal*. We can distinguish two types of search goals [29]. *The navigational goal* is a desire by the user to be taken to the home page of an institution or organization. To be considered navigational, the query must have a single authoritative Web site that the user already has in mind. For this reason, most queries consisting of names of companies, universities, or well-known organizations are considered navigational. Also for this reason, most queries for people - including celebrities - are not. *The informational goal* includes the desire to locate something in the real world, or simply to get a list of suggestions for further research. Informational queries are focused on the user goal of obtaining information about the query topic. This category includes goals for answering questions that the user has in mind. The second part includes the *domain search*. It is referred by a concept from the domain ontology which aims to define the topics of the search. The domain search aims to restrict the user search on a specific field. We have also modeled a *pivotal concept* for the chosen ontological module. This concept is the most significant description of the user's need or query. It is the most interesting concept from the domain ontology. We have included a *set of queries* classified by a record. This part list the set of queries similar to the user query. The last part is *the semantic signature* of the pivotal concept, which is a list of terms that appear frequently with the chosen concept.

- **Solution Modeling.** The case solution contains a *module vector*, used for filtering documents of the current search and the similar cases. This vector is represented as a n-tuple $(W_1C_1, W_2C_2, W_3C_3..., W_nC_n)$, where each tuple (W_i, C_i) represents a concept

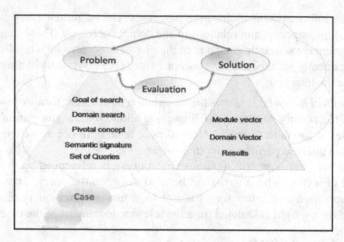

Fig. 2. Case structure

of the ontology module and its associated weight. The case solution also holds a *domain vector*, similar to the module vector, which contains the domain concepts and their weights. Finally, we have modeled a *results* part to collect the search results which are selected by the users. These results could be Web documents, ontological classes or text fragments, depending on the search goal.

In the example given above Table 1, we can see two use cases representing the semantic relation between two queries Q1 and Q2 that appear as two independent requests submitted by two users but, according to the domain knowledge, the result of one is the answer to the other. The query Q1 is "*What is an EMS in Computer Science?*" and Q2 is "*How to make an online meeting or workshop?*".

The difference between the two queries is that the first user knows exactly the concept he is searching for ("EMS") while the second user knows only the role of this concept (without knowing the fact that an EMS is a tool to make online meetings or workshops). The use of a traditional search engine to answer Q2 will lead to the difficulty of discovering that EMSs are used to make online meetings or workshops, as "EMS" does not appear as a term in the second query. But, according to the proposed system, the ontological index of cases can provide the second user with a similar case existing in the case base. By applying our proposal, the system has noticed that there is a common result between the two queries. However, by using the search engine, we wouldn't be able to find the same result. In fact, the retrieval of cases similar to query Q2 provides the most ranked recommended cases.

Procedures Applied to a Case. We present in the following paragraphs the different procedures that allow the enrichment of the cases.

- Search for Similar Cases. The submission of a query by a user is treated as a case search. The system starts by searching in the case base. Two requests are called similar if they belong to the same search area for the same search goal, are connected to the same ontological concept and include a similar set of terms (i.e., the terms are identical

Table 1. Comapaison of two similar cases

	Case1 modeling	Case2 Modeling(Similar)
Case Index	http://www.SemSearch.com/ Computer_Science/ontology ♯EMS.owl	http://www.SemSearch.com/ Computer_Science/ontology ♯online_meeting.owl
Problem	GS= navigational D=http://www.SemSearch.com/ Computer_Science PC: EMS TopicSign= meeting, conference, online, collaborative... Set_Q= (EMS, (EMS + electronic meeting system))	GS= navigational D=http://www.SemSearch.com/ Computer_Science PC: online_meeting TopicSign= meeting, online,tools,... Set_Q= (make online meeting, workshop making tools, interactive meeting tools,....)
Solution	Module Vector Domain Vector Set_response=http://en.wikipedia. org/wiki/Electronic_meeting _system	Module Vector Domain Vector Set_response= http://www.online- tech-tips.com /cool-websites/free-online-meeting-software/

to the keywords of the original query, or synonyms of them). If a similar case exists in the database, the system executes two processes.

The first is to collect all the relevant documents in the case base related to similar queries. The second is to collect all the documents from the Web related to the user query after treating and enriching the query with the semantic signature. After this document collecting phase, a classification process is triggered to classify all the selected documents and display them to the user. Then, these documents are evaluated by the user then an ontology learning tool analyzes them in order to enrich the ontological module with new concepts and new relationships with other modules. Finally, the system updates the case.

- **Cases update.** Once an existing case in the base case is selected, it is updated in two steps. The first one is the addition of new relevant documents to the case. After processing the user's query, a new class of terms is added to the semantic signature. The next step is to update the module vector and the domain vector, by recalculating the weights of the terms of the semantic signature related to the case. Finally, the system saves the new adapted case into the case base.

- **Insertion of New Cases.** This step involves the insertion of the case problem, which includes the addition of the search goal, the user-selected search field, the pivotal concept and the semantic signature (i.e., a list of terms obtained through the extraction of keywords from the user query). After inserting the case problem, the system returns a set of URLs from the search engine. The user mission is to mark those that he considers relevant. The addition of the module vector and the domain vector is made as soon as the user selection is finished.

Similarity Measures. The next step is crucial for the integration and update of the module and domain vectors.

- **Weighting Concepts by CF-ICF Method.** The CF-ICF measure is a good approximation to determine the importance of a concept in a document, particularly in a uniform size corpus. Here are some formulas to calculate CF-ICF.

CF: (Concept Frequency) is a value proportional to the occurrence and the frequency of a concept in the ontological module and so in the semantic signature.

$$CF_i = \frac{f_{ij}}{\sum_{j=1}^{N} f_{ij}^2} . \tag{1}$$

Where fij is a function normalized to reduce the differences between the values associated with the concepts of the document. It is defined by the following formula

$$f_{ij} = 0,5 + 0,5 \times \frac{cf_{ij}}{max_{c_i \in D_j}(cf_{ij})} . \tag{2}$$

where cf_{ij} is the number of occurrences of the concept c_i in document D_j.

The ICF (Inverse Concept Frequency) is defined as follows

$$ICF_i = Log(\frac{N}{n_i}) . \tag{3}$$

where N is the total number of concepts in the ontology and n_i is the number of documents that contain the concept c_i. The formula below (to obtain W_i) is the combination of CF and ICF:

$$W_i = (1 + Log(CF_i)) * ICF_i . \tag{4}$$

- **Calculation of Module Vector.** This involves the calculation of the weight of each term using the vector model, from the base of relevant documents related to a specific domain stored in the case base.

- **Calculation of Domain Vector.** This metric calculates the weights of all concepts of the domain ontology using the vector model, also from the basic documents stored in the case base.

It can be noticed that CBR is used for many tasks: reformulation of new queries on the basis of previous ones, proposal of recommendations that are represented by similar queries and their results, document classification and enrichment of ontological modules.

3.5   Filtering and Classification of Web Documents

The vector model of Salton [30] is used to retrieve the most relevant documents. We replace the terms by concepts to make it more suitable for our application.

More precisely, we filter the domain vector to eliminate documents outside the area, which do not belong to the same module; therefore, this vector can reduce the search noise. In this vector model each document is represented by a vector which has the following form: $Dj = (d_{1j}, d_{2j}, ..., d_{Nj})$. We can define d_{ij} as the weight of the concept c_i in the document D_j, being N the number of concepts that are in the semantic signature. The query is represented by the vector $Q = (q_1, q_2, ..., q_N)$, where q_i is the weight of the concept c_i in the semantic signature. The measure of similarity between a document and a query is calculated with the cosine formula:

$$Sim(D_j, Q) = \frac{\sum_{i=1}^{N} d_{ij} q_i}{\sqrt{\sum_{i=1}^{N} d_{ij}^2 \bullet \sum_{i=1}^{N} q_i^2}}. \qquad (5)$$

3.6 Ontology Enrichment from Web Documents

The final task of this system is the enrichment of the ontology fragments (associated to an old or new case) from the Web documents making up the solution of this case. Each document will be the input of the ontology learning phase of the process proposed in a previous work [31]. The body of the XML document corresponds to the natural language text and may contain information useful to enrich the ontology. In fact, it may contains terms and relations between them. Text mining techniques (syntactic patterns [8] and verb-based patterns) are used to discover new concepts and new relations between the concepts of the ontology fragment and the topic signature [32]. We chose to use lexico-syntactic patterns to identify semantic relations related to the pivotal concept.

A lexico-syntactic pattern describes a regular expression, consisting of words, with grammatical or semantic categories, and symbols to identify text fragments corresponding to this format. The application of such patterns requires text treatment using various tools (tokenizer, lemma, parser, etc.). The patterns exploit morpho-syntactic or semantic labels assigned by the software.

Our approach exploits document structure and forms new patterns by extracting semantic information from the ontological module. This module will be enriched by semantic relationships with new concepts or modules. It is possible to extract new terms which have not a stable relation with the ontology fragment. In this case, these terms are added to the topic signature of the case (instead of being added to the ontology) in order to be used in a next iteration.

We illustrate a simple scenario using the search system. We present an example in which the user wants to search for "a HP notebook". As it is shown in the last section, our system is composed of four modules. So, the first step consists in the treatment of the query. In figure 3 we present the first iteration, including the primitive query built by the system. The system begins by reformulating the user's query to refine the search. On the one hand, the system searches for similar cases to send their documents (if they exist) to the user, and inserts the module and domain vectors in the case base. Then, the search engine obtains a list of Web documents. This search tries to construct a primitive ontology. On the other hand, after selecting the relevant documents, the system uses text mining techniques to enrich the primitive ontology, by extracting concepts that are relevant to the ontological module from the selected Web documents; in this case the ontological module is the "notebook". After the completion of this scenario, new concepts and instances are added to the module and the query of the user is reformulated and enriched with related terms and concepts as represented in Figure 4.

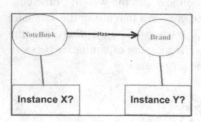
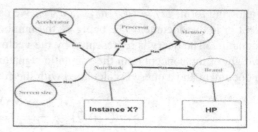

Fig. 3. An example: HP Notebook search   **Fig. 4.** Module enrichment

4   Evaluation

To evaluate the proposed system we computed the precision in the first part and the influence of CBR on ontology enrichment in the second part.

4.1   Search Results Evaluation

We measured the precision of our tool for retrieving Web documents. Several scenarios have been tested. The *first scenario* represents a classical search, which is a keyword search on Google. The *second scenario* represents the situation where there are similar cases in the database. The search is based on the vector model to filter the results. The *third scenario* represents the search for information based on the learning of cases, the query reformulation and the use of the vector model for classification.

The precision measures the rate of relevant documents that are manually assessed by a human domain expert. Its values are calculated for several queries. Indeed, we have calculated the precision for every single query and then we have made the average for several queries.

In Figure 5 we observe that the results show a significant improvement of the relevance of the returned information gradually as the concepts are used in experiments. In the second scenario, we notice a significant increase in the precision rate for the twenty first documents. This can show the impact of the CBR mechanism in our approach. Besides, we remark also a great improvement of the precision for more documents in the third scenario. This result shows the impact of query reformulation and document filtering in the approach. This demonstrates that the implemented ideas contribute significantly to improve the relevance of search results.

4.2   Ontology Enrichment Evaluation

As we have mentioned in section 3.4 a case is modeling a problem, a solution and an evaluation. The problem contains a set of queries. In this test we have fixed a number of ten queries per case and calculated the number of concepts added to the case base.

The evaluation of the relevance of the added concepts is done by an expert. For this reason, we have considered three scenarios. The first scenario presents the added concepts from WordNet. Indeed, we have used WordNet to reformulate the user's query

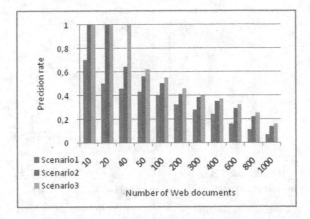

Fig. 5. An average precision for a system search

and disambiguate it. We have relayed on synonyms to disambiguate the meaning of the pivotal concept. Thus, we have added especially synonyms of query terms to the initial query. The second scenario presents the evaluation of the added concepts based on similar cases in the case base, in other words it represents added concepts from similar documents in the case base. This scenario shows the contribution of the CBR system for information retrieval. Finally, the last scenario presents added concepts that are extracted from relevant Web documents using a search engine.

In Figure 6 we can observe a significant improvement with the use of the CBR system and ontology enrichment from text. The relevance of the obtained results highlights the contribution of the semantic search based on case-based-reasoning and ontology enrichment.

In the first scenario, we use only WordNet. There are many relevant concepts in relation to the total of concepts, but the number of concepts is very low to describe the user's search (see Figure 7). In the second scenario we can extract concepts from similar cases. We can observe an increase in the number of relevant concepts in comparison with the first scenario, despite the decrease of the pertinent concepts average in relation to the total of concepts. This can highlight the contribution of CBR in our work (see Figure 8). The third scenario represents the concepts extracted from text (from relevant Web documents). We can notice the great number of concepts retrieved by the proposed system. This number of concepts, especially the relevant ones, demonstrates the contribution of ontology enrichment in our work (see Figure 9). Results highlight the contribution of CBR and the ontology enrichment. However, these results show some limitations, In fact, the documents returned for the user's query and the ontology enrichment are not synchronized, these two processes cannot unfold at the same time. The ontology enrichment process takes time because of the volume of Web documents. That is why we thought to use other ontology techniques based on Web metrics and data returned by the search engine to parallelize the search and the enrichment processes in our future work.

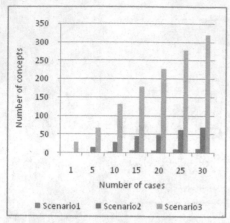

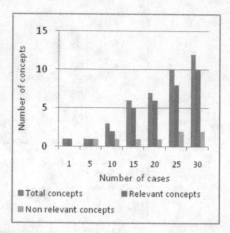

Fig. 6. Average number of added concepts for each scenario

Fig. 7. Concepts relevance for the first scenario

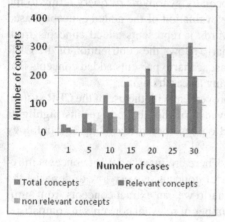

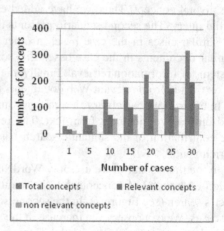

Fig. 8. Concepts relevance for the second scenario

Fig. 9. Concepts relevance for the third scenario

5   Conclusions and Future Work

The challenge addressed by this paper is to develop a system aiming to improve the contextualization of requests based on past users' queries and the enrichment of ontologies from the query context (documents selected by the user). In fact, since intelligent retrieval is one of the main applications of the CBR paradigm, the semantic formalization in CBR systems is becoming an increasing research area. In CBR systems, semantics are the main source of reasoning, similarity calculation and case adaptation.

The main contribution of our work is the ontology fragment building and enrichment within a semantic search engine and linking user queries with ontology fragments using the CBR. The ontology enrichment is based on our previous work that proposes to analyze the whole Web page to learn new concepts and relations [16]. Results obtained by the proposed system have been presented, showing an improvement on the precision of the Web search and ontology enrichment.

As future work, we plan to study especially ontology learning techniques based on Web metrics. On the other hand, we are integrating modularization which is a very important aspect in order to facilitate the enrichment, maintenance and the reuse of the ontologies. The main motivation of our future work is to adopt ontology learning techniques based on the Web to build modular ontologies by the composition of extracted ontology modules.

Acknowledgements. This work has been supported by the Spanish-Tunisian AECID project A/030058/10, "A framework for the integration of Ontology Learning and Semantic Search".

References

1. Watson, I.: Case-based reasoning is a methodology not a technology. Knowledge-Based Systems 12(5), 303–308 (1999)
2. Guarino, N.: Formal Ontology in Information Systems. In: 1st International Conference on Formal Ontology in Information Systems (FOIS 1998), Trento, Italy, pp. 3–15 (1998)
3. Charlet, J., Szulman, S., Pierra, G., Nadah, N., Teguiak, H.V., Aussenac-Gilles, N., et al.: DAFOE: amultimodel and multimethodplatform for building domain ontologies. In: Benslimane, D. (ed.) 2émes Journées Francophones sur les Ontologies, JFO 2008. ACM (2008)
4. Lassila, O., Swick, R.R.: Resource description framework (rdf) model and syntax specification. W3C Recommendation 22 (1999)
5. Fensel, D., Horrocks, I., van Harmelen, F., Decker, S., Erdmann, M., Klein, M.: Oil in a Nutshell. In: Dieng, R., Corby, O. (eds.) EKAW 2000. LNCS (LNAI), vol. 1937, pp. 1–16. Springer, Heidelberg (2000)
6. Karp, R., Chaudhri, V., Thomer, J.: An xml-based ontology exchange language (2002)
7. McGuinness, D.L., van Harmelen, F.: OWL Web Ontology Language Overview
8. Hearst, M.A.: Automated Discovery of WordNet Relations. In: Fellbaum, C. (ed.) WordNet: An Electronic Lexical Database, pp. 132–152. MIT Press (1998)
9. Morin, E.: Automatic acquisition of semantic relations between terms fromtechnical corpora. In: Proceedings of the Fifth International Congress on Terminology and Knowledge Engineering (TKE 1999), Vienna, pp. 268–278 (1999)
10. Faure, D., Poibeau, T.: First experiments of using semantic knowledge learned by ASIUM for information extraction task using INTEX. In: Staab, S., Maedche, A., Nedellec, C., Wiemer-Hastings, P. (eds.) Proceedings of the Workshop on Ontology Learning, 14th European Conference on Artificial Intelligence ECAI 2000, Berlin, Germany, pp. 7–12 (2000)
11. Hotho, A., Maedche, A., Staab, S.: Ontology-based text clustering. In: Proceedings of the IJCAI 2001 Workshop Text Learning: Beyond Supervision, Seattle (2001)
12. Alfonseca, E., Manandhar, S.: An unsupervised method for general named entity recognition and automated concept discovery. In: Proceedings of the1st International Conference on General WordNet, Mysore, India (2002)
13. Maedche, A., Staab, S.: Ontology Learning. In: Staab, S., Studer, R. (eds.) Handbook on Ontologies in Information Systems. Springer, Heidelberg (2003)
14. Kietz, J.U., Maedche, A., Volz, R.: A Method for Semi-AutomaticOntology Acquisition from a Corporate Intranet. In: Aussenac-Gilles, N., Biébow, B., Szulman, S. (eds.) EKAW 2000 Workshop on Ontologies and Texts, Juan-Les-Pins, France, pp. 37–50 (October 2000)

15. Sanchez, D., Moreno, A.: Learning non-taxonomic relationships from web documents for domain ontology construction. Data and Knowledge Engineering 63(3), 600–623 (2008)
16. Ben Mustapha, N., Baazaoui Zghal, H., Aufaure, M.A., Ben Ghezala, H.: Ontology learning from Web:survey and framework based on semantic search. In: Second International Conference on Web and Information Technologies, ICWIT 2009, Kerkena, Tunisia (2009)
17. Bergman, R., Schaaf, M.: On the Relation between Structural Case-Based Reasoning and Ontology-Based Knowledge management. In: Proc. of German Workshop On Experience Management (April 2003)
18. Salton, G., Buckley, C.: Term-weighting approaches in automatic text retrieval. Information Processing and Management 24(5), 513–523 (1988)
19. Aamodt, A., Plaza, E.: Case-based reasoning: foundational issues, methodological variations, and system approaches. AI Communications 7(1), 39–59 (1994)
20. Rissland, E.L., Daniels, J.J.: Using CBR to drive IR. In: IJCAI, pp. 400–407 (1995)
21. Minor, M., Staab, S. (eds.): Architecture-Based Integration of CBR-Components into KM-Systems. LNI, vol. 10. GI (2002)
22. Bergmann, R., Schaaf, M.: On the relations between structural case-based reasoning and ontologz-based knowledge management. In: Reimer, U., Abecker, A., Staab, S., Stumme, G. (eds.) Wissensmanagement. LNI, vol. 28, pp. 279–286. GI (2003)
23. Gao, J., Deng, G.: The research of applying domain ontology to case-based reasoning system. In: Proceedings of International Conference on Services Systems and Services Management, pp. 1113–1117 (2005)
24. Vehvilainen A., Olli Alm, E.H.: Combining case-based reasoning and semantic indexing in a question-answer service. In: 1st Asian Semantic Web Conference (2006)
25. Aktas, M., Pierce, M., Fox, G., Leake, D.: A web based conversational case-based recommender system for ontology aided metadata discovery. In: Proceedings of the Fifth IEEE/ACM International Workshop on Grid Computing (GRID 2004). IEEE Computer Society Press (2004)
26. Guan-yu, L., Li-ning, L. ,Shi-peng, L.: Design and realization of case-based ontology reasoning. In: 4th International Conference on Wireless Communications, Networking and Mobile Computing, WiCOM 2008 (2008)
27. Ben Mustapha, N., Baazaoui Zghal, H., Aufaure, M.A., Ben Ghezala, H.: Enhancing Semantic Search using Case-Based Modular Ontology. In: 25th Symposium on Applied Computing (SAC), track on "The Semantic Web and Applications" (SWA), Sierre, Switzerland, pp. 22–26 (2010)
28. Miller, G., Beckwith, R., Fellbaum, C., Gross, D., Miller, K.: Introduction to WordNet: An On-line Lexical Database. International Journal of Lexicography 3(4), 235–244 (1990)
29. Rose, D.E., Levinson, D.: Understanding User Goals in Web Search. In: International World Wide Web Conference, New York, USA, pp. 13–19 (2004)
30. Salton, G., Fox, E.A., Rocchio, J.: Extended Boolean information retrieval system. CACM 26(11), 1022–1036 (1983)
31. Baazaoui-Zghal, H., Aufaure, M.A., Ben Mustapha, N.A.: Model-Driven approach of ontological components for on-line semantic web information retrieval. Journal on Web Engineering, Special Issue on Engineering the Semantic Web 6(4), 309–336 (2008)
32. Agirre, E., Lacalle, O.: Publicly available topic signatures for all wordnet nominal senses. In: The 4rd International Conference on Language Resources and Evaluation (LREC), Lisbon, Portugal (2004)

Literature-Based Knowledge Discovery from Relationship Associations Based on a DL Ontology Created from MeSH

Steven B. Kraines[1], Weisen Guo[1], Daisuke Hoshiyama[2], Takaki Makino[3], Haruo Mizutani[4], Yoshihiro Okuda[5], Yo Shidahara[5], and Toshihisa Takagi[6]

[1] Future Center Initiative, The University of Tokyo
5-1-5 Kashiwa-no-ha, Kashiwa, Chiba, 277-8563, Japan
sk@fc.u-tokyo.ac.jp
weisen.guo@gmail.com
[2] Springer Japan KK
3-8-1 Nishi-Kanda, Chiyoda-ku, Tokyo 101-0065, Japan
daisuke.hoshiyama@springer.com
[3] Institute of Industrial Sciences, University of Tokyo
4-6-1 Komaba, Meguro-ku, Tokyo 153-8505 Japan
mak@sat.t.u-tokyo.ac.jp
[4] Department of Molecular and Cellular Biology,
Harvard University 52 Oxford St, Cambridge MA, 02138, USA
mizutani@fas.harvard.edu
[5] NalaPro Technologies, Inc.
4-12-16 Hongo, Bunkyo-ku, Tokyo 113-0033 Japan
{okuda,shidahara}@nalapro.com
[6] Department of Bioinformatics, School of Frontier Science, The University of Tokyo
5-1-5 Kashiwa-no-ha, Kashiwa, Chiba, 277-8568, Japan
tt@k.u-tokyo.ac.jp

Abstract. Literature-based knowledge discovery generates potential discoveries from associations between specific concepts that have been previously reported in the literature. However, because the associations are generally between individual concepts, the knowledge of specific relationships between those concepts is lost. A description logic (DL) ontology adds a set of logically defined relationship types, called properties, to a classification of concepts for a particular knowledge domain. Properties can represent specific relationships between instances of concepts used to describe the things studied by a particular researcher. These relationships form a "triple" consisting of a domain instance, a range instance, and the property specifying the way those instances are related. A "relationship association" is a pair of relationship triples where one of the instances from each relationship can be determined to be semantically equivalent. In this paper, we report our work to structure a subset of more than 1300 terms from the Medical Subject Headings (MeSH) controlled vocabulary into a DL ontology, and to use that DL ontology to create a corpus of A-Boxes, which we call "semantic statements", each of which describes one of 392 research articles that we selected from MEDLINE. Relationship associations were extracted from the corpus of semantic statements using a previously reported technique. Then, by making the assumption of the transitivity of association used in literature-based knowledge discovery, we generate hypothetical relationship associations by combining pairs of relationship

A. Fred et al. (Eds.): IC3K 2010, CCIS 272, pp. 87–106, 2013.
© Springer-Verlag Berlin Heidelberg 2013

associations. We then evaluate the "interestingness" of those candidate knowledge discoveries from a life science perspective.

Keywords: Relationship associations, Association rules, Semantic relationships, Semantic matching, Semantic web, Ontology, Logical inference, Life sciences, Literature-based knowledge discovery.

1   Introduction

Potentially interesting and valuable scientific discoveries can be made simply by following associations, such as co-occurrence, between terms describing particular concepts or entities that have been previously reported in the literature. For example, in the 1980's Don Swanson noted that many research articles mentioned "Raynaud's syndrome", which results in discoloration of extremities, together with medical terms such as "blood viscosity". Other articles mentioned the same medical terms together with "fish oil". However, no articles mentioned "fish oil" and "Raynaud's syndrome" together. He therefore proposed the new hypothesis that fish oil is effective for treating Raynaud's syndrome [1]. That hypothesis was later confirmed experimentally.

Following this pioneering discovery, just by examining the current literature, of the relationship between Raynaud's syndrome and fish oil, Swanson and other investigators made a few more interesting scientific discoveries by finding evidence in the existing literature for hitherto unreported associations between specific terms in the target domain [2], [3] ,[4], [5]. However, due to problems of polysemy and synonymy in natural language, the discovery process often produced a large number of false positives that had to be manually filtered out to find useful relationships.

In order to address the issues of term ambiguity in natural language, several research communities have established controlled vocabularies (CVs) that provide a one-to-one mapping between terms and concepts. One of the most well known CV is the Medical Subject Headings or MeSH terms. Currently, curators at the National Library of Medicine assign specific MeSH terms to research articles in life sciences that are stored in the MEDLINE repository. Because there is a controlled one-to-one matching between MeSH terms and the corresponding concepts from life sciences, term association can be replaced with actual concept association which should generate more semantically accurate discovery candidates. However, attempts to improve the accuracy of literature-based scientific discovery in life science by using the MeSH terms have been less successful than one might have hoped [6], [7], [8]. Some problems that have been noted include 1) the limited expressiveness of the MeSH vocabulary and 2) the inevitable mistakes in interpretation that are made by even the most careful curators.

Ontologies based on Description Logics (DL) extend the expressiveness of CVs in at least two important ways. First, in DL knowledge bases a distinction is made between ontology classes, which describe sets of semantically similar things, and instances of those classes, which represent actual things that are being described e.g. in a particular research article. Because ontology instances can be given arbitrary labels, this separation makes it possible to combine the precision of a CV with the flexibility of free text. For example, to represent a newly discovered protein "XYZ", an instance of the class **Protein** could be created and labeled "XYZ" (throughout this paper, we show class labels in bold, property labels in italics, and instance labels in quotes). Also, the DL instantiation mechanism makes it possible to describe multiple instances of a particular class, each having different attributes, and how they interact in the particular study being described. For example, one could describe the interactions between an adolescent and adult mouse each having different attributes.

The second important contribution of DL ontologies is a means for expressing unambiguously the specific relationships between the instances that have been created to represent the key concepts and entities in the resource described. These relationships are expressed by using special terms, called properties, that connect a domain instance to a range instance, forming a semantic triple that consists of a domain instance, a range instance, and a connecting property expressing a specific directed relationship between the two instances. The properties can be assigned logical characteristics, such as transitivity. Then DL reasoners can be used to infer additional relationships between instances that are implied by the stated properties [9].

Unfortunately, current text mining techniques cannot accurately extract semantic relationships between concepts from natural language text due to the complexity and ambiguity of natural language [10], [11]. Furthermore, annotations by third party curators suffer both from mistakes in interpretation and also the limited scalability of a small group of curators to the rate of research article publication [12], [13].

A third alternative that is receiving interest recently is to get the original authors of research articles to create computer-readable descriptors of the objects of their research [14], [15]. Several initiatives have been made to get the scientific community to create wiki entries for biological entities such as proteins or to create structured digital abstracts for research articles [12], [16], [17]. The descriptors are made "computer-readable" by using specific templates to mitigate the problem of natural language ambiguity. This enables search engines, text mining systems and perhaps even human readers to more accurately establish the relationships between the entities that are described [18]. Furthermore, this approach has the additional benefit of putting the responsibility of correctly describing a research article in the hands of the author, who is clearly the person who best knows the main points of the article. However, even in structured digital abstracts or wiki entries, the granularity of expression for most of the descriptive information is still at the sentence or paragraph level [12]. Consequently, computers still need to make sense of the sentences in the delimited entries in the digital abstracts [19], [20], which is notoriously difficult due to the complexity and ambiguity of natural language [21], [22].

We suggest that by drawing on new techniques and standards for semantic representation of knowledge in a computer-interpretable form, such as description logics, it should be possible to enable human researchers to author descriptions of their shared knowledge that are not just "computer-readable", but actually "computer-understandable". By "computer-understandable", we mean that computers can reason with the semantics of the descriptors in reference to shared mental models or conceptualizations of the knowledge domain, e.g. the ontologies, and that they can infer new "facts" or "assertions" in the form of relationships between concepts and/or entities that are implied but not explicitly stated. In order to test this idea, we have developed a system, called EKOSS for Expert Knowledge Ontology-based Semantic Search, that enables researchers to author computer understandable descriptors in the form of "semantic statements", which define the specific relationships between entities and concepts described by a research article [23]. The system provides a set of intuitive authoring tools that guide researchers who may not be experts in formal knowledge representation through the process of creating a semantic statement to represent their research work based on a shared DL ontology.

Here, we describe the process in which we developed a DL ontology from a subset of the MeSH vocabulary, and we present some statistics of the use of the ontology to create a corpus of semantic statements for 392 research articles that were chosen to represent the researchers in life sciences at the University of Tokyo. We then present an algorithm for discovering hypotheses based on associations between specific relationships, called "relationship associations". The relationship associations are mined from the semantic statements using a previously reported technique. We attempt to demonstrate the effectiveness of this approach by applying the algorithm to the corpus of semantic statements, and we discuss some of the hypothetical relationship associations that are discovered.

This paper is organized as follows. In Section 2, we describe the process of creating the UoT ontology by adding DL structure to a set of MeSH terms. In section 3, we describe the process of building the corpus of semantic statements based on the UoT ontology. In Section 4, we describe our algorithm for generating hypothetical relationship associations that represent new and potentially meaningful associations of specific relationships. In Section 5, we report the results of an experiment applying this algorithm to the corpus of semantic statements created previously. In Section 6, we review related work. In section 7, we finish with a discussion of the effectiveness of our approach and suggestions for future research.

2   Creating the UoT Ontology

In previous work to link a textbook used by undergraduate students at the University of Tokyo to research articles written by researchers at the same university, we have developed a DL ontology, called UoT for "University on Textbooks" [24]. The purpose of the DL ontology is to provide a formal knowledge representation language for positioning a research article in the "knowledge space" of the specific knowledge domain in a form that a computer can "understand" well enough to accurately determine the semantic similarity of different descriptors, e.g. in order to link the textbook and the articles. The UoT ontology was constructed by disambiguating the relationships between a subset of MeSH terms that were selected to cover the range both of the topics of the textbook and of 392 research articles selected from PubMed to represent the researchers in life sciences at the University of Tokyo.

Soualmia et al. reported initial efforts to add logical structure the entire MeSH CV [25]. However, they were only able to use a few heuristic methods to structure the terminology. Here, we focus our efforts on using various techniques to add logical structure to a relatively small subset of MeSH terms. This helps us to explore more thoroughly the possibilities for reframing the MeSH CV into a DL ontology that can function as a richly descriptive knowledge representation language.

We have implemented the ontology in OWL-DL [26]. The textbook is in Japanese, so we developed links from concepts in the UoT ontology to both English and Japanese terms. Thus the ontology also functions to link the natural languages of English and Japanese.

In the following subsections, we describe the details for the two main steps of the process of creating the UoT ontology: selecting the subset of MeSH terms to structure, and adding the logical structure to those MeSH terms using upper level classes and properties from other ontologies.

2.1   MeSH Term Selection Process

For the work reported here, we have focused on a small subset of the MeSH CV. Our hope is that the methods described in the next subsection could later be applied to the entire MeSH CV.

First, we used the 1997 version of the Japanese-English Life Science Dictionary (LSD) [27] and the UMLS (Unified Medical Language System) thesaurus to identify MeSH terms that match the 1078 Japanese terms in the index of the textbook. We identified a total of 883 MeSH terms (793 MeSH headings, 90 other MeSH terms) as possible matches with the terms in the index. Of those, 346 could be identified just by using the Japanese version of MeSH, 285 could be identified using the Japanese version of UMLS, and 252 were identified using the LSD. After manually filtering out false matches and choosing among candidate MeSH terms for each index term, we arrived at a list of 469 MeSH terms matching terms from the textbook index.

Simultaneously, we identified a subset of the 2024 MeSH terms (both major and minor) that had been assigned by PubMed to the 392 research articles selected for linking to the textbook. The subset to be added to the ontology was determined using the following conditions. First, only MeSH terms that were subsumed by one or more of the MeSH terms that had been mapped to the textbook were used – this eliminated about 700 terms. Second, the MeSH term must appear in at least 2 articles, which eliminated about 900 terms. We checked that at least one MeSH term from each article was included in the list of remaining terms.

The 297 MeSH terms that remained were added to the 469 MeSH terms from the textbook index, for a total of 766 MeSH terms mapped to either the textbook or the research articles. Next, we added all of the parent terms in the MeSH classification hierarchy, resulting in a grand total of 1360 MeSH terms. The total number of MeSH descriptors in 2011 was 26,142 [28], so our subset represents less than 10% of the entire MeSH CV. However, because our MeSH terms were selected based on the coverage of a general undergraduate textbook for life sciences and a set of research articles covering a wide range of topics, we believe that they represent much of the topic breadth of the MeSH CV and therefore the types of issues that would be involved in structuring the entire MeSH CV in a similar manner.

2.2   Adding Logical Structure to the MeSH Terms

In order to provide a higher-level logical structure for supporting logical reasoning to the set of 1360 MeSH terms, we added 45 upper level classes drawn from several popular upper-level ontologies. The most important upper level classes added to the basic MeSH categories are **Physical Objects, Phenomena** and **Abstract Classes** (we show classes in bold, properties in italics, and instances in quotes) – these correspond roughly to the ontological classifications used in the UMLS Semantic Network [29], ISO 15926 [30], SUMO [31] and GALEN [32].

We also added 51 abstract classes for defining roles of **Physical Objects** and **Phenomena**, 32 abstract classes for defining types of things, 189 classes for chemical elements, and 85 classes for general biological concepts such as chemical compounds and biological features. Most of these classes were reused from previous ontologies we had developed in other work [33]. The abstract classes were added to support faceted concept specification [34]. For example, the abstract class **Abnormal** can be used to specify the way in which a particular protein is reported to be abnormal in a research article. The other classes, such as the chemical elements and compounds, were added to increase the scope of the ontology. Thus the total number of classes in the ontology is 1762.

We next added a set of ontology properties (the OWL-DL objectProperty) for relating the concepts represented by the ontology classes. Based on reference to the upper-level ontologies described above, we compiled a list of 151 properties. These include domain specific properties such as *has homology* and *activates*, drawn from the UMLS semantic network [29] and GALEN [32], as well as more generic properties such as *has location* and *has part*, drawn mainly from SUMO [31] and the upper ontology based on ISO 15926 [30]. For ontologies that are not based on a description logic, such as the UMLS semantic network, some interpretation is required in order to determine how to represent the properties logically [35]. We have drawn on recent work in making these interpretations [36], [37], [38].

As discussed in the introduction, these properties play two important roles in making computer-aided knowledge sharing more intelligent. First, we can use them to define the specific types of relationships between the concepts expressed by classes in the UoT ontology. In particular, as described in the next paragraph, properties can be used to disambiguate "thesaurus type" subsumption relationships such as "related to", "narrower" and "broader". Second, the properties provide a means for connecting the specific entities in a semantic description of a research article or search query. In other words, they give us the "verbs" for making simple grammatical statements expressing knowledge in a computer understandable form.

The MeSH hierarchy is based on "thesaurus-type" subsumption relationships: the positioning of a term as a "child" of another term simply means that the child term somehow narrows the concept described by the parent term. The type of "narrowing" might be a set-theoretical *is a* relationship, but it can also be some other relationship such as composition, participation or location. For example, **Binding Sites, Antibody**, which is defined as "local surface sites on antibodies which react with antigen determinant sites on antigens", is positioned in the tree structure as a narrowing of **Antibodies, Binding Sites**, and **Antigen-Antibody Reactions** [28]. Clearly, the strict *is a* subsumption only holds for the relationship with **Binding Sites**. The relationship of **Binding Sites, Antibody** with **Antibodies** is compositional, and the relationship with **Antigen-Antibody Reactions** is locational.

We used the properties in the UoT ontology to disambiguate the subsumption relationships between MeSH terms in the MeSH term tree structure. For example, we define **Binding Sites, Antibody** in the DL ontology to be a subclass of **Binding Sites**, and we use existential restrictions (the OWL-DL "someValuesFrom" restriction) to specify that each instance of the class **Binding Sites, Antibody** is the part of an instance of the class **Antibodies** and the location of an instance of the class **Antigen-Antibody Reactions,** as shown in Figure 1.

Original version:
Binding Sites, Antibody *is a specialization of*
   Binding Sites,
   Antigen-Antibody Reactions, and
   Antibodies

Disambiguated version:
Binding Sites, Antibody *is a kind of*
   Binding Sites
   that *is the location of* some instances of **Antigen-Antibody Reactions** and
   that *is the structure part of* some instances of **Antibodies**

Fig. 1. Disambiguation of subsumption relationships in MeSH for the term "Binding Sites, Antibody"

In this way, we disambiguated the subsumption relationships for all of the MeSH terms that are subsumed by more than one parent term in the MeSH tree structure. The disambiguation step resulted in the definition of about 800 specific relationships between MeSH terms, such as the locational relationship between **Binding Sites, Antibody** and **Antigen-Antibody Reactions** described in the previous paragraph. In addition, we specified relationships between MeSH terms and the additional classes, such as roles and types, in order to add more structure to the ontology for supporting logical inference. For example, the class **Carrier Proteins** is specified in the ontology as being connected to some values of **Transport Roles** via the property *has role*.

A schematic diagram of the resultant knowledge model for the UoT ontology is shown in Figure 2. The figure shows the main upper classes that we have used to structure the MeSH terms, including **Process**, **Physical Object**, **Role** and **Characteristic**. In fact, the class **Process** is actually subsumed by a higher level class **Phenomena**, which subsumes a number of concepts from MeSH that we did not feel were actually processes in the context of the knowledge model that we constructed, such as **Acclimatization** and **Genetic Speciation**. **Regulation** is a class that reifies the regulation of a process by some physical object, such as an enzyme.

The main properties in the UoT ontology that can connect different upper level classes are also shown in Figure 2. Instances of the class **Process** can be connected to each other temporally with *occurs during, occurs before* and *occurs after*. They can also be connected compositionally with *process part of*, and by similarity with *has process homology*. Instances of **Physical Object** can be connected structurally with *has structure part, in contact* and *connects*, functionally with *interacts with* and *origin structure of*, and by similarity with *has structure homology*. Both instances of **Physical Object** and **Process** can be specified as members of instances of the respective family classes, which can turn can be members of other family class instances.

Physical Object instances can be described as active participants of particular **Process** instances by using *regulating agent of* or *actor of* and as passive participants by using *transported agent of, consumed agent of, produced agent of,* and *unaffected agent of*. In addition, process regulation can be reified using instances of the class **Regulation** linked to regulating **Physical Objects** with *regulation actor of* and regulated processes with *regulation of*. Reification of process regulation makes it possible to specify the manner in which the regulation occurs, e.g. the participation of cofactor molecules. Locations at specific **Physical Objects** can be specified for all instances of **Process**, **Physical Object** and **Regulation** using the *has location* property. **Physical Objects** and **Processes** can also be targets of **Investigative Techniques** using *analysis object of.*

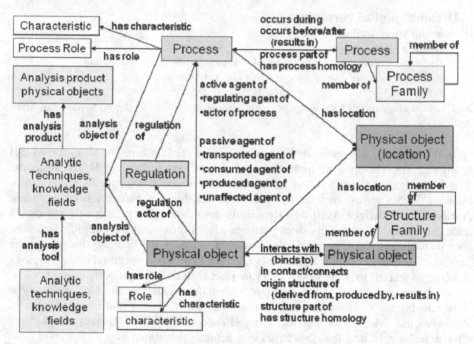

Fig. 2. Schematic diagram of the UoT ontology knowledge model showing top level classes and properties. Classes are shown in boxes. Physical objects, processes and analysis techniques are shown as green, yellow, and grey boxes. Directed arrows show properties that can connect a domain class to a range class. All boxes with the same name are interchangeable – e.g. all Physical Objects can be a passive agent of all Processes.

All properties are subsumed by the top level property *associated with*, which provides a way to describe an undetermined relationship between two instances. Therefore, any assertion of a specific relationship between two instances also produces an undetermined relationship usable in query matching. The logical characteristics and restrictions for all of the properties are specified using standard mechanisms provided by OWL-DL. In addition, as described in the Methods section, the OWL-DL restrictions "someValuesFrom" and "allValuesFrom" are used to define the allowable and descriptional usage of properties with specific classes.

3   Building the Corpus

We used the UoT ontology to create a corpus of semantic statements for a set of 392 research articles that were selected from MEDLINE for the UoT Project. The statements were created by curators having at least undergraduate degrees in life sciences, using the EKOSS (Expert Knowledge Ontology-based Semantic Search) system [23]. Each statement took about 3 to 4 hours to create and contains on average 26 class instances and 34 relationship triples of the form (domain instance, property, range instance). The entire corpus contains 13,283 semantic triples. An example of a complete semantic statement for the research article entitled "Oncogenic role of MPHOSPH1, a cancer-testis antigen specific to human bladder cancer" [39] is shown in Figure 3. Details on how to create semantic statements can be obtained at the EKOSS website at: www.ekoss.org.

The segment of the graph shown in Figure 3 that is circled corresponds to the following simple DL statement, where we use the notation of [9]:

- **Gene Expression**(colocalization)
- **Protein**(PRC1)
- *has produced agent*(colocalization, PRC1)

We can paraphrase this statement in natural language as follows:

"Colocalization" is a **Gene Expression** that *produces* a **Protein** called "PRC1."

As before, we show classes in bold font, properties in italics, and instance labels in quotes. The *is a* represents class instantiation, so "Colocalization" is an instance of the class **Gene Expression**.

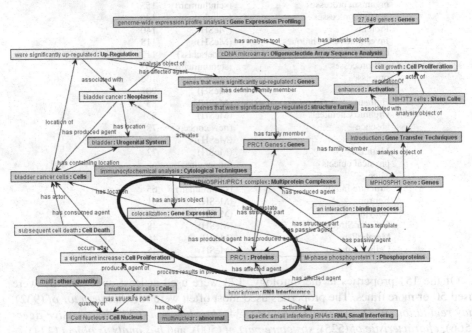

Fig. 3. Graph view of the semantic statement based on [39]. Boxes show instances of classes from the domain ontology. The text to the left of the colon in a box is the instance label, and the bold text to the right of the colon is the class name of that instance. Arrows show properties expressing the asserted relationships between instances. Colors are as described in Fig 2. The semantic relationship described in the text is circled.

Of the 1762 classes in the ontology, 906 were used in the corpus at least once, and 210 classes were used 10 or more times. Of the 906 classes used, 751 were from the MeSH CV. Of the 210 commonly used classes, 156 were from the MeSH CV. The top 30 classes are listed in Table 1. The table shows that the most commonly used classes from the UoT ontology were high level classes that had been imported from other ontologies rather than from the MeSH CV. However, there were several MeSH terms that were used more than 40 times in the corpus.

Table 1. Usage counts for the 30 most often used classes in the UoT ontology. Source is the CV or ontology from which the term was drawn: "MeSH" for MeSH, "scinthuman" for the previous version of the ontology "scinthuman", "upper class" for a newly introduced upper level class, and "new concept" for a newly introduced domain level class.

Class Name	(source)	Usage Count
Regulation	(scinthuman)	535
characteristics	(upper class)	376
molecular processes	(scinthuman)	360
Activation	(scinthuman)	274
status	(upper class)	269
Genes	(MeSH)	269
Inhibition	(scinthuman)	208
Gene Expression	(MeSH)	159
molecule parts	(scinthuman)	157
organism processes	(scinthuman)	155
binding processes	(scinthuman)	148
Proteins	(MeSH)	140
Investigative Techniques	(MeSH)	118
quantity	(upper class)	112
cell processes	(scinthuman)	107
processes	(upper class)	106
Cells	(MeSH)	90
Humans	(MeSH)	85
Organic Chemicals	(MeSH)	83
absence	(new concept)	78
Diseases	(MeSH)	76
Mammals	(MeSH)	72
physical objects	(upper class)	66
Mice	(MeSH)	66
structure family	(scinthuman)	65
Neurons	(MeSH)	61
Enzymes	(MeSH)	49
Cellular Structures	(MeSH)	48
Metabolism	(MeSH)	46
Cells, Cultured	(MeSH)	45

Of the 151 properties in the ontology, 123 were used at least once, and 61 were used 50 or more times. The properties used most often were *regulation actor of* (932), *has regulation* (927), *has structure part* (834), *has location* (657), *has passive agent* (566), *characteristic of* (525), *structure part of* (500), and *has analysis object* (474).

Finally, we extracted associations between specific relationships of concepts using the technique that we reported previously [40], [41]. A relationship association is analogous to concept association, such as that evidenced by term co-occurrence in article titles, except that instead of being between singleton concepts, the association is between semantic relationships of the form "A has specific directed relationship X with B." Therefore, a relationship association is a special kind of association rule that states "if concept A has relationship R1 with concept B, then it is likely that concept A has relationship R2 with concept C."

4   Generating New Hypothetical Relationship Associations

In order to generate new and potentially interesting scientific discoveries from the relationship associations described in the previous section, we use a modification of the ABC open discovery model developed by Swanson and his colleagues [2], [6]. We first choose a small number of the most potentially "interesting" relationship associations that were extracted from the semantic statements. Then, for each of the relationship associations, irrespective of the "interestingness" criteria, we create all of the possible A-C relationship associations from the (A-B, B-C) pairs where the B triples match. Finally, we check that they are indeed "new" discoveries by searching for a match for each of the A-C relationship associations in the corpus of semantic statements. We consider an A-C relationship association that did not match with any of the semantic statements to be a potential discovery.

In the following subsections, we briefly describe each of the three main steps in generating potential knowledge discoveries: 1) matching the B triples of A-B and B-C relationship associations, 2) generating A-C relationship associations, and 3) searching for a match for each of the A-C relationship associations to the full set of semantic statements in the corpus. More detailed descriptions are given in [42].

4.1   Matching B Triples

The basic assumption in Swanson's literature-based knowledge discovery model is that associations between concepts are transitive: if there is an association between concept A and B and between concept B and C, we can infer that there may be an association between concept A and C via the intermediary concept B. We consider two specific relationship triples to be associated if they are collocated in a particular semantic statement and there is an instance from each relationship triple that belongs to the same class, which we call the "connecting class". The defined classes for the two instances do not have to be the same; we only need to show that they are semantically equivalent. Furthermore, unlike the original Swanson ABC model, relationship associations support directionality in the form of "if Triple 1 occurs in a semantic statement, then it is likely that Triple 2 will occur" [41]. Therefore, we also include the inverses of the relationship associations in the B-C set, which doubles the size of the B-C set.

4.2   Generating A-C Relationship Associations

Next, we create a new A-C relationship association from each pair of matching A-B and B-C relationship associations by connecting the non-matching triples in the two relationship associations, the A and C triples, via the connecting class in each relationship association. This means that in addition to having a matching B triple, the A-B and B-C relationship associations must also have matching connecting classes. We can think of this matching criterion as follows: a relationship association is essentially an association of two typed relationships that apply to one entity, which is represented by the connecting class. If two relationship associations can be found that describe two typed relationships for the same "connecting class" and one of those typed relationships are the same, we can create a new relationship association that associates the two non-matching typed relationships.

For example, consider the A-B relationship association No 3 in Table 2, "if a **neoplasm process** involves a **cell** then the **cell** is likely to be the actor of a **cell proliferation process**." The connecting class of this relationship association is **cell**, so this relationship association can only be matched to a B-C relationship association that also has **cell** (or a subclass or superclass of **cell**) as the connecting class. Therefore, the B-C relationship association "if a **bone marrow cell** is involved in a **neoplasm process**, then the **bone marrow cell** is likely to contain an **oncogene protein**" can match because it has **bone marrow cell** as the connecting class, which is a subclass of **cell**. However, the B-C relationship association "if a **bone marrow cell** is involved in a **neoplasm process**, then the **neoplasm process** is likely to involve an **oncogene protein**" cannot match because it has **neoplasm process** as the connecting class. In the case where the connecting class in one relationship association is a subclass of the connecting class of the other we create two new relationship associations using each class. Both of these are considered to be potential scientific discoveries.

4.3   Matching A-C Relationship Associations to the Semantic Statement Corpus

We look for matches for each of the A-C relationship associations in the entire semantic statement corpus using the standard semantic search algorithm based on RacerPro that we have reported in previous papers [23]. Note that the use of logic and rules makes it possible to find matches to relationship associations that are only implied at a semantic level because the reasoner can infer relationships between instances that are implied but not explicitly stated in the semantic statement. Any A-C relationship association that is found to match with at least one semantic statement is discarded from the set of knowledge discovery candidates.

5   Case Study

We have applied the process described above to the corpus of 392 semantic statements that we created using the UoT ontology. Because the corpus is small, our goal is only to demonstrate the potential effectiveness of the approach of generating hypotheses from relationship associations that could be realized with a larger set of semantic statements. The following subsections detail the application to the semantic statement set of each of the steps of the process for generating knowledge discovery candidates.

5.1   Selecting the A-B Set

For the A-B set, we chose five of the 984 relationship associations that met the relevance criteria for "interestingness" that we specified in our previous work: the first criterion is that the first triple must occur in no more than 40 semantic statements, and second criterion is that the probability that the association query occurs when the first triple occurs must be twice the probability that the second triple occurs when the connecting class occurs [41]. The five relationship associations are shown in Table 2.

Table 2. The five relationship associations we extracted previously [41]. Each triple is shown in the form "domain class I property I range class". The conditional triple is separated from the consequent triple using ">". The connecting class is shown in bold type.

No.	Relationship association
1	Flagella I has structure part I **Cytoplasmic Structures** > physical objects I interacts with I **Cytoplasmic Structures**
2	**Cytoplasmic Structures** I has structure part I Microtubules > Chlamydomonas I has structure part I **Cytoplasmic Structures**
3	**Cells** I passive agent of I Neoplasms > Cell Proliferation I has active agent I **Cells**
4	**Gene Expression** I has passive agent I Receptors, Cell Surface > **Gene Expression** I has location I Neurons
5	**organism parts** I structure part of I Drosophila > Growth and Development I has passive agent I **organism parts**

5.2   Creating the B-C Set

As discussed in the previous section, we want to use as many relationship associations as possible for the B-C set, even ones that might not be so interesting. Therefore, we used all 4821 of the relationship associations extracted from the corpus of semantic statements together with their inverses, for a total of 9642 B-C relationship associations to match with the five A-B relationship associations shown in Table 2.

5.3   Creating the Candidate Discovery A-C Set

We created all of the A-C relationship associations that results from pairing each of the 9642 B-C relationship associations with each of the five A-B relationship associations, both from pairs where the A-B relationship association is first and from pairs where the B-C relationship association is first. The number of A-C relationship associations generated for each A-B varies from 18 to 29, with an average of 24. Therefore, on average, just 0.25 percent of the B-C relationship associations match with each A-B relationship association. We suggest that the small number of B-C relationship associations matching with each A-B relationship association together with the relatively small variance in the matches for each A-B relationship association may be indicative of the diversity of the triples making up the B-C relationship associations because each different A-B relationship association matched with at least 18 B-C relationship associations.

5.4   Matching the A-C Relationship Associations to the Corpus of Semantic Statements

On average, 53% of the A-C relationship associations were found to already exist in the initial set of semantic statements, which disqualifies them as knowledge discovery candidates. The remaining A-C relationship associations are potential "discoveries". However, as we noted earlier, the number of semantic statements is far too small to cover all of the semantic relationships that have been reported in the literature. We expect that with a larger corpus of semantic statements, many more of the A-C candidate relationship associations will be found to occur in the existing literature. In the following, we examine some of the knowledge discovery candidates that were generated.

One example of an A-C relationship association generated by the third A-B relationship association:

Cells | passive agent of | Neoplasms > Cell Proliferation | has active agent | **Cells**

and the B-C relationship association

Cell Proliferation | has active agent | **Cells, Cultured**
> Cell Differentiation | has passive agent | **Cells, Cultured**

that did not appear in any of the statements is:

Cells, Cultured | passive agent of | Neoplasms
> Cell Differentiation | has passive agent | **Cells, Cultured**

Here we express the relationship associations with the notation used in Table 2: "triple1 > triple2", where each triple is expressed as "domain class | property | range class" and the connecting class is shown in bold type.

We can interpret this relationship association to mean that if a researcher happens to be studying cells involved in neoplasm processes, then it might be interesting for that researcher to look at the cell differentiation processes of those cells.

An example resulting from the fourth A-B relationship association:

Gene Expression | has passive agent | Receptors, Cell Surface
> **Gene Expression** | has location | Neurons

combined with the B-C relationship association:

Gene Expression | has location | Neurons
> **Gene Expression** | has passive agent | Carboxy-Lyases

is the hypothetical relationship association:

Gene Expression | has passive agent | Receptors, Cell Surface
> **Gene Expression** | has passive agent | Carboxy-Lyases

The hypothesis generated here is that if a researcher is studying gene expression involving cell surface receptors, it might be interesting to look for carboxy-lyase enzymes also involved in the gene expression.

An example resulting from the fifth A-B relationship association:

organism parts | structure part of | Drosophila
> Growth and Development | has passive agent | **organism parts**

combined with the B-C relationship association:

Growth and Development | has passive agent | **Synapses**
> Gene Expression | has location | **Synapses**

is the hypothetical relationship association:

Synapses | structure part of | Drosophila
> Gene Expression | has location | **Synapses**

The resulting hypothesis is that if a researcher is studying the synapses of *Drosophila*, it might be interesting to look at the gene expression located at those synapses.

We hope that these three examples have provided a clear demonstration of the type of scientific hypotheses that can be generated using the approach of literature-based knowledge discovery from relationship associations. With a larger corpus of semantic statements, it should be possible to extract more interesting potential discoveries of new relationship associations and to check more thoroughly that those relationship associations do not already occur in the published literature. We are currently exploring ways to increase the size of the semantic statement corpus, e.g. by integrating the statement authoring tools into the scientific paper publication process.

6   Related Work

The goal of the work presented in this paper is to discover new knowledge or hypotheses from the literature. Several previous research studies have attempted to attain this goal as we mentioned earlier. However, there are only a few studies that look at knowledge discovery about specific relationships between concepts.

Natarajan et al. (2006) used a combination of microarray experiments and NLP methods for extracting specific gene and protein relationships, such as inhibits and phosphorylates, from full-text research articles, in order to discover gene interactions linked to the protein S1P and the invasivity phenotype. However, their sentence-based text mining results had to be manually checked, and the problem of gene name polysemy was noted as being particularly difficult to resolve. They also did not appear to use any kind of inference.

Hristovski et al. used the natural language processing tool, BioMedLEE, to extract relationships between genotypic and phenotypic concepts in research articles, expressed in the form of "associated with change" [43]. They also used another NLP system, SemRep, to extract semantic relationships in the form of "treats". They then used the extracted relationships to construct a "discovery pattern", which they defined as a "set of conditions to be satisfied for the discovery of new relations between concepts." The conditions are given by combinations of relations between concepts that were automatically extracted from articles on MEDLINE. Finally, they conducted a novelty check to find discovery patterns that actually do not occur in the medical literature. However, their approach suffers from the low accuracy of automatically extracted semantic relationships and the limited number of relationship types that could be handled.

Another technique for extracting and interconnecting knowledge at the relationship level is automatic text summarization based on relationship extraction. The CLEF (clinical e-sciences framework) project aims to generate summaries or "chronicles" of patient medical histories based on relationships that are extracted from individual medical records [44]. The authors indicate that inference is used in assembling individual events into chronicles, but it is not clear if the inference is done at the level of specific relationships between events and entities in the records. MIAKT (Medical Imaging and Advanced Knowledge Technologies) is another system for automatically summarizing knowledge in medical examination reports that focuses on image annotations [45].

7   Discussion and Future Directions

Literature-based knowledge discovery is a technique that can be used to assist researchers in making scientific hypotheses that are well-based in the existing literature but have not been reported by any previous articles. A "discovery", or more accurately a "potentially interesting hypothesis", is generated in the form of an association between a pair of key terms in the literature that have not actually appeared together in any article but that have each occurred multiple times in the literature with the same intermediary concepts terms.

Existing techniques for literature-based knowledge discovery only consider associations between singleton concepts. Because most scientific knowledge takes the form of specific binary relationships between concepts rather than just unnamed associations, hypotheses that are generated from the implied associations of pairs of relationship triples, consisting of two concept instances and a typed and directed relationship between them, are potentially more interesting and meaningful.

A well formulated "heavy weight" ontology based on a description logic (DL) can function as a formalized knowledge representation language for expressing descriptions of knowledge resources that contain not only lists of key concepts, but also explicit assertions of specific relationships between pairs of concepts. Using a DL reasoner, one can even infer relationships that have not been explicitly stated but that are implied by the asserted relationships.

In order to test the effectiveness of such an ontology to realize more accurate literature-based knowledge discovery, we have constructed a DL ontology from a subset of the MeSH CV. The "thesaurus-type" relationships specified between terms in the MeSH CV were disambiguated manually by human experts with only a limited amount of automatic preprocessing based on identification of multiple superclasses and use of simple regular expressions. We then used that DL ontology to create a corpus of 392 semantic statements describing research articles in life sciences.

Next, we described an algorithm that we have developed for generating potential discoveries in the form of relationship associations that are implied by the extracted relationship associations but that do not appear in any of the semantic statements in the corpus. A relationship association is analogous to concept association, such as that evidenced by term co-occurrence in article titles, except that instead of being between singleton concepts, the association is between relationship triples. We applied the algorithm to the relationship associations extracted previously from the 392 semantic statements [40], [41]. Each semantic statement contains an average of 34 properties, and the corpus contains more than 13,000 semantic triples, which is comparable to the size of other major corpora used for testing knowledge discovery applications. In fact, the number of triples that are logically entailed is easily more than 100,000. However, even this corpus is too small to provide a good guarantee that a new relationship association has not actually been reported in the literature. Therefore, the aim of this case study has been to provide a demonstration of the kind of knowledge discoveries that could be possible if more semantic statements become available. We were able to find several implied relationship associations that at least appear to be somewhat novel and of interest in life sciences.

There are two major conditions for producing interesting knowledge discoveries using relationship associations. First, the classes and properties in the ontology must be sufficiently detailed to be able to express meaningful relationship associations. Second, the corpus of semantic statements must be large enough to check that a potential discovery has not already been reported in the literature. Unfortunately, we only have 392 semantic statements to work with, which is insufficient to satisfy the second condition. The EKOSS system is based on the idea that if the task of authoring the semantic statements could be distributed over the entire scientific community, the problem of scalability would be solved [12], [17]. However, here we have a typical "chicken and egg" problem: in order to convince scientists to make the effort to create the semantic statements, we must show their utility, but in order to show the utility of the semantic statements, we need a certain minimum number of statements to work with. Still, we believe that our corpus of 392 semantic statements will be sufficient to indicate the kind of discovery process that might be possible with a larger corpus of statements, thereby helping to "jump-start" a virtuous cycle of creating and applying semantic statements representing research articles.

There are several areas in which to continue this research. First, it would be useful to expand the DL ontology to cover a larger part of the MeSH CV. Regular expressions have been used to resolve relationships between terms in the Gene Ontology (GO) [46]. That was possible in part due to the particular concept labeling convention in GO together with the highly specific focus of GO on genes and gene products. Unfortunately, the MeSH vocabulary, with its broader concept coverage, is less amendable to this kind of approach. An alternative might be to use our manually disambiguated results to train a machine algorithm for disambiguating similar relationships in MeSH by using grammatical expressions in the term definitions as features for machine learning. For example, application of a part-of-speech tagger, named entity recognition, and grammatical analysis to the definition of "Binding Sites, Antibody" can identify that the term refers to "sites" that are spatially related to "antibodies" and that are participants of some "reaction" process involving antigens. This might be enough to enable a computer algorithm to disambiguate the relationships between some MeSH terms. However, applications to other terms may be less effective (for example "Antigen-Antibody Reactions" does not have any definition).

Other future tasks include 1) establishing additional measures of "interestingness" for the generated relationship associations that mirror the measures that we developed in our previous work and 2) building a larger corpus of semantic statements. In order to facilitate the process of creating semantic statements and reduce the cognitive overhead for the human authors, we are developing semi-automatic methods, including incorporation of natural language processing and machine learning algorithms into the semantic statement authoring tools. Finally, we would like to investigate the possibility for integrating the semantic statement authoring approach into the research article publication process in order to leverage the potential for network effects in the scientific community [12], [17], [47].

Acknowledgements. The authors thank the President's Office of the University of Tokyo for funding support. Hideo Ogimura assisted in the creation of the semantic statements in the UoT corpus.

References

1. Swanson, D.R.: Fish oil, Raynaud's syndrome, and undiscovered public knowledge. Perspectives in Biology and Medicine 30, 7–18 (1986)
2. Swanson, D.R.: Somatomedin C. and Arginine: Implicit connections between mutually isolated literatures. Perspectives in Biology and Medicine 33(2), 157–179 (1990)
3. Weeber, M., Kors, J.A., Mons, B.: Online tools to support literature-based discovery in the life sciences. Briefings in Bioinformatics 6(3), 277–286 (2005)
4. Racunas, S.A., Shah, N.H., Albert, I., Fedoroff, N.V.: HyBrow: a prototype system for computer-aided hypothesis evaluation. Biofinformatics 20(suppl. 1), i257–i264 (2004)
5. Natarajan, J., Berrar, D., Dubitzky, W., Hack, C., Zhang, Y., DeSesa, C., Van Brocklyn, J.R., Bremer, E.G.: Text mining of full-text journal articles combined with gene expression analysis reveals a relationship between sphingosine-1-phosphate and invasiveness of a glioblastoma cell line. BMC Bioinformatics 7, 373 (2006)
6. Srinivasan, P.: Text Mining: Generating Hypotheses From MEDLINE. JASIST 55(5), 396–413 (2004)
7. van der Eijk, C.C., van Mulligen, E.M., Kors, J.A., Mons, B., van den Berg, J.: Constructing an associative concept space for literature-based discovery. JASIST 55(5), 436–444 (2004)
8. Yamamoto, Y., Takagi, T.: Biomedical knowledge navigation by literature clustering. Journal of Biomedical Informatics 40(2), 114–130 (2007)
9. Baader, F., Calvanese, D., McGuinness, D.L., Nardi, D., Patel-Schneider, P.F.: The Description Logic Handbook: Theory, Implementation, and Applications. Cambridge University Press, New York (2003)
10. Erhardt, R.A.-A., Schneider, R., Blaschke, C.: Status of text-mining techniques applied to biomedical text. Drug Discovery Today 11(7-8), 315–325 (2006)

11. Rinaldi, F., Schneider, G., Kaljurand, K., Hess, M., Romacker, M.: An environment for relation mining over richly annotated corpora: the case of GENIA. BMC Bioinformatics 7(suppl. 3), S3 (2006)
12. Ceol, A., Chatr-Aryamontri, A., Licata, L., Cesareni, G.: Linking Entries in Protein Interaction Database to Structured Text: the FEBS Letters Experiment. FEBS Letters 582(8), 1171–1177 (2008)
13. Rebholz-Schuhmann, D., Kirsch, H., Couto, F.: Facts from text–is text mining ready to deliver? PLoS Biol. 3(2), e65 (2005)
14. Gerstein, M., Seringhaus, M., Fields, S.: Structured digital abstract makes text mining easy. Nature 447, 142 (2007)
15. Seringhaus, M., Gerstein, M.: Manually structured digital abstracts: a scaffold for automatic text mining. FEBS Lett. 582, 1170 (2008)
16. Mons, B., et al.: Calling on a million minds for community annotation in WikiProteins. Genome Biol. 9(5), R89 (2008)
17. Pico, A.R., Kelder, T., van Iersel, M.P., Hanspers, K., Conklin, B.R., Evelo, C.: WikiPathways: Pathway Editing for the People. PLoS Biol. 6(6), e184+ (2008)
18. Hartley, J., Betts, L.: The effects of spacing and titles on judgments of the effectiveness of structured abstracts. JASIST 58(14), 2335–2340 (2007)
19. Cafarella, M.J., Re, C., Suciu, D., Etzioni, O.: Structured Querying of Web Text Data: A Technical Challenge. In: Proceedings of CIDR 2007 (2007)
20. O'donnell, M., Mellish, C., Oberlander, J., Knott, A.: ILEX: an architecture for a dynamic hypertext generation system. Nat. Lang. Eng. 7(3), 225–250 (2001)
21. Hunter, L., Cohen, K.B.: Biomedical language processing: what's beyond PubMed? Mol. Cell. 21, 589–594 (2006)
22. Natarajan, J., Berrar, D., Hack, C.J., Dublitzky, W.: Knowledge discovery in biology and biotechnology texts: A review of techniques, evaluation strategies, and applications. Critical Rev. in Biotech. 25, 31–52 (2005)
23. Kraines, S.B., Guo, W., Kemper, B., Nakamura, Y.: EKOSS: A Knowledge-User Centered Approach to Knowledge Sharing, Discovery, and Integration on the Semantic Web. In: Cruz, I., Decker, S., Allemang, D., Preist, C., Schwabe, D., Mika, P., Uschold, M., Aroyo, L.M. (eds.) ISWC 2006. LNCS, vol. 4273, pp. 833–846. Springer, Heidelberg (2006)
24. Kraines, S.B., Makino, T., Guo, W., Mizutani, H., Takagi, T.: Bridging the Knowledge Gap between Research and Education through Textbooks. In: Proc. 9th Intl Conference on Web Learning, Shanghai, China (2010)
25. Soualmia, L.F., Golbreich, C., Darmoni Soualmia, S.J.: Representing the MeSH in OWL: Towards a Semi-Automatic Migration. In: Proceedings of the KR 2004 Workshop on Formal Biomedical Knowledge Representation, Whistler, BC, Canada (2004)
26. OWL Web Ontology Language Overview,
    http://www.w3.org/TR/2004/REC-owl-features-20040210
27. Life Science Dictionary Project,
    http://lsd.pharm.kyoto-u.ac.jp/en/service/weblsd/index.html
28. U.S. National Library of Medicine,
    http://www.nlm.nih.gov/pubs/factsheets/mesh.html
29. McCray, A.T.: An upper-level ontology for the biomedical domain. Comparative and Functional Genomics 4, 80–84 (2003)
30. Batres, R., West, M., Leal, D., Price, D., Masaki, K., Shimada, Y., Fuchino, T., Naka, Y.: An upper ontology based on ISO 15926. Computers & Chemical Eng. 31, 519–534 (2007)
31. Niles, I., Pease, A.: Towards a Standard Upper Ontology. In: Welty, C., Smith, B. (eds.) Proc. 2nd Intl Conf. on Formal Ontology in Information Systems, Ogunquit, Maine (2001)

32. Rector, A., Bechhofer, S., Goble, C., Horrocks, I., Nowlan, W., Solomon, W.: The GRAIL concept modelling language for medical terminology. Artificial Intelligence in Medicine 9, 139–171 (1997)
33. Kraines, S.B., Iwasaki, W., Usuki, H., Yamamoto, Y.: A description logics ontology for biomolecular processes (poster). In: Bio-Ontologies SIG Workshop, Vienna, Austria (2007)
34. Yee, K.P., Swearingen, K., Li, K., Hearst, M.: Faceted metadata for image search and browsing. In: CHI 2003: Proceedings of the SIGCHI Conference on Human Factors in Computing Systems, pp. 401–408 (2003)
35. Kashyap, V., Borgila, A.: Representing the UMLS Semantic Network using OWL (Or "What's in a Semantic Web Link?"). In: Proceedings of the Second International Semantic Web Conference, Sanibel Island, Florida (2003).
36. Allemang, D., Hender, J.: Semantic Web for the Working Ontologist. Morgan Kaufmann, Burlington (2008)
37. Rector, A., Drummond, N., Horridge, M., Rogers, J., Knublauch, H., Stevens, R., Wang, H., Wroe, C.: OWL Pizzas: Practical Experience of Teaching OWL-DL: Common Errors & Common Patterns. In: Motta, E., Shadbolt, N.R., Stutt, A., Gibbins, N. (eds.) EKAW 2004. LNCS (LNAI), vol. 3257, pp. 63–81. Springer, Heidelberg (2004)
38. Smith, B., Ceusters, W., Klagges, B., Kohler, J., Kumar, A., Lomax, J., Mungall, C., Neuhaus, F., Rector, A.L., Rosse, C.: Relations in biomedical ontologies. Genome Biol. 6(5), R46 (2005)
39. Kanehira, M., Katagiri, T., Shimo, A., Takata, R., Shuin, T., Miki, T., Fujioka, T., Nakamura, Y.: Oncogenic role of MPHOSPH1, a cancer-testis antigen specific to human bladder cancer. Cancer Research 67, 3276–3285 (2007)
40. Guo, W., Kraines, S.B.: Discovering Relationship Associations in Life Sciences Using Ontology and Inference. In: Proceedings of 1st International Conference on Knowledge Discovery and Information Retrieval 2009, Madeira, Portugal, pp. 10–17 (2009)
41. Guo, W., Kraines, S.B.: Extracting Relationship Associations from Semantic Graphs in Life Sciences. In: Fred, A., Dietz, J.L.G., Liu, K., Filipe, J., et al. (eds.) IC3K 2009. CCIS, vol. 128, pp. 53–67. Springer, Heidelberg (2011)
42. Kraines, S.B., Guo, W., Hoshiyama, D., Mizutani, H., Takagi, T.: Generating Literature-Based Knowledge Discoveries in Life Sciences Using Relationship Associations. In: Proc. 2nd Intl. Conf. on Knowledge Discovery and Information Retrieval, Valencia, Spain (2010)
43. Hristovski, D., Friedman, C., Rindflesch, T.C., Peterlin, B.: Exploiting Semantic Relations for Literature-Based Discovery. In: AMIA Annu. Symp. Proc. 2006, pp. 349–353 (2006)
44. Taweel, A., Rector, A., Rogers, J.: A collaborative biomedical research system. Journal of Universal Computer Science 12, 80–98 (2006)
45. Bontcheva, K., Wilks, Y.: Automatic Report Generation from Ontologies: The MIAKT Approach. In: Meziane, F., Métais, E. (eds.) NLDB 2004. LNCS, vol. 3136, pp. 324–335. Springer, Heidelberg (2004)
46. Wroe, C.J., Stevens, R., Goble, C.A., Ashburner, M.: A methodology to migrate the Gene ontology to a description logic environment using DAML+OIL. In: Pacific Symposium on Biocomputing, vol. 8, pp. 624–635 (2003)
47. Berners-Lee, T., Hendler, J.: Publishing on the Semantic Web. Nature 410, 1023–1024 (2001)

Early Warning and Decision Support
in Critical Situations of Opinion Formation
within Online Social Networks

Carolin Kaiser, Sabine Schlick, and Freimut Bodendorf

Department of Information Systems, University of Erlangen-Nuremberg
90403 Nuremberg, Germany
{Carolin.Kaiser,Sabine.Schlick,
Freimut.Bodendorf}@wiso.uni-erlangen.de

Abstract. A growing number of people are exchanging their opinions in online social networks and influencing one another. Thus, companies should observe opinion formation concerning their products in order to identify risks at an early stage. By doing so counteractive measures can be initiated by marketing managers. A neuro fuzzy system detects critical situations in the process of opinion formation and issues warnings for the marketing managers. The system learns rules for identifying critical situations on the basis of the opinions of the network members, the influence of the opinion leaders and the structure of the network. The opinions and characteristics of the network are identified by text mining techniques and social network analysis. Simulations based on swarm intelligence are used to derive recommendations which help the marketing managers influencing the right opinion leaders to prevent the negative opinions from spreading. The approach is illustrated by an exemplary application.

Keywords: Opinion mining, Social network analysis, Neuro Fuzzy Model, Internet, Opinion spreading, Influence maximization, Recommendation.

1   Introduction

The number of people who are engaged in online social networks is increasing steadily. Within these networks, people are passing on information and evaluations of products. By discussing with each other they influence one another's opinions and purchasing behavior. It is important for companies to monitor the development of online opinions continously in order to detect risks at an early stage and to take preventive actions. Thus, the spread of negative opinions can be stopped and the compay's image can be saved from damage.

According to diffusion theory [32] not only the characteristics of a product but also the social network have a great impact on the spread of opinions. Opinion leaders are in a position to influence many members of the network. The structure of the network, i.e. the relationships among the network members, determines how fast opinions disseminate.

A new approach is being introduced which detects critical situations in the process of opinion formation and provides marketing managers with recommendations for counteractive actions by taking the overall social network into account. The opinions

A. Fred et al. (Eds.): IC3K 2010, CCIS 272, pp. 107–121, 2013.

of the networks members are first recognized by methods coming from text mining. The opinion leaders and the network structure are then characterized by key figures coming from social network analysis. Based on this information, a fuzzy perceptron learns rules which enable the discovery of critical situations and the warning of marketing managers. In critical situations, recommendations derived from simulations are made for selecting the opinion leaders who should be influenced in order to prevent the negative opinions from being spread.

2   Related Work

Opinion mining on the Internet has recently become a popular field of research. There are many papers which apply text mining to online discussions in order to reveal consumer opinions about products or product features ([15], [8], [29], [31]). Glance et. al [11] integrate several mining methods to enable online opinion tracking. Kim and Hovy [21] propose a similar approach for predicting election results by analyzing predictive opinions. All of these approaches only take a static view of online opinions.

Several papers focus on the dynamic evolution of online activities. Viermetz et al. [36] monitor the evolution of short term topics and long term trends. The system of Tong and Yager [34] automatically summarizes the development of opinions in online discussions in form of linguistic statements. Huang et al. [14] introduce a method for detecting and tracking the evolution of online communities. Choudhury et al. [6] extract and monitor key groups in blogs in order to study the dynamics of the whole community. These approaches deal with the dynamic evolution in the past.

Other approaches take online chatter as a basis for prediction. Gruhl et al. [13] use online postings to predict changes and peaks in Amazon's sales. Dahr and Chang [9] detect that user-generated content correlates with future music sales. Onishi and Manchada [28] arrive at the conclusion that blogging activity correlates with the sales of green tea, movie tickets and cell phone contracts. All three approaches do not predict the future behavior of Internet users but only the consequences of users' online activities.

There are also studies dealing with the prediction of online behavior. Choudhury, et al. [4] describe a method for predicting the communication flow in social networks. The work of Choudhury [5] allows the modeling and forecasting of activities in online groups. However, they do not identify opinion leaders and predict opinion formation.

Welser et al. [41], Chang et al. [3] as well as Gomez et al. [12] study social networks with regard to the different roles of their members (e.g. opinion leaders). However, the content of the conversation is not taken into consideration. Bodendorf and Kaiser [1] extract opinions by text mining and identify opinion leaders with the aid of social network analysis in order to analyze opinion formation.

None of the mentioned approaches focus on the evaluation of the situation as a whole and none of them consider all the influencing variables. Hence, former work did not enable the recognition of critical situations and the alerting of marketing managers. Besides, no recommendations for counteractive marketing measures were made.

Several researchers deal with influence maximization in social networks. They look for those persons in a network who must be convinced of an innovation so that a maximum number of followers is reached. For example, Domingos and Richardson [10], Kempe et al. [20] as well as Ma et al. [22] address this problem of optimization.

However, these approaches do not consider real behavior of network members [17] and, therefore, are not appropriate for recommending marketing actions in critical situations.

3   Approach

The aim of the approach is to detect critical situations during opinion formation in online social networks and to recommend appropriate marketing actions. Situations are considered as critical when negative opinions are on the verge of spreading and harming the company's image or sales volume. In these cases, marketing managers must be warned immediately in order to be able to take counteracting measures. Recommendations for ceasing the diffusion of negative opinions by influencing opinion leaders are needed.

The approach comprises four succeeding steps. In the first step, the opinions of all network members towards a product are identified by methods coming from text mining. Opinions are distinguished as positive, negative and neutral. In the second step, the opinion leaders and the network structure which have a great impact on the spread of opinions are determined by using key figures from social network analysis. In the third step, rules for discovering critical situations during opinion formation are revealed on the basis of the overall opinion of the network, the opinion and power of the opinion leaders as well as the network structure. With the aid of a fuzzy perceptron, linguistic rules are learned which can be easily understood by marketing managers. Rules learned from past situations can be employed to recognize future critical situations and to warn marketing managers at an early stage, i.e. before the spread of negative opinions. In case of critical situations, the fourth step would involve providing the marketing managers with recommendations for appropriate marketing actions. The recommendations result from simulations which show how opinions spread throughout the network if negative-oriented opinion leaders were convinced of positive opinions.

4   Data Collection

The presented approach is applied to the German Gaming community Gamestar.de for purposes of illustration and validation. The online platform Gamestar.de is provided by Europe's most popular magazine for computer games. Fans of computer games meet frequently on Gamestar.de to exchange opinions on many games within the discussion forum. 6596 postings submitted from Oct. 8th to Nov. 28th 2008 were extracted from threads discussing the games "Fallout 3", "Far Cry 2" and "Dead Space". For each of these three games, a sequence of time-dependent networks was generated by connecting those people with each other who have submitted postings directly before or after one another on one day.

5   Identification of Opinions

The identification of opinions aims at detecting the attitude of each user towards a product on the basis of his/her postings. Attitudes are assigned according to their polarity to the classes "positive", "negative" or "neutral".

The process of opinion formation consists of two phases [16]. First, the postings of the forum users are characterized by attributes. Second, the postings are classified according to their polarity on the basis of these attributes.

Statistical and linguistic attributes are used to describe the postings. For this reason, postings are decomposed into words. Unimportant stop words are removed. All remaining words are reduced to their word stem. The relative frequency of each word stem for each of the three classes is then calculated. Word stems which appear frequently in one class but rarely in the other two classes are chosen as attributes.

Several methods such as Hidden Marcov Models or Maximum Entropy enable the solving of classification tasks [40]. Support Vector Machines [7] are specially suited for text classification since they are able to process numerous attributes. A lot of papers (e.g. [29]) have empirically demonstrated the appropriateness of Support Vector Machines for text classification. Therefore, Support Vector Machines are employed for classifying the polarity of postings based on their attributes.

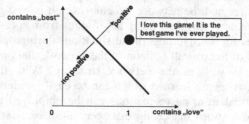

Fig. 1. Classification of opinions

In order to learn classification, training data consisting of the postings' attributes and manually assigned polarities are required. With the aid of this training data, Support Vector Machines learn the parameters of binary classification rules. In the case of three classes, three classification rules are learned: "positive" versus "not positive", "negative" versus "not negative" and "neutral" versus "not neutral". The final decision to which class a posting is assigned is based on a majority vote. In the simple case of just two attributes, the classification rule can be depicted as a straight line separating the postings into two classes. Figure 1 shows a line which classifies a posting as positive due to the word stems it contains.

After classification, the average opinion of each user is determined based on all the postings he/she has submitted to the discussion forum per day.

In order to validate this procedure, it was applied to the German Online Forum of Gamestar.de. 4010 postings from threads discussing the games "Dead Space", "Fallout 3" and "Far Cry 2" were manually classified as positive, negative and neutral. The validation is executed in form of a stratified 10-fold cross-validation. The data set is divided into 10 portions so that each portion contains the same amount of postings per class. Each of the portions is used once to test the rules learned on the basis of the other nine portions. After all ten test runs, the average performance in form of precision, recall and F-measure is calculated [40]. While precision measures the accuracy of the classification learned, recall determines its completeness. The F-measure is calculated as the harmonic mean of precision and recall.

Table 1 shows the results of the cross-validation. While the detection of negative and neutral opinions is very good, the detection of positive opinions is less successful. The examination of misclassification reveals that those postings with a positive introduction or positive conclusion but a neutral or negative statement in-between are often not recognized as positive. This problem can be solved by attaching more weight to the sentences at the beginning and the end of a posting.

Table 1. Results of opinion classification

Class	Precision	Recall	F-Measure
positive	62,96%	62,52%	62,74%
negative	86,35%	86,05%	86,20%
neutral	81,65%	81,07%	81,36%

6   Characterization of Social Network

6.1   Opinion Leaders

Opinion leaders are persons who have great influence on other people's opinion, attitude and behavior ([18]; [32]). While prior research in social psychology characterized opinion leaders on the basis of their personal attributes such as age and education, recent research defines opinion leaders on the basis of their social activities. Due to their central position and communicative behavior they play a leading role in opinion formation [35]. A small number of opinion leaders is sufficient to influence the opinions of many others in a network [19]. Persons are not split up in the classes "opinion leader" or "no opinion leader" but are characterized by the degree to which they affect others' opinions [32].

Social Network Analysis provides three key figures for measuring the degree of opinion leadership: degree centrality, closeness centrality and betweenness centrality ([39], [33]). The normalized values of these centrality key figures range from zero to one. While a value of one indicates maximum opinion leadership, a value of zero indicates minimum opinion leadership.

Degree centrality measures how many direct relationships a person has to other network members. It is calculated as the ratio of the number of a user's relationships to the number of all relationships in the network. Degree centrality specifies how often a person communicates directly with other persons in the network. Persons with high degree centrality are in a position to influence their local surroundings and can be considered as local opinion leaders.

In contrast to degree centrality, closeness centrality does not only take direct but also indirect communication relationships into account. Closeness centrality characterizes how close a person is to all other persons in the network. It is calculated as the inverse sum of the distances from each user to all other users within the network. Persons with high closeness centrality are able to influence the overall network due to their short distance to all other users in the network. Therefore, they can be considered as global opinion leaders.

Betweenness centrality describes how frequently a user can be found on the shortest connecting paths between all pairs of users. This key figure is determined as the fraction of the shortest paths which pass a user to all shortest paths within the network. Since a lot of communication flows via persons with high betweenness centrality they act as intermediaries and have the power of influencing the flow of information.

6.2   Network Structure

Besides opinion leaders, the structure of the network has an impact on opinion formation. The network structure can be characterized by the key figures centralization and density coming from the social network analysis ([39], [33]).

Centralization measures how the centralities of the network members differ from the centrality of the most central person. A strongly centralized network consists of only a few central opinion leaders and many peripheral users. In this case, the leaders' opinion can spread easily from the center to the periphery of the network [1].

Density specifies the connectivity of a network. It is calculated as the fraction of the number of relationships which exist in a network and the maximum number of relationships which are possible in a network. Density indicates the frequency of communication within the network. The higher the density of a network, the more opinions can be exchanged between the network members [1]. In a very dense network the opinion can disseminate quickly among the network members.

7   Discovery of Critical Situations

7.1   Objective

The objective is to discover critical situations automatically and to alert marketing managers as soon as such a situation arises in the process of opinion formation. The classification of situations depends on many different variables such as the overall opinion, the opinions of the opinion leaders or the structure of the network.

The approach attempts to fulfill two countervailing requirements. On the one hand, marketing managers should be in a position to easily comprehend why a situation is classified as critical. Consequently, the system must be able to process linguistic rules that can be formulated by the managers due to their expertise. On the other hand, interdependencies among influencing factors are very complex and make it difficult for marketing managers to define rules for detecting critical situations. For this reason, the system must enable supervised learning of such linguistic rules from data.

7.2   Method

With regard to the two requirements mentioned above, two methods coming from the discipline of soft computing come into consideration, i.e. artificial neural networks and fuzzy systems.Neural networks are capable of learning classification from data.

However, they are a black box. There is no way of understanding the rules behind the classifications [25]. In contrast, fuzzy systems cannot learn classification from data but they can process linguistic rules that are based on fuzzy sets [42]. Experts formulate such linguistic rules which are then employed for classification. This enables the understanding of classification results.

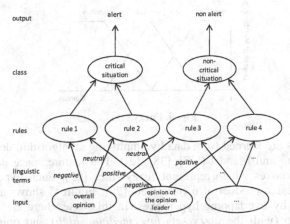

Fig. 2. Structure of a fuzzy perceptron

In order to combine the advantages and minimize the disadvantages of neuronal networks and fuzzy systems [25], a neuro fuzzy approach is applied in this work. Neuro fuzzy systems have the ability of learning linguistic rules from data. There are many different neuro fuzzy approaches. Here the NEFCLASS model (NEuro Fuzzy CLASSification) is chosen. This system is capable of learning fuzzy sets and fuzzy rules. Moreover, it can also deal with manually defined rules and optimize them [25]. The NEFCLASS model is a 3-layer fuzzy perceptron [23]. The input layer represents the input variables, the hidden layer the fuzzy rules and the output layer the two classes (critical situation and non-critical situation). The linguistic terms are represented by the weights between the input layer and the hidden layer [24]. Figure 2 illustrates the structure of a fuzzy perceptron.

The fuzzy perceptron depicted in figure 2 consists of four fuzzy rules. For example, rule 1 classifies a situation as critical if the input variables overall opinion and opinion of opinion leader take the value of the linguistic term negative (see figure 3).

Rule 1

If overall opinion is negative
and opinion of opinion leader is negative
then situation is critical

Fig. 3. Fuzzy rule

Fuzzy sets specify whether and to what degree the values of the input variables belong to linguistic terms. While in classical set theory objects either belong or do not belong to a set, in fuzzy set theory objects belong to a set with a certain degree of membership. A fuzzy set is a function which assigns a degree of membership for a linguistic term to each value of the input variable (see figure 4).

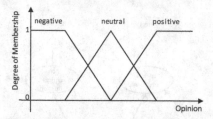

Fig. 4. Fuzzy sets

The fuzzy rules are learned from data by employing an algorithm derived from the algorithm of Wang and Mendel ([37], [38], [2]). The feature space is structured by overlapping hyperboxes which represent fuzzy rules [25]. Each hyperbox is an n-dimensional Cartesian product of n fuzzy sets [26]. Figure 5 shows a feature space that is structured by overlapping hyperboxes. In this case, there are two variables: opinion leadership (with the fuzzy sets *low, medium, high*) and opinion (with the fuzzy sets *negative, neutral, positive*).

The algorithm learns the fuzzy rules by running through the training data set twice. In the first run, all antecedents (*if* parts) of the rules are generated. For each pattern of the training data set, the combination of fuzzy sets which achieves the highest degree of membership is selected. In the second run, the rules are completed by determining the best consequent (*then* part) for each antecedent (*if* part). The resulting rules enable the classification of input patterns. However, there may still be some classification errors. Figure 5 (left side) exemplifies the classification of the input pattern (circles and triangles) after rule learning. One pattern is not classified (triangle) and one is misclassified (circle).

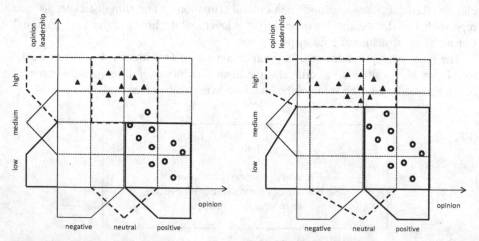

Fig. 5. Classification after rule learning (left side) and after fuzzy set modification (right side)

On the basis of the detected errors the shape and the position of the fuzzy sets are modified in order to improve classification [24]. Figure 5 (right side) shows the classification results after the process of modification. All patterns are classified. There are no misclassifications.

After modification of the fuzzy sets, the rule base is pruned by deleting variables or whole rules. Thus, the rule base is easier to interpret and can be applied to a broader range of cases [27].

7.3 Application

Training Sets. Since the NEFCLASS model is based on supervised learning, the training datasets must be classified manually before learning. Each day's snapshot of the discussion network is evaluated as a critical or non-critical situation.

Figure 6 (right side) shows a situation which is not critical. The local opinion leader and the intermediary is User 25 (degree centrality 0.57, betweenness centrality 0.48). The global opinion leader is User 26 (closeness centrality 0.65). The opinion of both of them is positive. The overall opinion is neutral. The centralization, i.e. the likelihood that the opinion of the opinion leader will diffuse, is medium (0.4). The density, i.e. the speed of diffusion, is small (0.18). Consequently, this situation is classified as non-critical.

In contrast, the situation on the left side of figure 6 is classified as critical. Local opinion leader and intermediary is User 8. Global opinion leader is User 9. The opinion of all opinion leaders is neutral. The overall opinion is slightly negative (-0.1). The likelihood that the opinion of the opinion leader will diffuse is high (centralization 0.47). The speed of diffusion is medium (density 0.36). The situation is critical since a neutral opinion can be interpreted as disinterest.

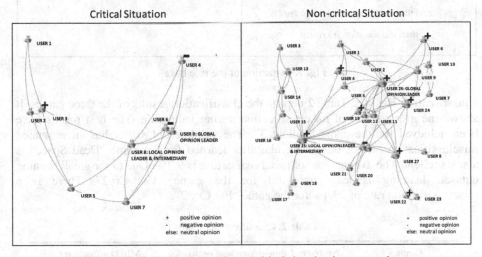

Fig. 6. Critical situation and non-critical situation in a network

Rule Bases. Based on the training datasets, the classification rules for all three games are learned. The resulting rule bases for all three games are clear and easily interpretable.

They consists of eight rules for the game "Dead Space", nine rules for the game "Fallout 3" and three rules for the game "Far Cry 2". Figure 7 shows an extraction of the rule base for the game "Dead Space".

The first rule classifies critical situations. It states that if the overall opinion is *negative* and the opinion of the local and the global opinion leader is *negative* as well, then the marketing manager must be alerted. It is also important that local opinion leadership and centralization, i.e. likelihood of opinion diffusion, are *high*. In these situations the opinion will remain negative in the future or even become more negative. As a consequence thereof marketing manager must take actions to influence opinion formation.

The second rule classifies situations that are not critical. The opinion of the global opinion leader is *positive*. The probability that the opinion of the opinion leader will diffuse is *low*. However, this is of no disadvantage since the overall opinion is already *positive*. There is no indication that the overall opinion will become negative in the future. For this reason the marketing manager there is no need for altering the marketing manager.

Rule 1	Rule 2
If the local opinion leadership is *high*	**If** the opinion of the global opinion leader is *positive*
and the opinion of the local opinion leader is *negative*	and the likelihood that the opinion of an opinion leader will diffuse is *low*
and the opinion of the global opinion leader is *negative*	and the overall opinion is *positive*
and the likelihood that the opinion of an opinion leader will diffuse is *high*	**then** *the situation is not critical.*
and the overall opinion is *negative*	
then *the situation is critical.*	

Fig. 7. Extraction of the rule base

Classification Results. Table 2 depicts the classification results of the three games. It shows the average rate of misclassification during validation. The best results have been achieved for the game "Fallout 3". The validated classifier has an estimated misclassification rate of 6.5%. The classifier learned for the game "Dead Space" is also excellent. The average rate of misclassification is 9.7%. Due to the small training dataset, learning is less successful for the game "Far Cry2". There is a misclassification rate of 29.4% for the game "Far Cry 2".

Table 2. Classification results

Game	Number of time-dependent networks	Misclassification rate
Dead Space	56	9.7%
Fallout 3	45	6.5%
Far Cry 2	33	29.4 %

8   Recommendations for Opinion Manipulation

If critical situations are detected, counteractive marketing actions can be taken to prevent negative opinions from spreading. One effective marketing strategy is influencing the opinion leaders. By addressing only a few opinion leaders, cascades of influence can be triggered [20]. Their central position in the social network can promote the rapid diffusion of influence. However, network effects are very complex and addressing the right opinion leaders poses a challenge to marketing.

Therefore, marketers are provided with recommendations for opinion manipulation. The recommendations are derived from simulations which demonstrate the effects achieved by targeting the opinion leaders [17]. Various simulations are executed in order to show how opinions spread throughout the network if different types of opinion leaders were convinced to change their opinion due to marketing measures. The results of the simulations are compared and the most successful marketing measure is recommended.

The simulations rely on the repeated application of an algorithm coming from swarm intelligence. The algorithm is employed to reveal the principles of opinion formation in the social network. With the aid of these principles, changes in opinion arising from targeting the opinion leaders are predicted.

The principles explain the opinion of each network member on the basis of the opinion of his neighborhood, the opinions of the opinion leaders in his surrounding and his own personality. The AntMiner algorithm [30], which is inspired by the foraging behavior of ants, is employed for detecting the principles of opinion formation. Digital ants search for the principles which explain opinion formation best by means of following and dropping pheromones. Figure 8 illustrates this procedure.

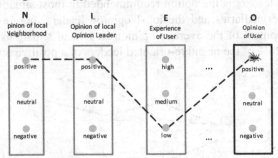

Fig. 8. Procedure for detecting principles of opinion formation

The application of this algorithm to gamestar.de yielded high accuracies ranging from 75% to 87% as depicted in table 3.

Table 3. Accuracies of rule detection

Game	Accuracy
Dead Space	87.85%
Far Cry 2	75.29%
Fallout 3	86.31%

The simulation of the spread of opinions is based on the detection and application of the principles of opinion formation (see Figure 9). At the beginning of each simulation, principles are revealed which explain opinion formation in the given social network. Next, the negative opinions of certain types of opinion leaders are converted into positive opinions. Within several subsequent iterations, masses of changed opinions are derived from this action. The principles of opinion formation are applied to all users adjacent to the changed users in order to predict their new opinion. Afterwards, new principles are revealed which explain opinion formation in the changed network and applied once again to the users adjacent to the changed users. This loop continues until there are no more users who are changing their opinion.

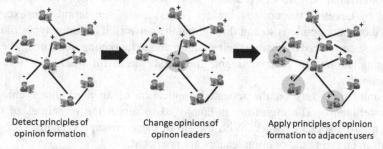

Detect principles of   Change opinions of   Apply principles of opinion
opinion formation   opinon leaders   formation to adjacent users

Fig. 9. Procedure for simulating opinion manipulation

The simulations were carried out in critical situations for gamestar.de. The effects on the average opinion of the social network achieved by converting the negative opinions of the local opinion leaders, the global opinion leaders and the intermediaries into positive opinions were measured. Simulations show that changing the opinions of the global opinion leaders is the option recommended in most situations followed by addressing the intermediaries and the local opinion leaders. Figure 10 exemplarily depicts the development of the average opinion in critical situations for "FarCry 2" when convincing 25% of the negative-oriented leaders of a positive opinion.

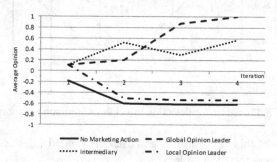

Fig. 10. Results of simulating opinion manipulation

9   Conclusions

The presented approach allows detecting critical situations and recommending appropriate marketing measures during opinion formation in online social networks by taking four steps. First, the opinions of all network members towards a product are recognized by methods coming from text mining. Second, the opinion leaders and the structure of the network are determined by key figures coming from social network analysis. Third, critical situations during opinion formation are spotted by a fuzzy perceptron on the basis of the opinions of the network members, the influence of the opinion leaders as well as the structure of the network. Choosing a neuro fuzzy approach enables learning linguistic rules which can be easily interpreted by marketing managers. These rules are learned from past situations and can be employed to judge future situations. When critical situations are detected, marketing managers are provided with recommendations for counteractive measures. The recommendations result from simulating the spread of opinions when targeting opinion leaders. This supports the marketing managers in their decisions as to which opinion leaders should be addressed. Opinion leaders could be convinced of a positive opinion by providing them with free product samples or by asking their advice about product improvements.

The application of the approach to the online social network gamestar.de yielded encouraging results. Future tasks will include the expansion of the database and the application of the approach to further social networks. In addition, the approach will be compared to similar approaches in order to improve validation.

References

1. Bodendorf, F., Kaiser, C.: Detecting Opinion Leaders and Trends in Online Social Networks. In: Proceedings of the 2nd Workshop on Social Web Search and Mining, Hong Kong (2009)
2. Borgelt, C., Klawonn, F., Kruse, R., Nauck, D.: Neuro-Fuzzy-Systeme: Von den Grundlagen künstlicher Neuronaler Netze zur Kopplung mit Fuzzy Systemen. In: Engl.: Neuro-Fuzzy-Systems: Foundations of the Combination of Neural Networks and Fuzzy-Systems, 3rd edn. Vieweg, Wiesbaden (2003)
3. Chang, C.L., Chen, D.Y., Chuang, T.R.: Browsing Newsgroups with a Social Network Analyzer. In: Proceedings of the Sixth International Conference on Information Visualization, London (2002)
4. Choudhury, M.D., Sundaram, H., John, A., Seligmann, D.D.: Contextual Prediction of Communication Flow in Social Networks. In: Proceedings of the IEEE/WIC/ACM international Conference on Web intelligence, pp. 57–65. IEEE Computer Society, Washington (2007)
5. Choudhury, M.D.: Modelling and Predicting Group Activity over Time in Online Social Media. In: Proceedings of the Twentieth ACM Conference on Hypertext and Hypermedia, Torino (2009)
6. Choudhury, M.D., Sundaram, H., John, A., Seligmann, D.D.: Which are the Representatative Groups in a Community? Extracting and Characterizing Key Groups in Blogs. In: ACM Student Research Competition, HyperText 2009 (2009)
7. Cortes, C., Vapnik, V.N.: Support Vector Networks. Machine Learning 20, 273–297 (1995)
8. Dave, K., Lawrence, S., Pennock, D.M.: Mining the peanut gallery: Opinion extraction and semantic classification of product reviews. In: Proceedings of the 12th International Conference on World Wide Web (2003)

9. Dhar, V., Chang, E.: Does Chatter Matter? The Impact of User-Generated Content on Music Sales. Technical Report, Leonard N. Stern School of Business, New York University (2007)
10. Domingos, P., Richardson, M.: Mining the network value of customers. In: Proceedings of the 7th ACM SIGKDD International Conference on Knowledge Discovery and Data Mining, pp. 57–66. ACM, New York (2001)
11. Glance, N., Hurst, M., Nigam, K., Siegler, M., Stockton, R., Tomokiyo, T.: Deriving Marketing Intelligence from Online Discussion. In: Proceedings of the Eleventh ACM SIGKDD International Conference on Knowledge Discovery in Data Mining, Chicago, pp. 419–428 (2005)
12. Gomez, V., Kaltenbrunner, A., Lopez, V.: Statistical Analysis of the Social Network and Discussion Threads in Slashdot. In: Proceedings of the International World Wide Web Conference. ACM Press, Beijing (2008)
13. Gruhl, D., Guha, R., Kumar, R., Novak, J., Tomkins, A.: The Predictive Power of Online Chatter. In: Proceedings of the Eleventh ACM SIGKDD International Conference on Knowledge Discovery in Data Mining, Chicago, pp. 78–87 (2005)
14. Huang, Y., Liu, S., Wang, Y.: Online Detecting and Tracking of the Evolution of User Communities. In: Third International Conference on Natural Computation, pp. 681–685 (2007)
15. Kaiser, C.: Combining Text Mining and Data Mining For Gaining Valuable Knowledge from Online Reviews. In: Isaías, P. (ed.) IADIS International Journal on WWW/Internet, vol. 6(2), pp. 63–78 (2009)
16. Kaiser, C., Bodendorf, F.: Opinion and Relationship Mining in Online Forums. In: Proceedings of the 2009 IEEE/WIC/ACM International Joint Conference on Web Intelligence and Intelligent Agent Technology, pp. 128–131. IEEE, Milan (2009)
17. Kaiser, C., Kröckel, J., Bodendorf, F.: Ant-Based Simulation of Opinion Spreading in Online Social Networks. In: Proceedings of the 2010 IEEE/WIC/ACM International Joint Conference on Web Intelligence and Intelligent Agent Technology, pp. 537–540. IEEE, Toronto (2010)
18. Katz, E., Lazarsfeld, P.F.: Personal influence, the part played by people in the flow of mass communication. Free Press, Glencoe (1955)
19. Keller, E.B., Berry, J.: The influentials. Free Press, New York (2003)
20. Kempe, D., Kleinberg, J., Tardos, E.: Maximizing the spread of influence in a social network. In: Proceedings of the ACM SIGKDD International Conference on Knowledge Discovery and Data Mining (2003)
21. Kim, S.-M., Hovy, E.: Crystal: Analysing Predictive Opinions on the Web. In: Proceedings of the 2007 Joint Conference on the Empirical Methods in Natural Language Processing and Computational Natural Language Learning, Prague, pp. 1056–1064 (2007)
22. Ma, H., Yang, H., Lyu, M.R., King, I.: Mining Social Networks Using Heat Diffusion Processes for Marketing Candidates Selection. In: Proceedings of the 17th ACM Conference on Information and Knowledge Management, Napa Valley (2008)
23. Nauck, D., Kruse, R.: A Fuzzy Perceptron as a Generic Model for Neuro-Fuzzy Approaches. In: Fuzzy Systeme 1994 (1994)
24. Nauck, D., Kruse, R.: NEFCLASS - A Neuro-Fuzzy Approach for the Classification of Data. In: George, K.M., Carrol, J.H., Deaton, E., Oppenheim, D., Hightower, J. (eds.) Applied Computing 1995: Proceedings of the 1995 ACM Symposium on Applied Computing, pp. 26–28. ACM Press, Nashville (1995)
25. Nauck, D., Klawonn, F., Kruse, R.: Foundations of neuro-fuzzy systems. John-Wiley & Sons, Chichester (1997)

26. Nauck, D., Kruse, R.: A neuro-fuzzy method to learn fuzzy classification rules from data. Fuzzy Sets and Systems 89, 277–288 (1997)
27. Nauck, U.: Design and Implementation of a Neuro-Fuzzy Data Analysis Tool in Java. Diploma Thesis, University of Braunschweig, Braunschweig (1999)
28. Onishi, H., Manchanda, P.: Marketing Activity, Blogging and Sales. Technical Report, Ross School of Business, University of Michigan (2009)
29. Pang, P., Lee, L., Vaithyanathan, S.: Thumbs up? Sentiment Classification using Machine Learning Techniques. In: Proceedings of the Conference on Empirical Methods in Natural Language Processing, pp. 79–86. ACM (2002)
30. Parpinelli, R., Lopes, H., Freitas, A.: Data Mining with an Ant Colony Optimization Algorithm. IEEE Transactions on Evolutionary Computing 6(4), 321–332 (2002)
31. Popescu, A.-M., Etzioni, O.: Extracting Product Features and Opinions from Reviews. In: Proceedings of Human Language Technology Conference and Conference on Empirical Methods in Natural Language Processing, pp. 339–346 (2005)
32. Rogers, E.: Diffusion of innovations, 5th edn. Free Press, New York (2003)
33. Scott, J.: Social Network Analysis – A Handbook. Sage, London (2000)
34. Tong, R.M., Yager, R.R.: Characterizing Attitudinal Behaviors in On-Line Open-Source. In: Proceedings of Association for the Advancement of Artificial Intelligence, Spring Symposium 2004, Atlanta (2004)
35. Valente, T.W.: Network Models of the Diffusion of Innovations. Hampton Press, Cresskill (1999)
36. Viermetz, M., Skubacz, M., Ziegler, C.-N., Seipel, D.: Tracking Topic Evolution in News Environments. In: 10th IEEE Conference on E-commerce Technology and the Fifth IEEE Conference on Enterprise Computing, E-Commerce and E-Services, pp. 215–220 (2005)
37. Wang, L.-X., Mendel, J.M.: Generating Rules by Learning from Examples. In: International Symposium on Intelligent Control, pp. 263–268. IEEE Press, Piscataway (1991)
38. Wang, L.-X., Mendel, J.M.: Generating Fuzzy Rules by Learning from Examples. IEEE Trans. Systems, Man, and Cybernetics 22(6), 1414–1427 (1992)
39. Wassermann, S., Faust, K.: Social Network Analysis – Methods and Applications. Cambridge University Press, Cambridge (1999)
40. Weiss, S., Indurkhya, N., Zhang, T., Damerau, F.: Text Mining – Predictive Methods for Analyzing unstructured Information. Springer, New York (2005)
41. Welser, H.T., Gleave, E., Fisher, D., Smith, M.: Visualizing the Signatures of Social Roles in Online Discussion Groups. Journal of Social Structure 8 (2007)
42. Zadeh, L.: Fuzzy Sets. Information and Control 8(3), 338–353 (1965)

A Connection between Extreme Learning Machine and Neural Network Kernel

Eli Parviainen and Jaakko Riihimäki

BECS, Aalto University, P.O. Box 12200, FI-00076 AALTO, Finland

Abstract. We study a connection between extreme learning machine (ELM) and neural network kernel (NNK). NNK is derived from a neural network with an infinite number of hidden units. We interpret ELM as an approximation to this infinite network. We show that ELM and NNK can, to certain extent, replace each other. ELM can be used to form a kernel, and NNK can be decomposed into feature vectors to be used in the hidden layer of ELM. The connection reveals possible importance of weight variance as a parameter of ELM. Based on our experiments, we recommend that model selection on ELM should consider not only the number of hidden units, as is the current practice, but also the variance of weights. We also study the interaction of variance and the number of hidden units, and discuss some properties of ELM, that may have been too strongly interpreted previously.

Keywords: Extreme learning machine, ELM, Neural network kernel.

1 Introduction

Extreme Learning Machine [1] (ELM) is a currently popular neural network architecture based on random projections. It has one hidden layer with random weights, and an output layer whose weights are determined analytically. Both training and prediction are fast compared with many other nonlinear methods.

We study a connection between ELM and neural network kernel (NNK). NNK is a kernel that is often used with Gaussian processes [2]. It is derived by assuming a neural network with an infinite number of hidden units, and integrating out the network weights. We interpret ELM as an approximation to the infinite network used in this derivation.

We show that ELM and NNK can, to certain extent, replace each other in computations. The output of the hidden layer of ELM can be used to form a kernel [3] for kernel classifiers; this observation has been the main inspiration for our work. We experimentally show that this kernel approaches NNK when the number of hidden units grows. On the other hand, we can decompose the NNK matrix into a possible set of feature vectors, and use these vectors instead of the hidden layer of ELM.

The connection between ELM and NNK leads us to question the current practice of parameterizing ELM. Usually, only the number of hidden units is carefully selected, and a fixed variance or range parameter is used for the distribution from which the random weights are drawn. In NNK, the weights are integrated out, so that the only parameter left is the variance of the weights. In ELM, the individual weights have little

A. Fred et al. (Eds.): IC3K 2010, CCIS 272, pp. 122–135, 2013.

meaning since they are random, but the variance of the distribution from which they are drawn should matter, just like it does in NNK. We study the effect of the weight variance parameter in ELM, finding that it affects the predictions. We therefore think that the model selection for ELM should consider both the variance and the number of hidden units.

Sect. 2 introduces ELM, NNK and the data sets we use. In Sect. 3 we motivate out work by noting how the random nature of ELM makes ELM resemble kernel methods, although it is usually thought as a neural network method. In Sect. 4 we discuss the connection between ELM and NNK in detail. Sect. 5 concentrates on the effects of variance parameter, and Sect. 6 studies the interaction of the weight variance and the number of hidden units. We finish by discussing our results and ELM in Sect. 7.

2 Methods and Data

2.1 Extreme Learning Machine

Extreme Learning Machine [1] is a feedforward neural network which consists of one hidden layer of sigmoid units, and one (linear) output layer. The key idea of ELM is the way the network weights are determined. The weights of the hidden layer are randomly chosen and not trained at all. The output weights are calculated analytically to predict a target variable from the hidden layer outputs. This makes training a single ELM network simple and fast. The price to pay is variation of results due to randomness in weights. For assessing average quality of results and their uncertainty, repeated runs are needed.

Several variants of ELM have been proposed, many of them incorporating some weak or indirect form of training in the first layer. For example, the least useful hidden units can be pruned [4], or the best of several ELM can be chosen for use in an evolutionary algorithm [5]. We use the plain ELM of [1] in this work.

The weights w_h and biases b_h of the hidden layer must be drawn from some continuous distribution [1]. Often a uniform distribution, spread over a fixed interval, is used. The width of the interval is not seen as a model parameter, but it is simply a constant that must be suitably fixed to guarantee that the sigmoid operation neither remains linear nor saturates to ± 1. In this work we use zero-mean Gaussian distributions, parameterized with variance σ^2. We use equal variance for all dimensions of input data. Gaussian was chosen because it is used in the derivation of the neural network kernel (Sect. 2.2).

ELM places few restrictions to the activation of hidden units. In this work we use the error function $f(z) = 2/\sqrt{\pi} \int_0^z \exp(-t^2)\mathrm{d}t$. Also this choice is motivated by its correspondence with the derivation of NNK. In [1] it is shown that, with a large enough number of hidden units, it is possible to achieve arbitrarily small training error, on the condition that the hidden unit sigmoids are infinitely differentiable. The error function fullfills this condition, so using it instead of the more common logistic or tanh sigmoids does not diminish the representational power of ELM.

The weights β of the output layer are determined by fitting a linear model to the outputs of the hidden layer. The outputs are collected into a matrix $\mathbf{F}$ with entries $[\mathbf{F}]_{ih} = f(w_h^T x_i + b_h)$, with i indexing the data points x_i and h indexing the hidden units. The weights β are found from

$$\mathbf{F}\beta = \mathbf{Y} , \tag{1}$$

where $\mathbf{Y}$ is the matrix or vector of target variables. We use ELM for binary classification, in which case $\mathbf{Y}$ is a vector with entries ± 1. A least squares estimate for β is obtained as

$$\beta = \mathbf{F}^\dagger \mathbf{Y} , \tag{2}$$

where $\mathbf{F}^\dagger$ denotes the Moore-Penrose pseudoinverse [6] of $\mathbf{F}$.

The actual classification happens by thresholding the network output at zero. To our knowledge nothing would prevent use of sigmoid outputs, so the outputs would be class probabilities instead of binary labels. In such case, the linear model from the hidden layer to the targets would be replaced by a generalized linear model [7]. Sigmoid outputs are common in other neural network models. For simplicity, we stick to the thresholding that is common for ELM.

2.2 Neural Network Kernel

Neural network kernel is derived in [8] by letting the number of hidden units in the hidden layer of a neural network go to infinity. A Gaussian prior is set to the hidden layer weights, which are then integrated out. The only parameters remaining after the integration are the variances of the weight priors. This leads to an analytical expression for the expected covariance between two feature space vectors,

$$k_{NN}(\boldsymbol{x}_i, \boldsymbol{x}_j) = \frac{2}{\pi} \sin^{-1} \left(\frac{2\tilde{\boldsymbol{x}}_i^T \Sigma \tilde{\boldsymbol{x}}_j}{\sqrt{(1 + 2\tilde{\boldsymbol{x}}_i^T \Sigma \tilde{\boldsymbol{x}}_i)(1 + 2\tilde{\boldsymbol{x}}_j^T \Sigma \tilde{\boldsymbol{x}}_j)}} \right) . \tag{3}$$

Above, $\tilde{\boldsymbol{x}}_i = [1 \; \boldsymbol{x}_i]$ is an augmented input vector and Σ is a diagonal matrix with variances of inputs. In this work all variances are assumed equal.

NNK also arises as a special case of a more general arc-cosine kernel [9]. NNK should not be confused with tanh-kernel $\tanh((x_i \cdot x_j) + b)$, which is sometimes called MLP kernel (from Multi-Layer Perceptron) because of its connection to neural networks [10].

2.3 Data Sets

For comparing ELM and NNK we use six binary classification data sets from UCI machine learning repository [11]: Arcene, US votes, WDBC, Pima, TicTacToe and Internet ads. Their properties are summarized in Table 1. For representative results, the data sets were chosen to have different sample sizes and different dimensionalities.

In addition, we use an artificial data set Saturni, illustrated in Fig. 1. It consists of four spheres in random locations in a three-dimensional space. Each sphere is surrounded by a ring, whose points have a different class label than the points of the sphere. The data set is highly nonlinear, but easy to classify with a nonlinear model.

3   ELM Falls between Neural Networks and Kernels

Extreme learning machine, although introduced as a fast method for training a neural network, is in some sense closer to a kernel method in its operation. A fully trained neural network has learned a mapping such that the weights contain information about the

Table 1. Data sets for the experiments in Sects. 4 and 5

name	# samples	# dims	data types
Arcene [12]	200	10000	continuous
US votes	435	16	binary
WDBC	569	30	continuous
Pima	768	8	continuous
Tic Tac Toe	958	27	categorical
Internet ads	2359	1558	continuous, binary

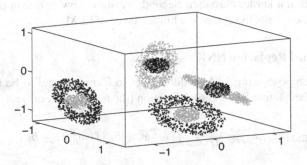

Fig. 1. The Saturni data set that is used in the experiments in Sect. 6

training data. ELM uses a fixed mapping from data to a feature space. This is similar to a kernel method, except that instead of some theoretically derived kernel, the mapping ELM uses is random.

The ability to learn features on data is an essential property of a fully trained neural network. Features should be good for predicting the target variable of a classification/regression task. In a network with one hidden and one output layer, the hidden layer learns the features, while the output layer learns a linear mapping. We can think of this as first non-linearly mapping the data into a feature space and then performing a linear regression/classification in that space.

ELM has no feature learning ability. It projects the input data into whatever feature space the randomly chosen weights happen to specify, and learns a linear mapping in that space. Parameters affecting the feature space representation of a data point are the type and number of neurons, and the variance of hidden layer weights. Training data can affect these parameters through model selection, but not directly through any training procedure.

This is similar to what a support vector machine does. A feature space representation for a data point is derived, using a kernel function with a few parameters, which are typically chosen by some model selection routine. Features are not learned from data, but dictated by the kernel. Weights for linear classification or regression are then learned in the feature space. The biggest difference is that where ELM explicitly generates the feature space vectors, in SVM or in another kernel method only similarities between feature space vectors are used.

Seeing ELM as a neural network method with kernel-like behavior raises a question about a connection between ELM and the neural network kernel. In the following section we will interpret ELM as an approximation to NNK.

4 ELM Feature Space Approximates NNK Feature Space

NNK is derived from a neural network with an infinite number of hidden units. We interpret ELM as an approximation to this infinite neural network. In this section to study the connection of NNK and ELM from two opposite points of view. First, we use ELM as a kernel in a kernel classifier. Second, we show how to decompose the NNK matrix and use the results to replace the hidden layer of ELM.

4.1 ELM Kernel Replacing NNK

Authors of [3] propose using ELM hidden layer to form a kernel to be used in SVM classification. They define ELM kernel function (the notation is ours) as

$$k_{ELM}(\boldsymbol{x}_i, \boldsymbol{x}_j) = \frac{1}{H} \sum_{h=1}^{H} f(\boldsymbol{w}_h^T \boldsymbol{x}_i + \boldsymbol{b}_h) f(\boldsymbol{w}_h^T \boldsymbol{x}_j + \boldsymbol{b}_h) , \qquad (4)$$

that is, the data is fed trough the ELM hidden layer to obtain the feature space vectors, and their covariance is then computed and scaled by the number of hidden units H.

When the number of hidden units grows, this kernel matrix approaches the NNK matrix. Fig. 2 shows the approach, measured by Frobenius norm, as function of H. Especially for small H the ELM kernel varies due to random weights, but it converges towards NNK.

Experiment. We use ELM kernel in a Gaussian Process classifier [2]. The experiment is implemented by modifying the GPstuff toolbox[1]. Expectation propagation [13] is used for Gaussian process inference. The data is split into train and test sets (50 % / 50 %). Zero-mean unit-variance normalization is used for the data. We are more interested in ELM behavior as function of hidden units than the prediction accuracy. We therefore only consider variation from the random weights (using 30 repetitions) and do not repeat over splits of data.

Results of the experiment are shown and compared to NNK results in Fig. 3. ELM kernel behavior in GP seems qualitatively similar to that observed in [3] for SVM. The classification accuracy first rises rapidly and then sets as a fixed level. Variation due to random weights remains, but NNK result stays inside the 95% interval of ELM.

4.2 NNK Replacing ELM Hidden Layer

We use NNK to replace the hidden layer computations in ELM. This is done by first forming the NNK matrix, and then deriving a possible set of explicit feature space vectors by matrix decomposition. This corresponds to using ELM with an infinite number of hidden units.

---

[1] http://www.lce.hut.fi/research/mm/gpstuff/

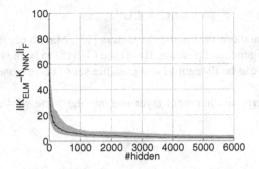

Fig. 2. ELM kernel (K_{ELM}) approaches neural network kernel (K_{NN}) in Frobenius norm when number of hidden unit grows. Mean by black dots, 95% interval by shading. Variation is caused by randomness in weights. WDBC data set.

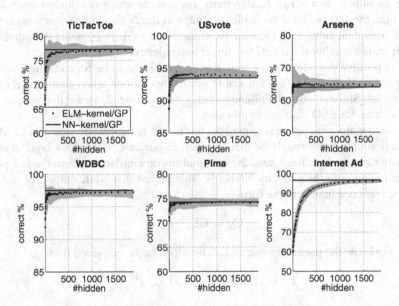

Fig. 3. Mean accuracy of GP classification when using ELM kernel (black dots and shading) versus NNK accuracy (horizontal line). The shading indicates variability due the repeated initializations of the ELM kernel (95 % interval).

Derivation. When using ELM, we only deal with vectorial data, with data space vectors transformed into feature space vectors by the hidden layer. Kernel methods rely on pairwise data, where only similarities from any point to all training points are considered. The kernel matrix specifies the pairwise similarities. In order to use pairwise information from the NNK instead of the ELM hidden layer, we must find a vectorial representation for the data. The problem of recovering points given their mutual relationships is old [14], and can be solved by a decomposition of the pairwise matrix.

NNK is derived as a covariance, and is therefore positive semidefinite (PSD). Any PSD matrix can be decomposed into a matrix and its Hermitian conjugate

$$\mathbf{C} = \mathbf{L}\mathbf{L}^H \ . \tag{5}$$

There are different methods for finding the factors [15]. Matlab `cholcov` implements a method based on eigendecomposition. If we take $\mathbf{C}$ in (5) to be the output of the NNK function (3), then $\mathbf{L}$ can be thought as one possible set of corresponding feature space vectors.

We use $\mathbf{L}$ to determine the output layer weights the same way we used the ELM features $\mathbf{F}$ in (2),

$$\boldsymbol{\beta} = \mathbf{L}^\dagger \mathbf{Y} \ . \tag{6}$$

The factors $\mathbf{L}$ are unique only up to a unitary transformation, but this is not a problem in ELM context, as the linear fitting of output weights is able to adapt to linear transformations.

With an infinite number of hidden units, the feature space is infinite-dimensional. Meanwhile, the data we have available is finite, and the N data points can span at most an N-dimensional subspace. Thus the maximum size of $\mathbf{L}$ is $N \times N$; the number of columns can be smaller if the data has linear dependencies.

If $\mathbf{C}$ is positive definite or close to it, a triangular $\mathbf{L}$ could be found using Cholesky decomposition, leading to fast and stable matrix operations when finding the output layer weights. Since positive definiteness cannot be guaranteed, we use the more general decomposition for PSD matrices in all cases.

The one remaining problem is to map the test points to the feature space. In ELM, the test data is simply fed through the hidden layer. In our case, the hidden layer does not physically exist, and we must base the calculations on similarities from the test points to the training points, as given by NNK (3). This means that NNK output for test data $\mathbf{C}_*$ is a covariance matrix of the form

$$\mathbf{C}_* = \mathbf{L}\mathbf{L}_*^H \ . \tag{7}$$

We already know the pseudoinverse of $\mathbf{L}$. Therefore $\mathbf{L}_*$ is recovered from

$$\mathbf{L}_* = (\mathbf{L}^\dagger \mathbf{C}_*)^H = (\mathbf{L}^\dagger \mathbf{L}\mathbf{L}_*^H)^H \ , \tag{8}$$

and the predictions for the test targets are computed as

$$\mathbf{Y}_* = \mathbf{L}_* \boldsymbol{\beta} \ . \tag{9}$$

Experiment. We compare ELM classification accuracy (as function of number of hidden units) to the accuracy obtained by replacing the ELM hidden layer with NNK. The latter variant of ELM is from now on referred to as NNK-ELM. The comparison is done for five different variances ($\sigma \in \{0.1, 0.325, 0.55, 0.775, 1\}$). The authors' Matlab implementation[2] is used.

The data is scaled to range $[-1, 1]$. Each data set is divided into 10 parts. Nine parts are used for training and one for testing, repeating this 10 times. This variation from data

---

[2] Available from http://www.becs.tkk.fi/~eiparvia/matlabcode.html

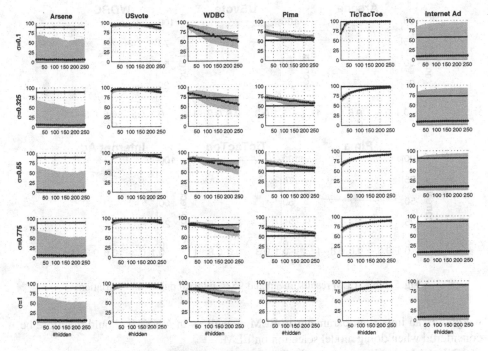

Fig. 4. ELM results (mean as black dots, 95 % interval as shading) for different values of σ. Mean of NNK results (horizontal line) are shown for comparison.

is shown in figures. ELM results have another source of variation, the random weights. This is handled by repeating the runs 10 times, each time drawing random weights, and averaging over results. The maximum number of hidden units is 250, to make sure to cover the sensible operating range of ELM (up to N hidden units) for all data sets.

The predictions given by the ordinary ELM are shown in in Fig. 4, and the mean predictions of NNK-ELM are included for comparison. We notice that when the variance is properly chosen, using NNK gives equal or better results than ELM for most data sets. Pima data set is an exception, ELM has some predictive power whereas NNK-ELM performs almost at the level of guessing.

5 Variance of Weights Affects ELM Predictions

An infinite network performing equally well or often better than ELM raises a question about meaningfulness of choosing model complexity based on hidden units only, as is traditionally done with ELM. NNK is parameterized by the variance of weights. The network has an infinite number of hidden units, and when the weights are integrated out, the values of the individual weights become meaningless. Also in ELM, the essential information about the weights is captured by their variance, since the individual, random weights are not meaningful as such.

We argue, and support the argument by experiments, that the variance of hidden layer

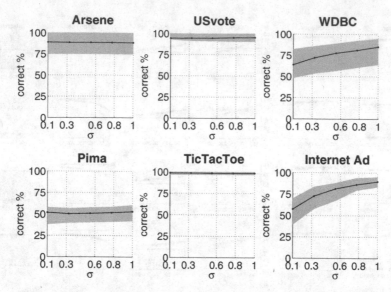

Fig. 5. NNK-ELM results, mean and 95 % interval due to data variation

weights is an important parameter in ELM as well as in NNK. It should therefore be considered when doing model selection on ELM.

NNK-ELM results are shown in Fig. 5, as function of σ. We notice that the choice of variance has a marked effect on two and some effect on other data sets, both for ordinary ELM and the NNK variant.

Fig. 6 summarizes the variance effects from Fig. 4. The mean predictions of ELM are shown as function of H. For TicTacToe and WDBC data sets the predictions are clearly affected by the variance parameter. For Internet ad data the overall effect of both H and σ is very small. In that scale, the smallest variance nonetheless gives results clearly different than larger values. Results for other data sets are not very sensitive to the variance values that were tried. For TicTacToe and Internet ads smaller variance gives better predictions, for WDBC the biggest one. Clearly no fixed sigma can be used for all data sets.

When thinking about the mechanism by which the variance parameter affects the results, differences between data sets are to be expected. Variance affects model complexity, and obviously different models fit different data sets. Variance and distribution of the data together determine the magnitude of values seen by the activation function. This is illustrated in Fig. 7. One-dimensional data points, spread over range [-1,1] (the x-axis), are given random weights drawn from a zero-mean Gaussian distribution and then fed through an error function sigmoid (denoted *erf* in the figure), repeating this 10000 times. Mean output and 95 % interval are depicted. On average, the sigmoid produces a zero response, but the distribution of responses is determined by the variance used. Small variance means mostly small weights, and linear operation. Large variance produces many large weights, which increase the proportion of large responses by the network, allowing nonlinear mappings.

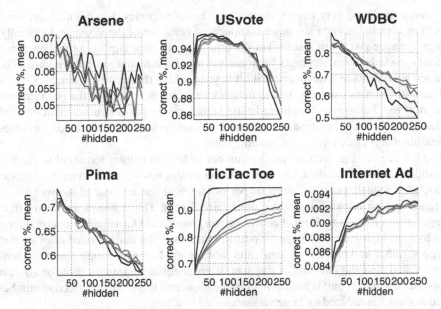

Fig. 6. Effect of variance on mean of ELM predictions. Darker shade indicates smaller variance.

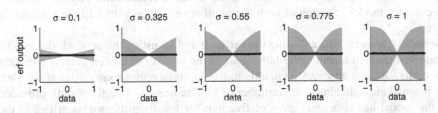

Fig. 7. Distributions of predictions of an error function sigmoid for different σ

6   Variance and the Number of Hidden Units Interact

In [16] we recognized the importance of weight variance as a parameter of ELM, and noted, that we now have two parameters that both affect the nonlinearity of the model. We did not, however, study their relative importance or joint effects. In this section we look at the interaction of the two parameters.

We use the artificial Saturni data set, from which we can easily draw repeated samples of different sizes. The size of the training set varies, and the test data always has 500 samples. The results are averaged over 10 ELM initializations and 10 samples drawn from the Saturni.

Fig. 8 shows the prediction accuracy of ELM for data sets of 250, 500, 750 and 1000 points, as a contour plot over a grid of weight variances and numbers of hidden units.

In the direction of hidden units, the plots seem to divide in three regions with qualitatively different behavior for the variance parameter. Left, where the number of hidden units is small, the variance has very little effect. Right, with a large number of hidden units, very small and very large variances give better accuracy than the medium-sized ones. The area in between has more complicated behavior of the variance. Generally, a large variance gives best results, but the area of optimal or almost optimal accuracy stretches towards lower variances for some values of the number of hidden units. The variance also seems to have an upper limit beyond which the accuracy decreases.

These differences can be related to network capacity in relation to the phenomenon to be modeled. Following complicated class boundaries requires a sufficient number of hidden units. On the other hand, if there are too many hidden units, the model becomes so flexible that it can overfit to the training data.

The left region is an area, where the number of hidden units is too small to capture the nonlinearities of the data. On the right, the model has started to overfit. This explains, why a small variance improves performance: it causes most of the weights to be small, making most of the hidden units fairly linear. This achieves a similar effect as regularizers have in traditional neural network models. Our tentative interpretation for why also high variances improve performance over the middle-sized ones is, that a large variance makes many hidden units saturate to -1 or +1. If such a unit happens to be located near the outskirts of the data, it gives equal outputs for most or all data points. Such a hidden unit is useless in classification, and therefore the effective number of hidden units is reduced by large variances.

The region of optimal predictions, between these two extremal areas, lies in a certain range of numbers of hidden units, but its exact location is determined by the variance. This calls for model selection of both the number of hidden units and the variance in this region.

Generally, it seems that adjusting the variance and adjusting the network size can be alternative means of reaching the same goal. If the number of hidden units is correctly chosen, the variance may vary a lot, although an optimal region exists. And, if the variance is small enough, almost any number of hidden units will perform well, provided that the model has enough degrees of freedom for handling the nonlinearities in the data.

7   Discussion

7.1   ELM and NNK

We introduced NNK-ELM as a way for studying the effect of infinite hidden units in ELM, but it can also find its use as a practical method.

Computational complexity of NNK-ELM corresponds to that of N-hidden unit ELM. The matrix decomposition required in NNK-ELM scales as $O(N^3)$. In practice ELM training is much faster, since the optimal number of hidden units is usually much less than N. However, the optimal value is usually found by model selection, necessitating several ELM runs. If the selection procedure also considers large ELM networks, choosing ELM over NNK-ELM does not necessarily save time. Prediction is faster with ELM than with NNK-ELM, if $H \ll N$.

Furthermore, NNK-ELM has only one parameter, the variance of the distribution of the hidden layer weights. If the number of data points is reasonably small, NNK-ELM can thus result in considerable time savings when doing model selection. A factor

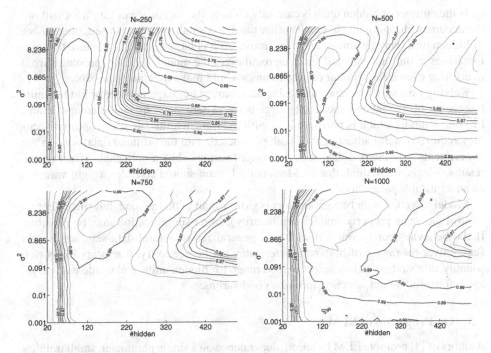

Fig. 8. ELM classification accuracy as function of the number of hidden units and of weight variance, for various N. Note the logarithmic scaling of the y-axis. The $N = 250$ plot has contour spacing of 0.02 throughout. In the other plots, contours are drawn at intervals of 0.1 below accuracy of 0.95 (this is to keep the narrow region of low accuracy a bit more readable), and at intervals of 0.005 above that.

adding to this is that, unlike ELM, NNK-ELM gives deterministic results, and only requires repetitions if variability due to training data is considered.

Use of ELM as a kernel, at least in a Gaussian process classifier, is likely to remain a curiosity. Classification performance seems to steadily increase as the number of hidden units grows, and, when considering the variation caused by ELM randomness, the performance does not exceed that of NNK. When an easy-to-compute, theoretically derived NNK function is available, we see no reason to favor a heuristical kernel, computation of which requires generating and storing random numbers and calculating an explicit mapping to a feature space.

7.2 Parameters of ELM

In our experiment with artificial data and large samples, repeating runs over both ELM initializations and data, it seemed that the weight variance and the number of hidden units may be alternative ways for controlling the nonlinearity of the model. This experiment probably captured the general principle of the interaction of the two parameters, although real data sets are usually not as easy to classify as the simple Saturni data.

If the number of hidden units is carefully chosen, the variance can vary a lot without the accuracy changing much. On the other hand, if a small variance is used, the number of hidden units may be almost arbitrary, provided it is large enough to capture the non-linearities in the data. A small variance regularizes the model, and may prevent it from overfitting even if the number of hidden units would make overfitting possible.

Relying on a small variance and using an arbitrary but large number of hidden units has the problem, that an unnecessary large network size slows down the computations. Finding a model which is parsimonious with the hidden units and gives optimal accuracy required careful adjustment of both parameters with the artificial data. Also in the experiments with the real data sets, the weight variance had a noticeable effect on the results. We therefore think that ELM model selection should consider weight variance as an adjustable parameter.

Model selection with two parameters is slower than with only one, but since the two control the same property (model nonlinearity), some simplification may be possible. If the behavior seen in our artificial data generalizes to natural data sets, a full grid search over the ranges of both parameters may not be necessary. It appears like a reasonably safe strategy to first select a good range for hidden units, and inside that range, to perform full model selection to find a good variance.

7.3   On Properties of ELM

Authors of [1] promote ELM by speed, dependence on a single parameter, small training error and good generalization performance. These claims have often been repeated by subsequent authors, but we have not come upon much discussion of them. Here we present some comments on these properties.

Training of a single ELM network is fast, provided the number of hidden units is small. *Speed of training* as the whole, however, depends also on the number of training runs. Model selection may require considerable number of repetitions, since all sensible parameter combinations should be considered. Further, due to the random nature of ELM, any runs must be repeated several times if average performance is to be assessed.

Complexity of model selection is determined by the *number of parameters*, and the ranges used for them. One parameter is the weight variance that we discussed above. Another is the number of hidden units. The only theoretically motivated upper limit for the number of hidden units to try is N (which is enough for zero training error). At that limit, computing pseudoinverse corresponds to ordinary inversion of an $N \times N$ matrix, with a complexity of $O(N^3)$. In practice, smaller upper limits are used.

Generally, *small training error* and good generalization may be contradictory goals. ELM has been proved [1] to be able to perfectly classify the training data if the number of hidden units equals or exceeds the number of data points. This behavior, though important in proving the computational power of ELM, is usually not desirable in modeling. A model should generalize, not exactly memorize the training data. This view is indirectly acknowledged in practical ELM work, where the number of hidden units is much smaller than N. This may prevent the ELM network from overfitting to the training data, a factor usually not discussed in ELM literature.

The *generalization ability* of ELM is attributed to the fact that computing output layer weights by pseudoinverse achieves a minimum norm solution. This is motivated by a work of Bartlett [17], who showed the generalization ability of a neural network to relate to small norm of weights. However, Bartlett's work considers the neural network as whole, not only the output layer. Although ELM minimizes the norm of output layer weights, the norm of the hidden layer weights depends on the variance parameter, and does not change in ELM training.

In the hidden layer, generalization ability is related to the operating point of the hidden unit activations, discussed in Sect. 5. A model with small hidden layer weights is nearly linear, and generalizes well. A highly non-linear model, produced by large weights, is more prone to overfitting. Therefore, conclusions about generalization ability of ELM should not be based on the output weights only.

References

1. Huang, G.B., Zhu, Q.Y., Siew, C.K.: Extreme learning machine: Theory and applications. Neurocomputing 70, 489–501 (2006)
2. Rasmussen, C.E., Williams, C.K.I.: Gaussian processes for machine learning. MIT Press (2006)
3. Frénay, B., Verleysen, M.: Using SVMs with randomised feature spaces: an extreme learning approach. In: Proc. of ESANN, pp. 315–320 (2010)
4. Miche, Y., Sorjamaa, A., Bas, P., Simula, O., Jutten, C., Lendasse, A.: OP-ELM: Optimally pruned extreme learning machine. IEEE Transactions on Neural Networks 21, 158–162 (2010)
5. Zhu, Q.Y., Qin, A.K., Suganthan, P.N., Huang, G.B.: Evolutionary extreme learning machine. Pattern Recognition 38, 1759–1763 (2005)
6. Penrose, R.: A generalized inverse for matrices. Mathematical Proceedings of the Cambridge Philosophical Society 51, 406–413 (1955)
7. McCullagh, P., Nelder, J.A.: Generalized linear models, 2nd edn. Monographs on statistics and applied probability, vol. 37. Chapman & Hall (1989)
8. Williams, C.K.I.: Computation with infinite neural networks. Neural Computation 10, 1203–1216 (1998)
9. Cho, Y., Saul, L.K.: Kernel methods for deep learning. In: Bengio, Y., Schuurmans, D., Lafferty, J., Williams, C., Culotta, A. (eds.) Proc. of NIPS, vol. 22, pp. 342–350 (2009)
10. Vert, J.P., Tsuda, K., Schölkopf, B.: A primer on kernel methods. In: Schölkopf, B., Tsuda, K., Vert, J.P. (eds.) Kernel Methods in Computational Biology, pp. 35–70. MIT Press (2004)
11. Asuncion, A., Newman, D.: UCI machine learning repository (2007)
12. Guyon, I., Gunn, S.R., Ben-Hur, A., Dror, G.: Result analysis of the NIPS 2003 feature selection challenge. In: Proc. of NIPS (2004)
13. Minka, T.: Expectation propagation for approximate Bayesian inference. In: Proc. of UAI (2001)
14. Young, G., Householder, A.S.: Discussion of a set of points in terms of their mutual distances. Psychometrika 3, 19–22 (1938)
15. Golub, G.H., Van Loan, C.F.: Matrix computations, 3rd edn. The Johns Hopkins University Press (1996)
16. Parviainen, E., Riihimäki, J., Miche, Y., Lendasse, A.: Interpreting Extreme Learning Machine as an approximation to an infinite neural network. In: Proc. of KDIR. INSTICC (2010)
17. Bartlett, P.L.: The sample complexity of pattern classification with neural networks: the size of the weights is more important than the size of the network. IEEE Transactions on Information Theory 44, 525–536 (1998)

Visually Summarizing Semantic Evolution in Document Streams with Topic Table

André Gohr[1], Myra Spiliopoulou[2], and Alexander Hinneburg[1]

[1] Martin Luther University, 06099 Halle Saale, Germany
[2] Otto-von-Guericke University, 39016 Magdeburg, Germany

Abstract. We propose a visualization technique for summarizing contents of document streams, such as news or scientific archives. The content of streaming documents change over time and so do themes the documents are about. Topic evolution is a relatively new research subject that encompasses the unsupervised discovery of thematic subjects in a document collection *and* the adaptation of these subjects as new documents arrive. While many powerful topic evolution methods exist, the combination of learning *and* visualization of the evolving topics has been less explored, although it is indispensable for understanding a dynamic document collection.

We propose Topic Table[1], a visualization technique that builds upon topic modeling for deriving a condensed representation of a document collection. Topic Table captures important and intuitively comprehensible aspects of a topic over time: the importance of the topic within the collection, the words characterizing this topic, the semantic changes of a topic from one timepoint to the next. As an example, we visualize content of the NIPS proceedings from 1987 to 1999.

Keywords: Visualization, Topic modeling, Evolving topics, Summarizing dynamic document collections, Stream analysis.

1 Introduction

Electronic document collections proliferate and grow continually. Research on summarizing document collections [7,13,14] aims at helping readers to keep up with changing content in growing collections. Topic learning is successfully pursued with probabilistic topic models like PLSA [12] or LDA [3] that derive a small number of groups of words that appear frequently together in documents. These groups of words are statistically represented as distributions over words called topics, of which a few most likely words are listed. Humans can often interpret such groups of words as themes. This interpretation task becomes more tedious in the dynamic case, when multiple batches of documents arrive over time. For each batch, a new topic model is derived and each topic model outputs a number K of topics (word distributions). Thus, in the end, the human user is required to look at the top words of a number of topics, which is K times number of batches in total, to read the summarization of the topic evolution of a growing document collection. To ease that task, we propose a new visualization method called "Topic Table" that visualizes inferred topics in combination with main pieces of information

---

[1] R script at http://users.informatik.uni-halle.de/~hinnebur

A. Fred et al. (Eds.): IC3K 2010, CCIS 272, pp. 136–150, 2013.

	1881	1890	1900	1910
Atomic Physics	force energy motion differ light measure magnet direct matter result	motion force magnet energy measure differ direct line result light	magnet electric measure force theory system motion line point differ	force magnet theory electric atom system measure line energy body
Neuroscience	brain move- ment action right eye hand left muscle nerve sound	movement eye right hand brain left action mus- cle sound ex- periment	brain eye movement right left hand nerve vision sound muscle	movement brain sound nerve active muscle left eye right nervous

Fig. 1. Box scheme for two topics (rows) over time (columns). Example topics were inferred from the Science corpus and are taken from the manuscript on dynamic topic models [2].

about these topics. This work is an extension of the study [9], where we had focused on topics associated with a tag in a social tagging system.

For a document collection Topic Table first derives from topics document prototypes that summarize the contents of documents. These topics may be learned by any dynamic topic modeling method; we use AdaptivePLSA [8], where a series of topic models is learned, *and* the topics learned at each timepoint are semantically linked to those appearing in the next timepoint. To visually present the evolving topics, Topic Table deals with following design challenges: (i) to use the canvas efficiently and (ii) display the dominant information (main document prototypes) while retaining less dominant information.

By visually presenting these evolving topics to the reader, Topic Table makes it possible to inspect topics and discover hidden structures. To this purpose, we extend the box scheme that is often used in topic modeling literature to present topics. An example of the conventional box scheme is depicted in Figure 1: each row corresponds to one evolving topic over time. This constraint requires the model to match topics of directly subsequent batches. For each topic, the most likely words are listed in a box. Columns correspond to points in time: vertically stacked boxes represent topics that are active in the batch of documents arrived at the same time. Beside the most likely words, other pieces of information that might be helpful for the reader are often neglected.

Topic Table delivers at a glance four additional pieces of information in an intuitive manner to the observer, namely i) new words not seen in the word lists of topics before, ii) the relative strength of topics, iii) an indication on how much a topic changes/evolves from one point in time to the next one, as well as iv) similarity among all topics at different points in time. All these pieces of information might assist the reader in identifying and assessing contents of the document stream; e.g. the strength of topics might be especially helpful for readers if they are interested in the relative prevalence of topics in the studied document collection. As Topic Table takes only word lists and inter-topic-distances derived from arbitrary sets of topics, it might be combined with any topic model that can infer topics and their evolution in dynamic document collections. This general applicability of Topic Table makes it a promising visualization tool for topic evolution.

The remainder of the paper is structured as follows. We discuss related work in the next section. Then, we formally introduce document streams in Section 3 and explain document prototypes summarizing document contents, a concept that allows to use Topic Table in combination with many different methods for summarizing document collections. Further on, we explain how to derive document prototypes using topic models and discuss the options to prevent the problem of label switching between topics inferred at timepoints. Next, in Section 4, we present details of Topic Table, and we present a case study (Section 5), in which we visualize the contents of research articles published at NIPS conferences from 1987 to 1999. The last Section 6 concludes the manuscript.

2   Related Work

Topic Table combines probabilistic topic learning with visualization over an evolving document collection. Dynamic topic modeling methods like [17,2,21,1,20,5,8,15] express a topic as a constellation of characteristic words, and adapt topics (and their word descriptions) over time. Some methods allow for words to become obsolete and irrelevant while others emerge [1,5,8]. Capturing terminological evolution is indispensable for visualizing the semantic evolution, because that evolution is inevitably associated with the increased importance of some words that were irrelevant or unknown in the past [8].

Topic modeling methods are often accompanied by some visualization aids, but these are often rudimentary. For example, [2] lists the most likely words for topics at each timepoint, and further plot the probability of certain words for a topic at different timepoints. In this way, they give hints about how a topic evolves. Mei et al. [16] and Boyd-Graber et al. [4] point out that human inspection is facilitated by choosing the most descriptive words for a topic rather than the most likely ones.

Statistical approaches for summarizing streaming document collections and visualization are combined in MemeTracker [15], TopicMaps [18], and Topics Over Time [21], while TimeFall [7] uses Minimum Description Length to match clusters over an evolving graph before visualizing them.

TimeFall [7] has been designed for visualization of clusters in evolving social networks, but the clusters it depicts are actually *communities of words* visualized as boxes. The authors describe an example visualization as follows: "In Figure 1 each box represents a community of words a user profile topic. Each topic is thus described by a set of keywords that are characteristic for the corresponding topic. Each line (horizontal group of topics) represents a time step, and arrows between the topics from the adjacent time steps represent the evolution of the topics. Notice the splits and merges of the topics over time." (cf. [7], page 1). The size and weight of each topic are printed inside the box, together with the characteristic keywords, making the visualization a bit cumbersome for inspection, since all boxes are of same size.

MemeTracker [15] studies phrases and their frequency as they occur in news channels like blogs or information media portals. For visualizing these frequencies, Meme-Tracker uses a ThemeRiver approach [11], which applies a river metaphor to visualize changes in document contents over time. For predefined phrases ThemeRiver [11] visualizes their document frequencies[2] in streaming document collections at several timepoints. For each phrase, the width of a corresponding flow (from left to right) is mapped to the respective document frequency over time. Flows are plotted in different colors to make perceptual discrimination possible and are combined on top of each other to a river that flows from the right of the canvas to the left through time. Space of the canvas is wasted whenever the width of the ThemeRiver is small and integration of text into narrow curved flows is difficult. ThemeRiver could be adapted to visualize the changing strength of topics through time as it is done for Topics over Time [21], but combining it with additional pieces of information is difficult due to space constraints; e.g. the ThemeRiver visualization used in [21] has not enough space to print the topic headline of each topics.

Other sophisticated approaches for visualizing document collections are Topic Map [18] and Topic Model Browser[3]. In contrast to Topic Table that aims at presenting topics and their evolution through time, Topic Map and Topic Browser summarize static collections. Topic Map uses topic modeling for deriving topic distributions for documents and map these into the 2-dimensional space. This approach is meant as help for observing similar documents in collections like search results. Topic Browser is an interactive visualization for presenting word distributions of topics, topic strengths and overviews of topic (headline and most likely words). It extends the simple box design by allowing the observer to interactively change among different websites each one presenting distinct pieces of information (topic strengths, word lists, topic distributions etc.).

3 Document Prototypes from Streams of Documents

The aim is to provide readers of streaming documents with a summary of contents that might be inspected for clarification. Therefore, we introduce the concept of document prototypes that are – in combination with additional pieces of information – presented to the reader by the developed visualization technique Topic Table. For a collection of documents, document prototypes abstract from these documents and give a condensed description of the summarized documents. Consequently, the number of prototypes is chosen to be substantial smaller than the number of documents and thus inspecting prototypes is more efficient than inspecting all documents. For summarizing documents' content, prototypes should give the reader hints for deducing themes the documents are about. Thus, a document prototype is a list of words that often co-occur in documents of the summarized collection. To allow for a fast perception by the reader, these lists should consist of a few words only, e.g. 5 up to about 25.

The concept of document prototypes makes it possible to separate the visualization technique from the inference method used to derive document prototypes. In general, arbitrary dimension reduction approaches are suited for learning prototypes; examples include principal component analysis, non-negative matrix factorization, and topic modeling [18]. Following [18], we use probabilistic topic modeling of documents for learning topics, which capture patterns of words that often co-occur. Because topics are

---

[2] Number of documents containing the word.

[3] http://www.cs.princeton.edu/ blei/topicmodeling.html

distributions over *all* words of the vocabulary [12,3] they are less suitable to summarize document collections. Instead, we derive a document prototype from each inferred topic as follows: the prototype consists of the N_{top} most likely words of that topic.

3.1 Streams of Documents

Before we discuss some subtle points of inferring prototypes and their evolution over a streaming document collection, we formally define such a stream and explain, how we deduce batches of subsequent documents from such a stream.

A stream of documents is an ordered sequence of documents defined by a sequence of their document IDs $D = \langle d_1, \ldots, d_N \rangle$. Ordering could be induced e.g. by time stamps of documents. The content of documents in a sequence reflect topics that occur in this stream. Examples of document streams are news articles over time, or conference proceedings over several years.

To visually summarizing the stream of documents D, we use a sliding window. This window might cover l successive documents [10]. The window covering l documents typically shifts by one document at a time, i.e. the least recent document within the sliding window is forgotten when a new document arrives. Such a fine-grained analysis is impractical for our purposes, because semantics of contents do not change when a single document arrives. We rather slide the window by l_{new} documents, i.e. the window slides to a new position after l_{new} new documents have been arrived. The sliding window at position i covers the following partial sequence of document IDs $D^i = \langle d_{r(i)}, \ldots, d_{r(i)+l} \rangle$ of D with $r(i) = 1 + (i - 1)l_{new}$. We call these partial sequences also batches in the following. Figure 2 shows an example with 22 documents arrived so far. The sliding window covers $l = 7$ documents and it slides by $l_{new} = 5$ documents. The sliding window at each of its four positions covers a certain subsequence D^1, D^2, D^3 and D^4 of the stream D of documents.

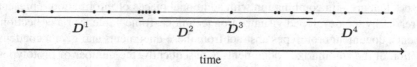

time

Fig. 2. Stream of documents (small black dots). Positions of the sliding window of length $l = 7$ are represented by horizontal lines. The sliding window shifts forward by $l_{new} = 5$ new documents and it covers at its four positions the following document subsequences D^1, D^2, D^3 and D^4.

Alternatively, the window could be defined to cover all documents arriving during a certain time period, e.g. one month, and it might be shifted by some constant amount of time again defining a sequence of document batches. Decision on the kind of the sliding window depends on the specific problem to be solved.

3.2 Evolving Document Prototypes

We derive document prototypes for each batch of documents of the document stream. By inspecting the prototypes for subsequent partial document collections, the reader gets an overview of the contents in each of these batches and how the contents change over time.

A subtle point of inferring the evolution of prototypes over time is that, at subsequent points in time, inferred prototypes have to be semantically linked to each other. This is a prerequisite of Topic Table because it presents in one row all prototypes at different points in time that are snapshots of the same topic through time. In other words, the problem here is to prevent label switching of topics of subsequent models.

Because we suggest to derive prototypes from topics learned by topic models, we identify three different general approaches for coupling topics of subsequent models. The model at timepoint i is denoted by ζ^i and each model learns K topics.

Post-hoc coupling means that models are learned independently of each other. Afterward, topics of subsequent models that are most similar to each other are matched. For example, say topic $\bar{t}$ of model ζ^{i+1} is most similar to topic $\hat{t}$ of model ζ^i compared to all topics of the latter model. Then topic $\bar{t}$ is identified as the successor of topic $\hat{t}$. This approach depends on the similarity measurement among topics. An example of this approach is the work of Mei and Zhai [17].

Coupling by initialization is taken by AdaptivePLSA [8], which we use for inferring topics through time for this work, and the similar approach proposed by Chen and Chou [5]. Each topic model ζ^i is learned by an iterative inference algorithm namely the EM algorithm. The EM algorithm is a gradient method that finds local optima; it updates model parameters in each step so that they are better fitted to the data. As such, this algorithm depends on the start points of model parameters. Coupling by initialization means that for learning the model ζ^i we set the start points of these model parameters equal to the already learned parameters of the preceding model ζ^{i-1}. Consequently, the kth topic at timepoint $i + 1$ evolves from the kth topic at timepoint i.

Coupling by priors is used by the dynamic topic model [2]. Information about topics of successive models is transferred via priors for model parameters (multinomial distributions that define topics). A state space model makes the prior for topics of the later model ζ^{i+1} statistically dependent on the prior of the preceding model ζ^i.

We present a short introduction into topic modeling in the Appendix by exemplarily specifying probabilistic latent semantic analysis (PLSA) [12]. In our experiments, we use PLSA and its extension AdaptivePLSA [8] to streams of documents.

4   Visualizing Document Prototypes

So far we have discussed how we construct successive batches of documents $\langle D^1, \ldots, D^{\bar{N}}, \ldots \rangle$ from a document stream D. The contents of documents of these batches indicate emerging or abandoned themes of documents of stream D. For this work we use AdaptivePLSA [8,9] that learns a sequence of PLSA topic models $\zeta^1, \ldots, \zeta^{\bar{N}}, \ldots$ for the batches. We assume that $\bar{N}$ batches and models have been defined and learned so far. AdaptivePLSA uses coupling by inference (see Section 3.2) for inferring evolving topics while preventing label switching of topics among subsequent timepoints. Each topic model infers K topics. From each topic we derive a document prototype that is the most likely words of this topic.

Topic Table visualizes these comprehensible document prototypes and additional pieces of information for inspection by the reader. Studying document prototypes derived from topics over time might reveal changing aspects that documents of the stream D are about. Topic Table builds upon the table-like structure of the simple box design.

For K learned topics at $\bar{N}$ points in time, Topic Table has K rows and $\bar{N}$ columns. The cell (k, i) in row k and column i corresponds to the kth prototype (derived from the kth topic) of PLSA model ζ^i; rows correspond to document prototypes and columns correspond to snapshots of these prototypes over time. By arranging the kth document prototypes of all models in the kth row, Topic Table visually establishes a correspondence among them. This correspondence stems from the fact that topics of successive PLSA models are coupled by the learning procedure AdaptivePLSA; the kth topic of model ζ^{i+1} evolves from the kth topic of model ζ^i. Inspecting snapshots of prototypes along the kth row eases deduction of how they change over time.

Topic Table arranges four pieces of information. From background to foreground, these pieces are i) how much does a topic change between successive batches (timepoints), ii) how similar are different prototypes at different points in time, iii) how prominent a topic and the derived prototype are in document stream D during a certain period of time, and iv) the corresponding document prototypes. Figure 3 depicts how these pieces of information are visually presented by Topic Table.

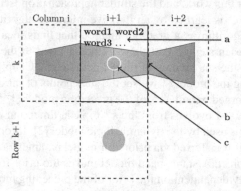

Fig. 3. The cell in row k and column $(i + 1)$ of Topic Table corresponds to the kth topic extracted from the documents D^{i+1} of the stream D. Features of Topic Table are (a) the N_{top} most likely words per topic that define the corresponding document prototype (bold face words are new and not part of the previous prototype), and (b) the river in the background of each row has a width at each border between the $(i + 1)^{\text{th}}$ and $(i + 2)^{\text{th}}$ column that is proportional to the similarity between the kth topic at $(i + 1)^{\text{th}}$ and $(i + 2)^{\text{th}}$ position of the sliding window, (c) the radius of the background circle is non-linearly proportional to the probability/strength of the corresponding topic/derived prototype relative to the strength of other topics in the same column (during the same time period), and (d) the color of the background river indicates similarities among all document prototypes at all timepoints: similar prototypes have a similar color of their background. A colored version of this figure is available online(footnote 1, page 136).

Firstly, Topic Table visually depicts how topics change between two successive points in time. To this end, Topic Table uses the metaphor of a *river* that "flows through time" and associate each evolving topic with a river along the corresponding row. Narrow parts of the river represent watergates that strongly separate what comes before and what afterward. These watergates indicate transitions at which the corresponding topic changes much. Topic Table visualizes the rivers as colored straps along each row, which correspond to the evolution of one topic over time. The width of each river changes between successive cells to indicate watergates. Successive cells, say (k, i) and $(k, i + 1)$, correspond to the kth topic of model ζ^i and ζ^{i+1}, respectively. These topics are multinomial distributions, which can be represented by two vectors ω_k^i and ω_k^{i+1}. The entries are probabilities of words of the respective vocabularies of documents of D^i and D^{i+1}. The more similar these two vectors are the more stable the corresponding topic is and the less it evolves. For measuring similarity, Topic Table uses cosine similarity, which is equal to 1 if both vectors point into the same direction and equal to 0 if the vectors are orthogonal to each other. Topic Table establishes the connection between the cosine similarity of successive topics ω_k^i and ω_k^{i+1} and the width of the background river at the transition of cell (k, i) to cell $(k, i + 1)$ as follows. When the similarity is maximal (equal to 1), then the width of the river is 0.7 times the height of the cell; this allows for a better perception of the background rivers. Further on, the width of the river is proportional to the fifths power of the similarity $\cos^5(\omega_k^i, \omega_k^{i+1})$; this emphasizes small differences among values close to 1. Vocabularies of D^i and D^{i+1} are likely to be different. To compute cosine similarity between the vectors ω_k^i and ω_k^{i+1}, which might be defined in different spaces, we embed them into the joint space defined by the union of both vocabularies. Because the rivers are drawn along a row they combine cells of one row and thereby strengthen the perception of rows and hence conveys the evolution of the document prototypes through time.

Secondly, Topic Table visualizes the similarity between different topics at different points in time by coloring the background rivers: similar topics and corresponding document prototypes get the same background color. Topic Table uses multi-dimensional scaling (MDS) for projecting at once all $\bar{N} \times K$ topics ω_k^i into a 2D plane of the 3D RGB color space. For computing MDS, Topic Table uses the cosine distance matrix, which contains pairwise cosine distances (equal to $1-$ cosine similarity) for all topics. MDS projects high dimensional points somewhere into the 2D plane. Topic Table linearly transforms the resulting 2D coordinates such that all points lie in the intervals $[0, 255] \times [0, 255]$. These 2D coordinates $((x, y)$ for each topic) are mapped directly to corresponding RGB colors R$= x$, G$= 161$, B$= y$.

A third useful information is the relative strength of the learned topics and corresponding document prototypes. This kind of information is helpful in two respects. First, a user might want to study only the strongest document prototypes at each timepoint. Second, a user might want to inspect at a glance temporal patterns; how the strength of a topic changes over time. Each PLSA model ζ^i allows to derive the probabilities of each extracted topic $p^i(z{=}k)$. All these probabilities sum to one $1 = \sum_{k=1}^K p^i(z{=}k)$ for the ith studied time period or in other words for each partial collection D^i. A large probability indicates a topic that is prevalent in documents D^i. Topic Table visualizes these probabilities by circles in the center of each cell (see (c) in Figure 3); the probability is mapped to the area of the circle. We follow [19] and map the probability of a topic to the circle area in a nonlinear way to enhance human perception of differences in the quantity; the radius of the circle visualizing the probability of a topic k is direct proportional to $p^i(z{=}k)^{5/7}/\sqrt{2\pi}$. The circles are depicted in the background of the cell on top of the background river. Studying all circles of a column top-down gives a fast impression about what topics are the most dominant ones at a certain time period. Inspecting the circles along a row allows to deduce how the relative strength of the corresponding topic changes over time. Because circles are positioned in the center of the cells they enhance the perception of the structure of Topic Table.

Last, Topic Table shows the document prototypes for each topic consisting of the most likely words. Topic Table lists these words in the foreground of the corresponding cells. The number of words that constitute a document prototype is not fixed. Common choices are ten to twenty words. An experienced user may need only ten-word prototypes ($N_{top} = 10$) to deduce what they are about. Consequently, Topic Table lets the user decide how many words should constitute the document prototypes. As an additional help for perceiving how prototypes change through time, Topic Table highlights words of the kth prototypes at timepoint $i + 1$ that are not part of the previous k^{th} prototype at timepoint i.

By displaying from background to foreground background rivers, circles and words on top of each other, Topic Table uses the canvas efficiently. And because Topic Table uses always one cell per document prototype, Topic Table does not suppress but visually retain less dominant prototypes.

5   Visually Summarizing Contents of NIPS Documents from 1987 to 1999

We exemplarily apply AdaptivePLSA to infer topics over time from NIPS documents from 1987 to 1999 and visualize these topics by Topic Table for visually summarizing NIPS proceedings.

5.1   Data Preparation and Parameter Setting

We use NIPS data[4] provided by Sam Roweis. Applying standard text preprocessing, i.e. Porter stemming and removal of English stop words, and removing the most frequent 50 words of the vocabulary we end up with 1740 documents, a vocabulary of 8621 words and about 1.7 million word tokens. Sorted by year of publishing, all these documents form a stream of NIPS documents.

Three main parameters influence learning with AdaptivePLSA and visualization with Topic Table. First, the length of the sliding window determines the number of documents a particular PLSA model is trained on. We define the sliding window such that it covers all NIPS documents of three successive years; this has the consequence that the number of documents described by Topic Table varies among different columns (see bottom of Figure 4). Thereby we also specify implicitly that a PLSA model is computed at a time scale of three years; the difference in time between the last and least covered document. A second parameter is how fast the sliding window moves over the document stream of NIPS documents. Without making further assumptions it can be meaningfully varied between one and three years; it would also be possible to let it jump by more than three years yielding very distinct snapshots out of the whole studied period of time from 1987 to 1999. We let the sliding window move by two years to its next position. As a result, successive batches of documents overlap by one year

---

[4] www.cs.nyu.edu/~roweis/data.html, file nips12raw_str602.mat

what strengthens the evolution of the kth topic of model ζ^i into the kth topics of the next model ζ^{i+1}. As a short annotation for each position of the sliding window we use the last covered year (see top of Figure 4), e.g. $[87, 88, 89]$ is denoted by 1989 (D^1), $[89, 90, 91]$ (1991; D^2), $[91, 92, 93]$ (1993; D^3) and so on. Third, the number K of hidden topics over time learned by each PLSA model affects the roughness of the summary. A reasonable choice is $K \ll N$ with $N = \min\{|D^i| \mid 1 \le i \le \bar{N}\}$ if a rough summary of the contents of documents is desired. Consequently, we set K equal to 5, and we show the top 15 words per topic.

5.2   Topic Table for NIPS Documents from 1987 to 1999

We present the resulting Topic Table in Figure 4. At a first glance, we observe that each background river has its own particular color. We conclude that the kth topics at different timepoints are relatively close to each other but that these are relatively dissimilar to other topics with index $\hat{k} \ne k$.

Closer inspecting Topic 1, we find in 1989 terms like *circuit* and *hopfield* that stand for specific architectures of neural networks. Terms like *implement*, *chip* and *matrix*, which we find from 1989 through 1993, might stand for research dealing with development of specific hardware using concepts of neural networks. From 1993 to 1999, we find terms like *theorem*, *theory*, *minimum,complex*, *loss*, *converg*, and *bound* that seem to represent research on complexity and learning theorems of neural networks.

The background river of Topic 2 first is relatively slim at the beginning and gets broader later on; the second topic changes more during transitions from 1989 to 1991 and from 1991 to 1993 as during later transitions. At the beginning we find terms like *cortex*, *simul*, *synaps*, and *oscillat* that deal with biological perspectives on neural networks. Later, in 1991 and 1993, we find *visu, field, eye, object, locat*, and *region*. These terms indicate research on image preprocessing using neural networks. Later on, terms like *motion*, *movement*, *activity* might represent research on processing of dynamic images, e.g. recorded by cameras. This topic further develops until 1999 and gets aspects of controlling robots (*head, orientat, motor, spaty*, and *task*) and learning to respond (*human, respons*).

Topic 3 is in general about signal processing. It starts in 1989 and 1991 with biological terms like *synaps, fire, spike, signal*. These terms seem to stand for biological research on how synapses process signals. Later from 1991 through 1999 terms like *nois, filter, channel*, and *sourc* occur that indicate research on implementing filters for signal processing using neural networks. The continuous change of the color of the background river from pink/peach in 1989 to lavender in 1999 visually emphasizes the change of the third topic from a biological perspective in signal processing toward a technical perspective.

The terms *class, classify*, or *recognit* are always among the top 2 words of Topic 2 from 1989 to 1999. We conclude that Topic 4 is about pattern recognition using neural networks. Subtopics are how decision rules are encoded by the structure of neural networks (*rule, code, architecture* in 1989 and 1991), speech recognition (*speech, word* from 1989 to 1999), segmentation tasks (*segment* in 1991 and 1993), different designs of neural networks for classification (*tree, architecture, combin, hmm* from 1995 to 1999), and face recognition (*face, cluster* in 1999).

	1989	1991	1993	1995	1997	1999
Topic 5	map propag equat back converg energy grady solut requir local simul term visu rate decis	**predict task** grady local propag back **dynam node** rate solut equat **minim** **step** term **optimal**	predict opti- mal task term local grady **select** equat **converg** step solut dynam minim **trajectory measur**	optimal pre- dict step con- verg term **nois** task **sampl** local equat **gaussian** minim mixtur **cluster** grady	gaussian op- timal predict sampl step nois **density** term **compon ma- trix** mixtur local likeli- **hood** converg **condit**	gaussian sampl density mixtur step **pol- icy** compon likelihood **prior action** nois matrix **bayesian** condit optimal
Topic 4	net classify recognit node speech mem- ory signal represent class rule bit code sequ architectur region	recognit clas- sify speech **word** rule net **classif** architectur class **level** **segment** **task** sequ **rate predict**	recognit class classify word rule speech classif net **structur** level segment **represent** **charact** sequ architectur	recognit clas- sify class classif rep- resent word sequ speech tree node repres struc- tur **context** net charact	classify recog- nit class tree classif word represent ex- **pery combin** sequ task label rate decis hmm	class tree clas- sify recognit classif **cluster** word **similar** face label select **pre- dict** expery **measur node**
Topic 3	fire respons synapt potenty rate spike object signal activity frequ effect prop- erty threshold increas tempor	object signal respons fire frequ spike synapt **nois** activity **visu** rate **stimulu** potenty **view** effect	signal respons spike **dynam** fire frequ ac- tivity synapt **oscillat** object nois **tempor** rate potenty **synaps**	signal spike ac- tivity respons fire synapt frequ oscillat rate **correl** channel fil- ter synaps **simul** tempor	signal spike frequ respons filter channel fire **circuit** rate tempor correl **stimulu** synapt **nois** analog	signal spike frequ rate synapt **sourc** circuit respons nois channel fire **compon** **synaps** stim- ulu **activity**
Topic 2	synaps activity simul structur represent local cortic cortex role repres field fig re- spons oscillat present	**map** field **visu eye re- cept** respons cortex simul activity repres local **posit ve- loc develop motion**	visu map posit eye field **direct move- ment** motion activity veloc repres **object locat region target**	map direct field visu posit object motion **motor** eye **head represent** movement **dy- nam orientat** locat	visu object di- rect map mo- tion field posit orientat **spaty** eye **view** se- lect head cen- ter locat	visu object field direct **respons** map **task local** motion spaty orientat posit **human** locat **represent**
Topic 1	circuit memory matrix chip fig analog operat hopfield implement equat stabl solut threshold store dynam	chip imple- ment analog circuit mem- ory threshold **bound bit** matrix equat operat **node size poly- nomy term**	implement bound circuit analog chip threshold **the- orem** bit size node polynomy **complex net** memory **defin**	node bound im- plement chip **dynam** ana- log **distanc** threshold com- plex **operat** **perceptron** size bit net **fix**	bound node **defin** dynam **loss matrix** equat class size **grady** **theorem** fix **solut** term **converg**	bound solut **kernel** theo- rem node defin equat term **support the- ory minim** matrix obtain loss machin
	286	388	415	436	455	452

Fig. 4. Topic Table for NIPS data from 1987 to 1999. Numbers at the bottom indicate the number of summarized documents. We recommend to use Adobe Reader for viewing Topic Tables because other PDF viewer might inadequately visualize shadings. A colored version of this figure is available online(footnote 1, page 136).

Topic 5 is mainly about learning approaches. Terms like *back propag*, *optim*, *converg*, *energy*, *grady*, and *trajectory* from 1989 to 1993 seem to stand for back propagation learning of neural networks and aspects thereof. Later on we find terms indicating learning (Gaussian) mixture models: *mixtur*, *cluster*, *likelihood*, *compon*, *sampl density* that we find from 1995 to 1999. This aspect later, in 1999, evolves into Bayesian learning: *bayesian*, and *prior*.

Inspecting the gray circles across all cells, which for each column represent the relative strength of the corresponding topic and document prototype among all topics of this column, we find that Topic 4 and Topic 5 are the most prevalent topics from 1989 to 1999. The majority of papers about neural networks will refer to learning principles like back propagation; back propagation is indeed the most prevalent approach of training neural networks. Although artificial neural networks are capable of approximating almost each function of input variables to one or several output variables, and hence are used e.g. also for regression, classification tasks are the most dominant field of applied neural networks.

Colors of the background rivers are distinct and indicate in agreement with studied document prototypes that the inferred five ($K = 5$) topics over time are relatively distinct.

6   Conclusions

We propose Topic Table, a visualization technique for studying content of streaming document collections and its evolution. Topic Table visualizes document prototypes meant as summarizing documents. For deducing prototypes, unsupervised learners like topic models are suited; a prototype is the set of a few most likely words per topic. Consequently, we set up Topic Table so that it might be used in combination with many different topic models. In addition to prototypes, Topic Table enriches the visualization with four additional pieces of information that are helpful for deducing knowledge about document contents and its evolution. These pieces are: similarity among all topics at all timepoints, indication on how much one topic changes between successive points in time, strengths of topics, and highlighting of new emerging words of prototypes.

We apply Topic Table to visually summarize document contents over time of the NIPS conference proceedings from 1987 to 1999. We learn five topics and visualize derived document prototypes by Topic Table. We find different aspects like image processing e.g. artificial neural networks, biological neural network, mixture modeling, Bayesian learning, support vector machines. Topics on learning artificial neural network and applications on classification, regression and prediction are the most prevalent topics during the studied time period. From the visualized indication on how much a topic changes between successive timepoints, we conclude that the topics do not change substantially between successive timepoints. A reason could be the small number of inferred topics; these concentrate on major strong subjects of the NIPS proceedings, which do not change much over time. This finding is visually supported by different background colors that are assigned to the prototypes over time. We find that each prototypes does not change its background color over time but different prototypes have particular different colors.

Because of its capability of being combined with many topic models, Topic Table is a promising tool for visualizing and summarizing document contents changing over time.

References

1. AlSumait, L., Barbara, D., Domeniconi, C.: On-line LDA: adaptive topic models for mining text streams with applications to topic detection and tracking. In: ICDM (2008)
2. Blei, D., Lafferty, J.: Dynamic topic models. In: ICML (2006)
3. Blei, D.M., Ng, A.Y., Jordan, M.I.: Latent Dirichlet Allocation. J. Mach. Learn. Res. 3, 993–1022 (2003)
4. Boyd-Graber, J., Chang, J., Gerrish, S., Wang, C., Blei, D.: Reading tea leaves: How humans interpret topic models. In: Neural Information Processing Systems, NIPS (2009)
5. Chou, T.-C., Chen, M.C.: Using incremental PLSI for threshold-resilient online event analysis. IEEE Trans. on Knowl. and Data Eng. 20(3), 289–299 (2008)
6. Dempster, A.P., Laird, N.M., Rubin, D.B.: Maximum likelihood from incomplete data via the EM algorithm. Journal of the Royal Statistical Society, Series B (Methodological) 39(1), 1–38 (1977)
7. Ferlez, J., Faloutsos, C., Leskovec, J., Mladenic, D., Grobelnik, M.: Monitoring network evolution using MDL. In: Proceedings of IEEE Int. Conf. on Data Engineering (ICDE 2008). IEEE (2008)
8. Gohr, A., Hinneburg, A., Schult, R., Spiliopoulou, M.: Topic evolution in a stream of documents. In: SIAM Data Mining Conf. (SDM 2009), Reno, NV, pp. 378–385 (April-May 2009)
9. Gohr, A., Spiliopoulou, M., Hinneburg, A.: Visually summarizing the evolution of documents under a social tag. In: International Conf. on Knowledge Discovery and Information Retrieval (KDIR 2010), Valencia, Spain, pp. 85–94. SciTePress Digital Library (October 2010)
10. Guha, S., Meyerson, A., Mishra, N., Motwani, R., O'Callaghan, L.: Clustering data streams: Theory and practice. IEEE Trans. of Knowlende and Data Eng. 15(3), 515–528 (2003)
11. Havre, S., Hetzler, E., Whitney, P., Nowell, L.: ThemeRiver: Visualizing thematic changes in large document collections. IEEE Trans. Visualization and Computer Graphics 8(1), 9–20 (2002)
12. Hofmann, T.: Unsupervised learning by probabilistic latent semantic analysis. Machine Learning 42(1), 177–196 (2001)
13. Ipeirotis, P., Ntoulas, A., Cho, J., Gravano, L.: Modeling and managing content changes in text databases. In: Proceedings of the IEEE Int. Conf. on Data Engineering, ICDE 2005 (2005)
14. Jin, W., Srihari, R.K., Ho, H.H., Wu, X.: Improving knowledge discovery in document collections through combining text retrieval and link analysis techniques. In: Proceedings of the 2007 Seventh IEEE International Conference on Data Mining, pp. 193–202. IEEE Computer Society, Washington, DC, USA (2007)
15. Leskovec, J., Backstrom, L., Kleinberg, J.: Meme-tracking and the dynamics of the news cycle. In: Proceedings of the 15th ACM SIGKDD International Conference on Knowledge Discovery and Data Mining, KDD 2009, pp. 497–506. ACM, New York (2009)
16. Mei, Q., Shen, X., Zhai, C.: Automatic labeling of multinomial topic models. In: KDD, pp. 490–499 (2007)
17. Mei, Q., Zhai, C.: Discovering evolutionary theme patterns from text: an exploration of temporal text mining. In: SIGKDD, pp. 198–207. ACM, New York (2005)
18. Newman, D., Baldwin, T., Cavedon, L., Huang, E., Karimi, S., Martinez, D., Scholer, F., Zobel, J.: Visualizing search results and document collections using topic maps. Web Semantics: Science, Services and Agents on the World Wide Web 8(2-3), 169–175 (2010); Bridging the Gap–Data Mining and Social Network Analysis for Integrating Semantic Web and Web 2.0; The Future of Knowledge Dissemination: The Elsevier Grand Challenge for the Life Sciences

19. Cleveland, W.S.: The Elements of Graphing Data. Hobart Press, Summit (1985/1994)
20. Wang, C., Blei, D., Heckerman, D.: Continuous Time Dynamic Topic Models. In: Proceedings of ICML (2008)
21. Wang, X., McCallum, A.: Topics over time: a non-markov continuous-time model of topical trends. In: SIGKDD, pp. 424–433. ACM (2006)

Appendix: Topic Modeling

For batches of documents $D^1, D^2, \ldots$ we use AdaptivePLSA [8,9] for learning a sequence of topic models $\zeta^1, \zeta^2, \ldots$; one model for each batch. Each of the topic models ζ^i is a model as defined by probabilistic latent semantic analysis (PLSA) [12]. We briefly review PLSA here as an example of a topic model. With many other topic models, like latent Dirichlet allocation [3], PLSA has in common that it models topics as multinomial distributions over words of the vocabulary.

Documents of a batch D^i (in the following we omit superscript i for ease of presentation) are represented by co-occurrences $X = \{(d, w)_j\}_{1 \leq j \leq |X|}$ of document IDs d and word IDs $1 \leq w \leq |V|$. We denote the vocabulary of this batch by V. For each word in a document d the representation contains one co-occurrence (d, w) of the corresponding document ID and word ID. Several occurrences of the same word are encoded by several distinct co-occurrences.

PLSA infers K hidden topics for documents X. Topics are denoted by the unobserved variable z which takes values $1 \leq z \leq K$. Each topic is represented by a multinomial distribution $p(w|z)$ over word IDs $1 \leq w \leq |V|$. Documents are assumed to be linear combinations of these topics; the word distribution of each document is a weighted sum of the topics $p(w|d) = \sum_{z=1}^{K} p(z|d)p(w|z)$. Topics (word distributions $p(w|z)$ for $1 \leq z \leq K$) and their document specific weights are unknown and have to be inferred from data. The parameters of a PLSA model ζ are:

- **document probabilities** which form a vector δ with elements $\delta_d = p(d), d \in D^i$,
- **mixture weights** which form a matrix θ with elements $\theta_d = (\theta_{1d}, \ldots, \theta_{Kd})$ and $\theta_{kd} = p(z = k|d), d \in D^i, 1 \leq k \leq K$, and
- **topics** which form a second matrix ω with elements $\omega_{kw} = p(w|z = k), 1 \leq w \leq |V|, 1 \leq k \leq K$.

The likelihood of the data X given a learned PLSA model ζ is

$$p(X|\zeta) = \prod_{j=1}^{|X|} p((d, w)_j|\zeta)$$

$$p((d, w)_j|\zeta) = p(d_j) \sum_{z=1}^{K} p(w_j|z)p(z|d_j) = \delta_j \sum_{z=1}^{K} \omega_{zw}\theta_{zd}$$

The last line follows because words and documents are assumed to be conditionally independent if the hidden topic, from which the word was sampled, is known.

Informally, estimating the parameters of a PLSA model for some given data means to find topics and mixture weights such that the empiric distribution of words and documents $p(w, d)$ for the training documents are approximated as best as possible. The expectation maximization algorithm (EM) [6] is used for parameter estimated because it allows to estimate model parameters even in presence of hidden (unobserved) variables like topic variable z. The EM algorithm is an iterative method for finding local maxima of the likelihood (or posterior of the model parameters). More details about PLSA can be found elsewhere [12].

A Clinical Application of Feature Selection: Quantitative Evaluation of the Locomotor Function

Luca Palmerini[1], Laura Rocchi[1], Sabato Mellone[1],
Franco Valzania[2], and Lorenzo Chiari[1]

[1] Biomedical Engineering Unit, DEIS, University of Bologna
Viale Risorgimento 2, 40136 Bologna, Italy
{luca.palmerini,l.rocchi,sabato.mellone,lorenzo.chiari}@unibo.it
[2] Department of Neuroscience, University of Modena and Reggio Emilia
via Pietro Giardini 1355, 41126 Baggiovara (MO), Italy
f.valzania@ausl.mo.it

Abstract. Evaluation of the locomotor function is important for several clinical applications (e.g. fall risk of the elderly, characterization of a disease with motor complications). We consider the Timed Up and Go test which is widely used to evaluate the locomotor function in Parkinson's Disease (PD). Twenty PD and twenty age-matched control subjects performed an instrumented version of the test, where wearable accelerometers were used to gather quantitative information. Several measures were extracted from the acceleration signals; the aim is to find, by means of a feature selection, the best set that can discriminate between healthy and PD subjects. A wrapper feature selection was implemented with an exhaustive search for subsets from 1 to 3 features. A nested leave-one-out cross validation (LOOCV) was implemented, to limit a possible selection bias. With the selected features a good accuracy is obtained (7.5% of misclassification rate) in the classification between PD and healthy subjects.

Keywords: Feature selection, Clinical, parkinson's disease, Accelerometer, Selection bias, Nested cross validation.

1 Introduction

Evaluation of the locomotor function is important for several clinical applications (e.g. fall risk of the elderly, characterization of a disease with motor complications). The Timed Up and Go (TUG) is a widely used clinical test to assess balance, mobility and fall risk in Parkinson's disease (PD). The traditional clinical outcome of this test is its duration, measured by a stopwatch. Since this single measure cannot provide insight on subtle differences in test performances, instrumented Timed Up and Go tests (iTUG) have been recently proposed [1], [2]. These studies demonstrated the potential of using inertial sensors to quantify TUG performance. As stated in [2], quantitative evaluation is especially important for early stages of PD when balance and gait problems are not clinically evident but may be detected by instrumented analysis. The aim of this study is to find, by means of a feature selection process, the

A. Fred et al. (Eds.): IC3K 2010, CCIS 272, pp. 151–157, 2013.

best set of quantitative measures that can allow an objective evaluation of gait function in PD and could be considered as possible early biomarkers of the disease. Feature selection has recently been used in the field of Parkinson's disease to quantify the performance of a PD subject [3]; in the mentioned study the quantitative data came from force/torque sensors.

2   Methods

We examined twenty early-mild PD subjects OFF medication (Hoehn & Yahr ≤ 3, 62±7 years old, 12 males and 8 females) and twenty healthy age-matched control subjects (CTRL, 64±6 years old, 7 males and 13 females). The OFF condition in PD subjects was obtained by a levodopa washout of at least 18 hours and a dopamine agonist washout of at least 36 hours. Subjects wore a tri-axial accelerometer, McRoberts© Dynaport Micromod, on the lower back at L5 level. They performed three TUG trials (single task, ST) and three TUG trials with a concurrent cognitive task (dual task, DT), which consisted in counting audibly backwards from 100 by 3s. The TUG trial consisted of rising from a chair, walking 7m at preferred speed, turning around, returning and sitting down again. A schematic representation of the task is shown in Fig. 1.

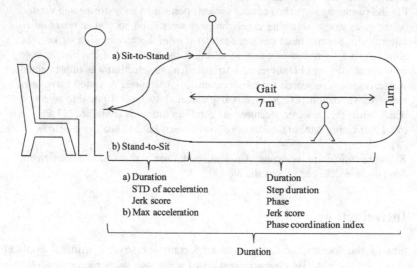

Fig. 1. Timed Up and Go Test and extracted measures

It has to be noted that a modified version of the TUG was used in this study, with a Gait section of 7m (instead of 3) to get more accurate measures of the various steps.

Several temporal (including total duration of the test), coordination and smoothness measures were extracted from the acceleration signals in different sections of the TUG. In Fig. 1 the main measures are reported.

Considering the Gait section, each stride (from one heel strike to the consecutive heel strike of the same leg) defines one gait cycle. The phase is determined by the ratio between the duration of the first step of the gait cycle and the entire duration of the gait cycle: a factor of 360 is used to transform the variable into degrees (360 degrees would correspond to the entire gait cycle) [4]. Among the other measures, phase coordination index measures the symmetry of gait [4] and jerk score (for both Sit-to-Stand and Gait sections) can be seen as an index of movement smoothness.

In the Gait section, jerk score and step duration were computed for each step; for the following analysis their averages across all the steps were considered, together with measures of variability between different steps (standard deviation, STD, and coefficient of variation, CV). Similarly, phase was computed for each gait cycle but only its average and variability measures were considered.

Jerk score (for both Sit-to-stand and Gait sections), Root Mean Square (RMS), and max value of acceleration, were computed along two orthogonal axes of the accelerometer: the first aligned with the direction of gait progression and coincident with the biomechanical anteroposterior (AP) axis of the body; the second in the left/right direction and coincident with the biomechanical mediolateral (ML) axis of the body.

For each measure, both in ST and in DT, we computed the mean value across the three repeated trials for the following analyses.

2.1   Feature Selection

The total number of measures (features) is higher (56, 28 for ST and 28 for DT) than the available samples (40 subjects). Therefore feature selection is necessary to avoid overfitting and to improve the performance of the classifiers. To select, from all the available features, the subset which has the best discriminative ability, a "wrapper" feature selection [6] was implemented: the objective function was the predictive accuracy of a given classifier on the training set. We used the following classifiers: linear and quadratic discriminant analysis (LDA and QDA, respectively), Mahalanobis classifier (MC), logistic regression (LR), K-nearest neighbours (KNN, K=1) and linear support vector machines (SVM). An exhaustive search among subsets of cardinality from one to three was implemented; the limit of three was chosen to permit a clinical interpretation of the result (it would be difficult to associate too many features with different aspects of the disease). Subsets of different cardinalities were considered separately.

The adopted procedure is similar to the one proposed by [3] where an exhaustive search of subsets of three features was performed. Still, in the present study, feature selection bias was also considered.

Since feature selection is part of the tuning design of the classifier, it needs to be performed on the training set, in order to avoid a possible bias (selection bias) in the final evaluation of the accuracy of the classifier [5]. The most common solution to this problem is to use a nested cross validation procedure [6]: the internal feature selection step is repeated for each training set resulting from the external cross validation. In this study, because of the small sample size (40), a leave-one-out cross validation (LOOCV) was implemented both for the feature selection steps and for the final evaluation of the classifier.

As it can be seen in Fig. 2, the external cross validation used for estimation of the accuracy of the classifier ($LOOCV_{ext}$) splits the dataset in 40 different training and testing sets (TR_i, TS_i $1 \leq i \leq 40$); for each TR_i, a different feature selection step was performed (FS_i, $1 \leq i \leq 40$). The objective function (predictive accuracy) of each feature selection was evaluated by an internal LOOCV ($LOOCV_{int}$). After each FS_i, a list of optimal subsets of features was generated: there was generally more than one subset with the same highest $LOOCV_{int}$ accuracy (more than one optimal subset). In the nested procedure TS_i should be classified from the classifier built with a single subset chosen by FS_i; in this study, since more than one optimal subset was found, it was not possible to make a unique choice. Moreover different FS_i's led to different lists of optimal subsets. So we decided to extract the subset which was selected as optimal more frequently over all the FS_i's (overall optimal subset, see Fig. 2). The number of times a certain subset was selected as optimal (selection times) can be seen as an index of how that subset is robust to changes in the training set, and therefore to selection bias. Eventually, the accuracy of the classifier (misclassification rate, MR) was computed by $LOOCV_{ext}$ for the overall optimal subset (see Fig. 2).

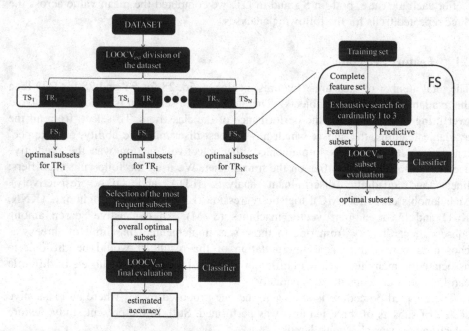

Fig. 2. Feature selection procedure

3   Results and Discussion

In Table 1 the results of the feature selection procedure for subsets of 3 measures are reported; the estimated accuracy is presented together with the *selection times* (the number of times a subset was selected as optimal among the 40 different feature selection procedures). Subsets of 3 measures were preferred since subsets of lower cardinality led to higher misclassification rates. It can be seen that a good misclassification rate could be achieved (7.5%-10%) by all the classifiers. As discussed in section 2, estimates of misclassification rates of subsets with higher *selection times* should be considered as more reliable, regarding selection bias, with respect to estimates with lower *selection times*.

The best subset from this point of view is the subset selected by the KNN classifier which is exclusively made of measures related to the jerk score in different sections of the TUG. It can then be seen that the smoothness of the movement during Gait and Sit-to-Stand is very important in discriminating between control and PD subjects.

Table 1. Results of the feature selection procedure

Class.	Overall optimal subsets	Task	Selection times /40	MR
LDA	RMS of AP acceleration during Sit-to-Stand	single task	32	7.5%
	Max AP acceleration during Stand-to-Sit	dual task		
	STD of the phase during Gait	dual task		
QDA	RMS of ML acceleration during Sit-to-Stand	single task	25	7.5%
	Max AP acceleration during Stand-to-Sit	single task		
	CV of the step duration during Gait	dual task		
LR	Jerk score of AP acceleration during Sit-to-Stand	single task	28	7.5%
	Jerk score of ML acceleration during Gait	single task		
	STD of the step duration during Gait	dual task		
KNN	Jerk score of AP acceleration during Sit-to-Stand	single task	36	7.5%
	Jerk score of AP acceleration during Gait	dual task		
	CV of the jerk score of ML acceleration during Gait	dual task		
MC	Jerk score of AP acceleration during Sit-to-Stand	single task	32	10%
	Jerk score of ML acceleration during Gait	single task		
	max AP acceleration during Stand-to-Sit	dual task		
SVM	Jerk score of ML acceleration during Sit-to-Stand	single task	25	7.5%
	CV of the jerk score of ML acceleration during Gait	single task		
	max AP acceleration during Stand-to-Sit	dual task		

Considering the overall optimal subsets from all the classifiers, the procedure always selected a measure related with the sit-to-stand and one or two measures related with the gait phase. In four subsets there is also a measure extracted during Stand-to-Sit. It should also be remarked that every subset presented in Table 1 is made of both single and dual task related measures.

These measures improve the discrimination power between CTRL and PD with respect to the traditional TUG duration (the best misclassification rate that can be obtained by using this single measure with the reported classifiers, in ST or in DT, is 35%), which interestingly was not selected in any of the overall optimal subsets. Moreover TUG duration alone was not significantly different between the two groups (as in [1] and [2]) and therefore it could not discriminate between CTRL and early-mild PD. Instead, considering various quantitative measures related to different parts of the TUG (see Table 1), allowed us to obtain good accuracy in the classification of PD subjects.

This accuracy would not have been obtained without feature selection; considering all the features altogether, the number of features is higher than the number of samples. In this case LDA, QDA and MC cannot be used because it is not possible to estimate the covariance matrix; similarly, in LR the model is over-parameterized and some coefficients of the logistic model are not identifiable. So the only classifiers that can be used without feature selection are KNN and SVM which, using all the features, have a MR of 52% and 20%, respectively; this reflects the importance of performing feature selection in this kind of datasets.

Furthermore it has to be noted that, even if our relatively small sample size limits the power of our data mining perspective, a nested cross validation was applied to limit the possible feature selection bias. Since it was not possible to follow the typical nested procedure (because several different combinations of features were selected as optimal), a value was derived which can be seen as an index of the reliability of the estimation of the misclassification rate.

4   Conclusions

The main result achieved by this work is that a set of few quantitative measures, derived from a clinical test for locomotor evaluation, can discriminate with a good accuracy between early-mild PD and CTRL subjects.

Further experiments should be made on new subjects to have an independent data set and validate these findings; in particular, the selected optimal measures could be tested on PD subjects in an earlier stage of their disease in order to check if they could also be used as early biomarkers of PD. On the other hand it should be investigated whether the presented measures remain valid and maintain their superiority over TUG duration for later stages of the disease. In fact, even if the presented subsets are optimal for classifying early-mild PD, there is no guarantee that they would be optimal to monitor the disease progression or to detect changes in gait patterns after a particular medical treatment; in this context, the next step will be a follow-up of the study with the same subjects.

Another future goal will be to assess if and how the TUG carried out under DT can add discriminative power with respect to the ST alone (as suggested by this study), since this would have important implications on the experimental design.

Acknowledgements. The authors wish to thank Luca Codeluppi, MD, and Valentina Fioravanti, MD, from the Department of Neuroscience, University of Modena and Reggio Emilia, Modena, Italy, for clinical supervision and assistance in data.

References

1. Weiss, A., Herman, T., Plotnik, M., Brozgol, M., et al.: Can an accelerometer enhance the utility of the Timed Up & Go Test when evaluating patients with Parkinson's disease? Med. Eng. & Phys. 32(2), 119–125 (2010)
2. Zampieri, C., Salarian, A., Carlson-Kuhta, P., Aminian, K., et al.: The instrumented timed up and go test: potential outcome measure for disease modifying therapies in Parkinson's disease. J. of Neurol., Neurosurg. & Psychiatry 81(2), 171–176 (2009)
3. Brewer, B.R., Pradhan, S., Carvell, G., Delitto, A.: Feature selection for classification based on fine motor signs of Parkinson's disease. In: 31st Annual International Conference of the IEEE Engineering in Medicine and Biology Society (2009)
4. Plotnik, M., Giladi, N., Hausdorff, J.M.: A new measure for quantifying the bilateral coordination of human gait: effects of aging and Parkinson's disease. Exp. Brain Res. 181(4), 561–570 (2007)
5. Simon, R., Radmacher, M.D., Dobbin, K., McShane, L.M.: J. Natl. Cancer Inst. 95(1), 14–18 (2003)
6. Kohavi, R., John, G.H.: Wrappers for Feature Subset Selection. Art. Intel. 97(1-2), 273–324 (1997)

Inference Based Query Expansion Using User's Real Time Implicit Feedback

Sanasam Ranbir Singh[1], Hema A. Murthy[2], and Timothy A. Gonsalves[3]

[1] Department of Computer Science and Engineering, Indian Institute of Technology, Guwahati
Guwahati-781039, Assam, India
ranbir@iitg.ernet.in

[2] Department of Computer Science and Engineering, Indian Institute of Technology, Madras
Guwahati-781039, Assam, India
hema@iitm.ac.in

[3] School of Computing and Electrical Engineering, Indian Institute of Technology, Mandi
Guwahati-781039, Assam, India
tag@iitmandi.ac.in

Abstract. Query expansion is a commonly used technique to address the problem of short and under-specified search queries in information retrieval. Traditional query expansion frameworks return static results, whereas user's information needs is dynamics in nature. User's search goal, even for the same query, may be different at different instances. This often leads to poor coherence between traditional query expansion and user's search goal resulting poor retrieval performance. In this study, we observe that user's search pattern is influenced by his/her recent searches in many search instances. We further propose a query expansion framework which explores user's real time implicit feedback provided at the time of search to determine user's search context and identify relevant query expansion terms. From extensive experiments, it is evident that the proposed query expansion framework adapts to the changing needs of user's information need.

1 Introduction

The task of query expansion [18] is the process of supplementing a search query with additional related terms or phrases to increase the chances of capturing more relevant documents. Traditional query expansion frameworks return static results, whereas user's information needs is dynamics in nature. User's search goal, even for the same query, may be different at different instances. This often leads to poor coherence between traditional query expansion and user's search goal resulting poor retrieval performance. For instance, if the query `jaguar` be expanded as the terms {`auto`, `car`, `model`, `cat`, `jungle`,...} and user is looking for documents related to `car`, then the expansion terms such as `cat` and `jungle` are not relevant to user's search goal. Therefore, it is important for the query expansion system to support dynamic expansion adapting to the change in user's information needs.

Possibly, the simplest way to determine user's search goal is to ask users for explicit inputs at the time of search. Unfortunately, majority of the users are reluctant to

A. Fred et al. (Eds.): IC3K 2010, CCIS 272, pp. 158–172, 2013.

provide any explicit feedback [5]. The retrieval system has to learn user's preferences automatically without any explicit feedback from the users. Query log is a commonly used resource to determine user's preferences automatically without incurring any extra overhead to the users [12,1,10]. However, such studies are not flexible enough to capture the changing needs of users over time. If we want to model the complete dynamics of user's preferences from query log, we will need an extremely large query log and huge computational resources. Moreover, user may always explore new search areas. This makes the task of modelling user's search dynamics an extremely difficult and expensive problem.

Further, user's information needs at the time of query submission relates to the activities at the time of submitting query. There is a notion of importance for capturing user's activities (at the time of submitting query) and inferring user's information needs in real time. In this paper, we study a framework to expand user's search query dynamically based on user's implicit feedback provided at the time of search. It is evident from the analysis that, in many instances, user's implicit feedback provided at the time of search provides sufficient clues to determine *what user wants*. Just an example, if the query jaguar is submitted immediately after the query national animals, it is very likely that user is looking for the information related to animal. Such a small feedback can provide a very strong clue to determine user's search preferences. This is the main motivation of this paper. The activities which drive a query request to a search engine may include all other Web or non-Web activities such as off-line documents read on Desktop, documents read on printed copies, conversation with a friend, any other Web activities, navigational queries or, information found through query chain etc. However, this paper focuses only on Web activities alone.

The rest of the paper is organized as follows. We first define formal problem statement in Section 2. In Section 3, we then discuss background materials. In Section 4, we present few observations of query log analysis which inspire the proposed framework. In Section 5, we discuss our proposed query expansion framework. Section 7 present experimental observations. The paper concludes in Section 8.

2 Problem Statement

Let q be a query and $\mathcal{E}^{(q)} = \{f_{q,1}, f_{q,2}, f_{q,3}, ...\}$ be the set of expansion terms for the query q returned by a traditional query expansion mechanism. In general, many of these expansion terms are not relevant to user's search goal. Now, the task is to identify the expansion terms in $\mathcal{E}^{(q)}$ which are relevant to user's search goal by exploiting user's implicit feedback provided by the user at the time of search.

3 Background Materials

3.1 Notations and Definitions

Vector Space Model. We use the vector space model [16] to represent a query or a document. A document d or a query q is represented by a *term vector* of the form $\mathbf{d} = \{w_1^{(d)}, w_2^{(d)}, ..., w_m^{(d)}\}$ or $\mathbf{q} = \{w_1^{(q)}, w_2^{(q)}, ..., w_m^{(q)}\}$, where $w_i^{(d)}$ and $w_i^{(q)}$ are the weights assigned to the ith element of the set d and q respectively.

Cosine Similarity. If $\mathbf{v}_i$ and $\mathbf{v}_j$ are two arbitrary vectors, we use cosine similarity to define the similarity between the two vectors. Empirically, cosine similarity can be expressed as follows.

$$sim(\mathbf{v}_i, \mathbf{v}_j) = \frac{\sum_{k=0}^{m} w_{ik} \cdot w_{jk}}{\sqrt{\sum_{k=0}^{m} w_{ik}^2} \cdot \sqrt{\sum_{k=0}^{m} w_{jk}^2}} \tag{1}$$

Kullback-Leibler Divergence (KLD). Given two probability distributions p_i and p_j of a random variable, the distance between p_i and p_j can be defined by Kullback-Leibler divergence as follows.

$$KLD(p_i \| p_j) = p_i \cdot \log\left(\frac{p_i}{p_j}\right) \tag{2}$$

Real Time Implicit Feedback (RTIF). In this paper, we differentiate two types of implicit feedback; *history* and *active*. The active implicit feedback is the feedback provided by user at the time of search. We also refer to it by *real time implicit feedback* in this paper. A query session has been defined differently in different studies [9,8]. This paper considers the definition discussed in [9] and defines as a sequence of query events submitted by a user within a pre-defined time frame. Any feedback provided before the current query session is considered history.

3.2 Background on QE

Global analysis [11,13] is one of the first QE techniques where a thesaurus is built by examining word occurrences and their relationships. It builds a set of statistical term relationships which are then used to select expansion terms. Although, global analysis techniques are relatively robust, it consumes a considerable amount of computational resources to estimate corpus-wide statistics. Local analysis techniques use only few top ranked documents that are retrieved through an initial ranking by the original query. Thus, it focuses only on the given query and its relevant documents (pseudo relevant). A number of studies including the ones in [18,19,2,6] indicate that local analysis is effective, and, in some cases, outperforms global analysis. In the study [21], authors explore query log to determine relevant document terms for a given query term by exploring clickthrough records. Further, ontology based query expansion such as Wikipedia, Wordnet are reported in the studies [22,23]. However, the above studies do not address the problem of poor coherence between expansion terms and user's search goal.

4   Few Motivating Observations

4.1   Query Log vs Academic Research

After AOL incident in August 2006[1], no query logs are available publicly (not even for academic research). Obtaining query log from commercial search engines had always

---

[1] http://www.nytimes.com/2006/08/09/technology/09aol.html

Table 1. Characteristics of the clicked-through log dataset

Source	Proxy logs
Search Engine	Google
Observation Periods	3 months
# of users	3182
# of query instances	1,810,596
% of clicked queries	53.2%

been a very difficult task for academic research communities. One alternative is to use organizational local proxy logs. From proxy logs, we can extract *in-house* click-through information such as *user's id, time of search, query, click documents* and *the rank of the clicked documents*.

In this study, we use a large proxy log of three months. We extract the queries submitted by the users to google search engine and users' clicked responses to the results. Table 1 shows the characteristics of the click-through query log extracted from the three-months long proxy logs. To prove that In-House query log has similar characteristics with that of server side query log, we also analyze AOL query log. The analysis described in this paper is strictly anonymous; data was never used to identify any identity.

Constructing Query Session. For every user recorded in query log, we extract sequence of queries submitted by the user. Figure 1 shows a pictorial representation of the procedure to construct query sessions. The upper arrows ↑ represent the arrival of query events. Each session is defined by the tuple $\Gamma = <t_{e_{1f}}, uid, E, \delta>$. Just before the arrival of first query from the user u, the first query session has an empty record i.e., $\Gamma = <\phi, u, \phi, \delta>$. When user u submits his/her first query q, Γ is updated as $\Gamma = <t_{e_{1f}}, u, E, \delta>$, where $E = \{e_1\}$, $e_1 = <t_{e_{1f}}, t_{e_{1l}}, q_1, \phi>$, $q_1 = q$ and $t_{e_{1f}} = t_{e_{1l}}$. The down arrows ↓ in the Figure 1 represent the clicked events. As user clicks on the results for the query q_1, e_1 gets updated as $e_1 = <t_{e_{1f}}, t_{e_{1l}}, q_1, \mathcal{D}^{(q_1)}>$ where $\mathcal{D}^{(q_1)}$ is the set of clicked documents and $t_{e_{1l}}$ is the time of the last click.

When the second query q is submitted by the user u, it forms the second event $e_2 = <t_{e_{2f}}, t_{e_{2l}}, q_2, \phi>$, where $q_2 = q$ and $t_{e_{2f}} = t_{e_{2l}}$. If $t_{e_{2f}} - t_{e_{1l}} \leq \delta$, then e_2 is inserted into Γ and E is updated as $E = \{e_1, e_2\}$. If $t_{e_{2f}} - t_{e_{1l}} > \delta$, then e_2 can not be fitted in current query session Γ. In such a case, e_2 generates a new query session with e_2 as its first event i.e., e_2 becomes e_1 and $E = \{e_1\}$ in the new query session. We, then, shift the current session Γ to the newly formed session. In this way, we scan the entire query sequence submitted by the user u and generate the query sessions.

4.2 Exploring Recent Queries

To form the basis of the proposed framework, we analyze the similarity of the user's search patterns during a short period of time defined by a query session. The average similarity between queries submitted during a query session (defined by $\delta = 30min$) is estimated using cosine similarity defined in Equation (1). Figure 2.(a) shows that almost 55% of the consecutive queries have non-zero similarity (58% for AOL).

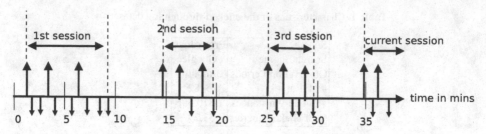

Fig. 1. Pictorial representation of the query sessions

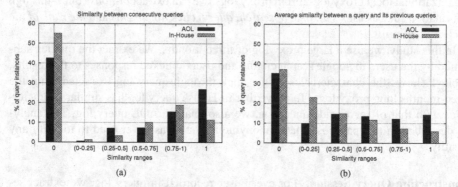

Fig. 2. Similarity between the queries in a query session

Further in Figure 2.(b), we report the average similarity between a query and its previous queries in a session. Almost 65% of the queries have similarity larger than 0. It suggests that majority of the queries in a session share common search context. Further, two queries with similar search context may have similarity 0. For example, the queries madagascar and die hard 2. Although, both the queries means movies, their similarity is 0. Therefore, the plots in Figure 2 represent the lower bound.

Remarks: The above observations show that, in many instances, queries in a session often share common search context. This motivates us to explore user's real time implicit feedback to determine user's search context.

5   Proposed QE Framework

To realize the effect of real time implicit feedback on query expansion, we systematically build a framework as shown in Figure 3. It has five major components.

1. Baseline retrieval systems. It retrieves a set of documents which are relevant with user's query and provides the top most R relevant documents to query expansion unit.
2. Baseline query expansion. Using the documents provided by the IR system, it determines a list of expansion terms which are related to the query submitted by the user. In this study, we use a KLD (see Equation 2) based QE as discussed in [3] as baseline QE (Algorithm 1).

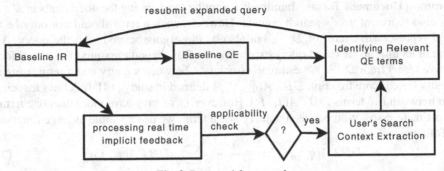

Fig. 3. Proposed framework

Algorithm 1. Conventional QE through local analysis

1: run original query q and retrieve relevant documents
2: select top n documents as local set R
3: extracted all terms t from local set R
4: **for all** terms $t \in R$ **do**
5:     calculate KLD
6: **end for**
7: rank terms t based on their KLD weight
8: add top $|E|$ terms to original query q
9: run expanded query q and rank documents using PL2

---

3. `Processing real time implicit feedback`. It constructs query session using the procedure discussed in Section 4.1.
4. `Applicability Check`. Some query session may not have enough evidences of sharing common search context. This unit verifies whether the newly submitted query shares common search context with that of the other queries in the session.
5. `Determining Search context`. It determines user's search context by exploiting the implicit feedback provided by the users in the current query session. It then identifies the relevant expansion terms.

5.1 Determining User's Search Context

Let $\Gamma = <t_{e_{1f}}, u, E, \delta>$ be the current query session as defined in Session 4.1, where E is the sequence of n query events. Let $\mathcal{Q}^{(\Gamma)}$ and $\mathcal{D}^{(\Gamma)}$ be the set of queries and visited documents respectively present in E. Let q_{n+1} be a new query submitted by the user u and $\mathcal{E}^{(q_{n+1})} = \{f_{q_{n+1},1}, f_{q_{n+1},2}, f_{q_{n+1},3}, ...\}$ be the set of expansion terms extracted using Algorithm 1 for the query q_{n+1}. Now the task is to identify relevant terms with that of user's search goal.

Common Query Terms. It exploits the list of previous queries $\mathcal{Q}^{(\Gamma)}$ submitted by the user in the current query session Γ) and determines the popular query terms using a function $qf(f, \mathcal{Q}^{(\Gamma)})$ which is the number of queries in $\mathcal{Q}^{(\Gamma)}$ containing the term f. We consider a term f popular if its frequency is greater than a threshold i.e., $qf(f, \mathcal{Q}^{(\Gamma)}) \geq \Theta_{\mathcal{Q}}$. In this study, majority of the query sessions are short and the term frequencies are small. Therefore, we set threshold to $\Theta_{\mathcal{Q}} = 1$.

Common Document Terms. Intuitively a popular term among the documents in $\mathcal{D}^{(\Gamma)}$ can also represent user's search context. However, such a term should not only be a good representative term of $\mathcal{D}^{(\Gamma)}$, but also be closely associated with the query. As done in local analysis based query expansion, KLD is a good measure to extract informative terms from $\mathcal{D}^{(\Gamma)}$. We estimate association between a query and a term using a density based score function $DBTA(q_{n+1}, f)$ defined in study [14]. It defines association between two terms $DBTA(f_i, f_j)$. However, q_{n+1} may have more than one term. To estimate association between a query and a term, we use a simple average function as follows:

$$DBTA(q_{n+1}, f) = \frac{1}{|q_{n+1}|} \sum_{f_i \in q_{n+1}} DBTA(f_i, f) \tag{3}$$

where $|q_{n+1}|$ is the number of terms in q_{n+1}.

Harmonic mean [17] is a popular measure to merge the goodness of two estimators. Therefore, the values of KLD and DBTA are combined using harmonic mean between the two. However, the two values are at different scales: KLD scales between $-\infty$ to $+\infty$ and DBTA scales between 0 to 1. To make the two estimators coherent to each other, the estimators are further normalized to the scale of 0 and 1 using the following equation.

$$\text{normalize}(g) = \frac{g - \min_g}{\max_g - \min_g} \tag{4}$$

where g is an arbitrary function. Now, the harmonic mean score between the two can be defined as follows:

$$\text{score}^{\mathcal{P}^{(\mathcal{D})}}(f) = \frac{2 \cdot KLD^{(\mathcal{D}^{(\Gamma)})}(f) \cdot DBTA(q_{n+1}, f)}{KLD^{(\mathcal{D}^{(\Gamma)})}(f) + DBTA(q_{n+1}, f)} \tag{5}$$

If an expansion terms $f \in \mathcal{E}^{(q_{n+1})}$ has a score greater than a threshold $\Theta_{\mathcal{P}^{(\mathcal{D})}}$ i.e., $\text{score}^{\mathcal{P}^{(\mathcal{D})}}(f) \geq \Theta_{\mathcal{P}^{(\mathcal{D})}}$, then the term f is selected. In this study, the threshold value is set to an arbitrary value 0.5. It is because intuitively the normalized average may cover the upper half of the term collections.

Expansion Terms of Previous Queries. Let $e_i = < t_{e_{if}}, t_{e_{il}}, q_i, \mathcal{D}_c^{(q_i)} >$ be a query event in E, where $i \neq n + 1$ and $\mathcal{E}^{(q_i)}$ be the expansion terms of the query q_i. If an expansion term $f \in \mathcal{E}^{(q_i)}$ is also present in any document $d \in \mathcal{D}_c^{(q_i)}$, then it is selected. The set of such terms is denoted by $\mathcal{P}_i^{(\mathcal{E})}$ and is formally defined as follows:

$$\mathcal{P}_i^{(\mathcal{E})} = \{f | f \in \mathcal{E}^{(q_i)} \text{ and } \exists d \in \mathcal{D}_c^{(q_i)} \text{ s.t. } f \in d\} \tag{6}$$

We assume that the visited documents against a query are relevant to user's information need of that query. Therefore, this set represents the set of expansion terms of previous queries in the same query session which are actually relevant to user's search goal. For all the queries in $\mathcal{Q}^{(\Gamma)}$, Equation (6) is repeated and all $\mathcal{P}_i^{(\mathcal{E})}$ are merged i.e., $\mathcal{P}^{(\mathcal{E})} = \cup \mathcal{P}_i^{(\mathcal{E})}$. An expansion term $f \in \mathcal{E}^{(q_{n+1})}$ is assumed to be relevant to user's current search context, if $f \in \mathcal{P}^{(\mathcal{E})}$.

Synonyms of Query Terms. There are publicly available tools like Wordnet[2], WordWeb[3] which can provide synonyms of a given term. Such expert knowledge can be used effectively to select the expansion terms.

Let $P^{(S)}$ be the list of synonyms[4] for all the query terms in $Q^{(\Gamma)}$ extracted using Wordnet. If an expansion terms $f \in \mathcal{E}^{(q_n+1)}$ has an score greater than a threshold Θ_{dbta} i.e., $\text{score}^{P^{(S)}}(f) \geq \Theta_{dbta}$, then the term f is considered to be relevant to user's search goal.

$$\text{score}^{P^{(S)}}(f) = \begin{cases} DBTA(f, f'), & \text{if } f \in P^{(S)} \text{ and} \\ & \exists f' \in P^{(\mathcal{E})} \text{ s.t.} \\ & DBTA(f, f') \geq \Theta_{dbta} \\ 0, & \text{Otherwise} \end{cases} \qquad (7)$$

In this study, the threshold Θ_{dbta} is set to an arbitrary value i.e., the average value of $DBTA(f, f')$ over the corpus. However, more sophisticated procedure to set threshold value will be to study the distribution of positive and negative associations.

Category Specific Terms. Another important information that can be extracted from implicit feedback is dominant class labels in $\mathcal{D}^{(\Gamma)}$. The relevant expansion terms should have close association with the dominant class labels. In the study [15], the authors studied a measure known as *within class popularity* and it is observes that WCP provides better assocition as compared to other estimators such as *mutual information*, *chi-square* [20]. In this study, we use the same measure WCP to estimate association between a term and class. If C be the set of global class labels and $C^{(\Gamma)}$ be the set of dominant class labels of the current query session Γ. We select a term $f \in \mathcal{E}^{(q_n+1)}$ if $\exists c \in C^{(\Gamma)}$ such that

$$c = \max_{\forall c_i \in C} \{wcp(f, c_i)\} \qquad (8)$$

Mining More Context Terms. Let $\mathcal{E}_{rtif}^{(q_n+1)}$ be the set of relevant expansion terms thus obtained from the above sections. Still there may be terms in $\mathcal{E}^{(q_n+1)}$ which are not included in $\mathcal{E}_{rtif}^{(q_n+1)}$, but closely related to some terms in $\mathcal{E}_{rtif}^{(q_n+1)}$. Intuitively, such missing terms are also related to the context of user's search goal. Therefore, we further determine missing terms as follows:

- for all terms $t \in \mathcal{E}^{(q_n+1)}$ and $t \notin \mathcal{E}_{rtif}^{(q_n+1)}$: if $\exists t' \in \mathcal{E}_{rtif}^{(q_n+1)}$ s.t. $DBTA(t, t') > \Theta_{dbta}$, then insert the term t in $\mathcal{E}_{rtif}^{(q_n+1)}$.

Now, we consider the terms in $\mathcal{E}_{rtif}^{(q_n+1)}$ as the expansion terms related to the context of user's search goal.

---

[2] http://wordnet.princeton.edu

[3] http://wordweb.info/free/

[4] We apply the Wordnet command wn auto synsn to get list of synonyms. We pass the output of this command to a script. This script processes the output and returns the list of synonyms.

5.2 Applicability Check

The above procedures to identify relevant expansion terms will return good results if the newly submitted query q_{n+1} indeed has the same search preference as that of other queries in E. But this condition is not always true. In some query sessions, there may not be enough evidences of having common search context.

Therefore, it is important to perform an applicability check before applying the above procedures. For every newly submitted query q_{n+1}, we perform an applicability check. We estimate average cosine similarity among the expanded terms of all queries in the session. If the average similarity of a current session is above a user-defined threshold Θ_{sim}, then it is assumed that the queries in the current query session share common search context.

6 Evaluation Methodology

To evaluate the proposed framework we define three metrics – (i)*quality*: to measure the quality of the expansion terms, (ii) *precision@k*: to measure retrieval effectiveness and (iii) *dynamics*: to measure the capability of adapting to the changing needs of the user.

The best evidence to verify the quality of the expanded terms or retrieval effectiveness of a system is to cross check with the documents actually visited by the user for the subjected query. Let q be an arbitrary query and $\mathcal{D}_c^{(q)}$ be the set of documents actually visited by the user for q. Now, given an IR system and a query expansion system, let $\mathcal{E}^{(q)}$ be the set of expansion terms for the query q. Then, the quality of the expansion terms is defined as follows:

$$quality = \frac{|\rho(\mathcal{E}^{(q)}, \mathcal{D}_c^{(q)})|}{|\mathcal{E}^{(q)}|} \tag{9}$$

where $\rho(\mathcal{E}^{(q)}, \mathcal{D}_c^{(q)})$ is the matching terms between $\mathcal{E}^{(q)}$ and $\mathcal{D}_c^{(q)}$ i.e.,

$$\rho(\mathcal{E}^{(q)}, \mathcal{D}_c^{(q)}) = \{f | f \in \mathcal{E}^{(q)}, \exists d \in \mathcal{D}_c^{(q)} \text{ s.t. } f \in d\}$$

Let $\mathcal{D}_n^{(q)}$ be the set of top n documents retrieved by the IR system. To define retrieval effectiveness, we determine the number of documents in $\mathcal{D}_n^{(q)}$ which are closely related to the documents in $\mathcal{D}_c^{(q)}$. We use cosine similarity (see Equation (1)) to define the closeness between two documents. Let $\mathcal{D}_r^{(q)}$ be a set of documents in $\mathcal{D}_n^{(q)}$ for which the cosine similarity with at least one of the document in $\mathcal{D}_c^{(q)}$ is above a threshold Θ_{sim} i.e.,

$$\mathcal{D}_r^{(q)} = \{d_i | d_i \in \mathcal{D}_n^{(q)}, \exists d_j \in \mathcal{D}_c^{(q)} \text{ s.t. } sim(d_i, d_j) \geq \Theta_{sim}\}$$

In this study we define $\mathcal{D}_r^{(q)}$ with the threshold value $\Theta_{sim} = 0.375$. In our dataset, the majority of the co-click documents have cosine similarity in the range of $[0.25,5)$. We have considered the middle point as the threshold value. Now we use the *precision@k* to measure the retrieval effectiveness and define it as follows:

$$precision@k = \frac{\mathcal{D}_r^{(q)}}{k} \tag{10}$$

Last we define the dynamics in query expansion. For a query, the system is expected to return different expansion terms for different search goals. Let $\mathcal{E}_i^{(q)}$ and $\mathcal{E}^{(q)}{}_j$ be the set of expansion terms for a query q at two different instances i and j. Then we define the dynamics between the two instances as follows:

$$\delta^{(q)}(i,j) = 1 - sim(\mathcal{E}_i^{(q)}, \mathcal{E}_j^{(q)}) \tag{11}$$

If there are n instances of the query q then we estimate the average dynamics as follows

$$E(\delta^{(q)}(i,j)) = \frac{n(n-1)}{2} \sum_{i \neq j} \delta^{(q)}(i,j) \tag{12}$$

Now, we are interested to investigate two forms of dynamics among the instances with – (i) same goal and (ii) different goals. We expect that a QE system which can adapt to the changing needs of the user should have low value for former case and high value for latter case.

7 Performance of the Proposed Framework

We build two baseline retrieval systems (i) an IR system which indexes around 1.6 million documents using PL2 normalization [7], denoted by *LIR*, and (ii) a meta-search interface which receives queries from the users and submit it to Google search engine, denoted by *GIR*. On top of these systems, we have incorporated the proposed framework.

To verify the performance of the proposed framework, we have used the In-House query log discussed in Section 4.1. We have extracted few experimental queries and their corresponding click-through information from this query log. First we discuss the procedure to extract our experimental queries.

7.1 Experimental Queries

A total of 35 queries are selected to conduct the experiments. All these queries are extracted from the In-House query log. Top most popular *non-navigational* queries [4] of length 1 and 2 words are selected. The *entropy* is a commonly used measure to analyse a probability distribution of a random variable. In this study, we also use an entropy based measure to study the distribution of the visited documents and identify navigational queries.

Let $\mathcal{D}_c^{(q,u)}$ be the set documents visited by a user u for the query q in the entire query log. Then, we define an entropy of the query q for the user u as follows:

$$H(q) = - \sum_{d_i \in \mathcal{D}_c^{(q,u)}} Pr(d_i|q,u).\log Pr(d_i|q,u) \tag{13}$$

where $Pr(d_i|q,u)$ is the conditional probability that user u visits the document d_i given the query q. If $H(q)$ is very closed to zero, the query q is considered as a navigation query.

Table 2. List of the 35 queries. #Γ indicates number of query sessions for each query and #$\mathcal{Z}$ indicates the number different search context.

query	#Γ	#$\mathcal{Z}$	query	#Γ	#$\mathcal{Z}$	query	#Γ	#$\mathcal{Z}$	query	#Γ	#$\mathcal{Z}$
blast	15	1	books	18	4	chennai	18	3	coupling	10	2
crunchy munch	38	1	indian	14	2	games	59	1	jaguar	3	2
kate winslet	23	2	mallu	38	1	milk	15	2	namitha	22	1
nick	20	1	rahaman	2	1	passport	38	2	roadies	10	1
statics	36	4	times	5	2	science	16	2	scholar	16	3
simulation	3	1	smile pink	2	1	tutorial	11	6	reader	11	3
ticket	38	3	crank	10	1	engineering village	12	1	maps	15	4
nature	28	2	reshma	15	1	savita	2	1	dragger	11	2
sigma	11	2	spy cam	10	1	java	17	2			

Table 2 shows the list of 35 selected queries. This table also shows the number of query sessions for each of the individual queries and denoted by "#". A total of 612 query sessions are found for these 35 queries. A query may have different search goals at different times. We manually verify and mark all these 612 instances. While verifying we broadly differentiate the goals (e.g. "java programming" and "java island" are two different goals, however "java swing" and "core jave" have same goal). Table 2 also shows the number of different search goals for individual query (denoted by "#$\mathcal{Z}$"). It shows that 20 out of 35 (i.e., 57.1%) queries have varying search preferences at different times.

7.2   Quality of Expansion Terms

We examine top 20 expansion terms of all 35 queries. If an expansion term predicted by a system is found in corresponding visited document, then, we assume that the term is indeed relevant to the search preference. Table 3 shows the average *quality* of the expansion terms over all 35 queries. There is a significant improvement in quality. On an average there is an improvement from 0.287 to 0.536 (86.7% improvement) on local IR system. For the Google meta search, there is an improvement of 70.8% from 0.329 to 0.562.

Table 3. Average quality of the top 20 expansion terms over 35 queries given in Table 2

Baseline		Proposed	
LIR	GIR	LIR	GIR
0.287	0.329	0.536(+86.7%)	0.562(+70.8%)

7.3   Retrieval Effectiveness

Now, we compare the retrieval effectiveness of the proposed expansion mechanism with the baseline expansion mechanism. We use the precision at k measure (defined in Equation (10)) to estimate retrieval effectiveness. In Table 4, we compare the retrieval performance of the baseline system and the proposed system in terms of the average of

Table 4. Precision@k returned by different systems using top 20 expansion terms

top k	Baseline		Proposed	
	LIR	GIR	LIR	GIR
10	0.221	0.462	0.749	0.763
20	0.157	0.373	0.679	0.710
30	0.113	0.210	0.592	0.652
40	0.082	0.153	0.472	0.594
50	0.052	0.127	0.407	0.551

the precision at k for all 612 query instances. If a query has no visited documents, we simply ignore them. Note that, the set of visited documents $\mathcal{D}_c^{(q)}$ is obtained from the query log whereas the set $\mathcal{D}_n^{(q)}$ is obtained from the experimental retrieval system after simulating the query sequence. Table 4 clearly shows that our proposed framework outperforms the baseline systems for both the local IR system and Google results.

7.4   Component Wise Effectiveness

In the section 5.1, we define different components that contribute to the expansion terms. In this section, we study the effect of each component separately. Table 5 shows the quality of the expansion terms returned by each component (considering the top 20 expansion terms). In the table, $P^{(\mathcal{Q})}$ denotes set of expansion terms based on query terms (Section 5.1), $P^{(\mathcal{D})}$ denotes the document terms (Section 5.1), $P^{(\mathcal{E})}$ denotes combine expansion terms of previously submitted queries (Section 5.1), $P^{(\mathcal{S}^{\mathcal{R}})}$ denotes word sense (Section 5.1)and $P^{(\mathcal{C})}$ denotes class specific terms (Section 5.1). We observe that expansion terms extracted using $P^{(\mathcal{D})}$ and $P^{(\mathcal{E})}$ contribute the most. This observation is true for both the local retrieval system and Google results. The summation of the percentages in each row is more than 100%. It is because, there are overlapping terms among the components.

Table 5. Average quality of individual components over 35 queries given in Table 2

	$P^{(\mathcal{Q})}$	$P^{(\mathcal{D})}$	$P^{(\mathcal{E})}$	$P^{(\mathcal{S}^{\mathcal{R}})}$	$P^{(\mathcal{C})}$
LIR	8.3%	39.8%	37.9%	4.6%	12.1%
GIR	8.8%	43.3%	39.2%	6.9%	8.4%

7.5   Retrieval Efficiency

Though the proposed framework provides better retrieval effectiveness, it has an inherent efficiency problem. Apart from the time required for query expansion (Algorithm 1), the proposed framework needs computational time for determining context for user's search goal. Table 6 shows the efficiency of different retrieval systems. It clearly shows that the proposed framework has poor efficiency. It can be noted that the computational overhead is an order of magnitude higher than that of general expansion and without expansion.

Table 6. Average retrieval efficiency of different expansion system in seconds

Baseline IR		Baseline QE		Proposed QE	
LIR	GIR	LIR	GIR	LIR	GIR
1.028	0.731	3.961	3.205	14.518	14.149

However, the focus of this paper is to prove that queries can be expanded dynamically by exploiting the real time implicit feedback provided by the users at the time of search. It is obvious that there will be additional computational overhead to process the expansion in real time. The implementation of the experimental systems are not optimal. Though the computational overhead reported in Table 6 is high, with efficient programming and hardware supports we believe that the overhead can be reduced to reasonable level.

7.6   Dynamics: Adapting to the Changing Needs

Table 7 shows the average of the average dynamics of different systems over the entire experimental queries. It clearly shows that the baseline system has a dynamics of zero in all cases. It indicates that baseline systems always return the same expansion terms irrespective of user's search goal. Whereas the proposed framework has a small dynamics among the instances of the same query with same goal and high dynamics among the query instances of the same query with different goals. It indicates that the proposed framework is able to adapt to the changing needs of the users and generate expansion terms dynamically.

Table 7. Average of average dynamics over the entire experimental queries

	Baseline QE		Proposed QE	
Goal	LIR	GIR	LIR	GIR
Same	0	0	0.304	0.294
Different	0	0	0.752	0.749

8   Conclusions

In this paper, we explore user's real time implicit feedback to analyse user's search pattern during a short period of time. From the analysis of user's click-through query log, we observe two important search patterns – user's information need is often influence by his/her recent searches and user's searches over a short period of time often confine to 1 or 2 categories. In many cases, the implicit feedback provided by the user at the time of search have enough clues of what user wants. We explore query expansion to show that the information submitted at the time of search can be used effectively to enhance search retrieval performance. We proposed a query expansion framework, which explores recently submitted query space. From various experiments, we observed that the proposed framework provides better relevant terms compared to the baseline query expansion mechanisms. Most importantly, it can dynamically adapt to the changing needs of the user.

References

1. Agichtein, E., Brill, E., Dumais, S., Ragno, R.: Learning user interaction models for predicting web search result preferences. In: Proceedings of the 29th Annual International ACM SIGIR Conference on Research and Development in Information Retrieval, pp. 3–10. ACM (2006)
2. Attar, R., Fraenkel, A.S.: Local feedback in full-text retrieval systems. Journal of ACM 24(3), 397–417 (1977)
3. Billerbeck, B., Scholer, F., Williams, H.E., Zobel, J.: Query expansion using associated queries. In: Proceedings of the Twelfth International Conference on Information and Knowledge Management. ACM (2003)
4. Broder, A.: A taxonomy of web search. SIGIR Forum 36(2), 3–10 (2002)
5. Carroll, J.M., Rosson, M.B.: Paradox of the active user. In: Interfacing Thought: Cognitive Aspects of Human-Computer Interaction, pp. 80–111. MIT Press (1987)
6. Croft, W.B., Harper, D.J.: Using probabilistic model of document retrieval without relevance information. Journal of Documentation 35, 285–295 (1979)
7. He, B., Ounis, I.: Term Frequency Normalisation Tuning for BM25 and DFR Models. In: Losada, D.E., Fernández-Luna, J.M. (eds.) ECIR 2005. LNCS, vol. 3408, pp. 200–214. Springer, Heidelberg (2005)
8. Jaime, T., Eytan, A., Rosie, J., Michael, A.S.P.: Information re-retrieval: Repeat queries in yahoo's logs. In: Proceedings of the 30th Annual International ACM SIGIR Conference on Research and Development in Information Retrieval, pp. 151–158. ACM (2007)
9. Jansen, B.J., Spink, A., Saracevic, T.: Real life, real users, and real needs: a study and analysis of user queries on the web. Information Processing and Management 36(2), 207–227 (2000)
10. Joachims, T.: Optimizing search engines using clickthrough data. In: Proceedings of the ACM Conference on Knowledge Discovery and Data Mining, pp. 133–142. ACM (2002)
11. Jones, S.: Automatic keyword classification for information retrieval. Butterworths, London (1971)
12. Kelly, D., Teevan, J.: Implicit feedback for inferring user preference: A bibliography. SIGIR Forum 32(2), 18–28 (2003)
13. Qiu, Y., Frei, H.: Concept based query expansion. In: Proceeding of the 16th International ACM SIGIR Conference on Research and Development in Information Retrieval, pp. 151–158. ACM (1993)
14. Ranbir, S.S., Murthy, H.A., Gonsalves, T.A.: Effect of word density on measuring words association. ACM Compute. 1–8 (2008)
15. Ranbir, S.S., Murthy, H.A., Gonsalves, T.A.: Feature selection for text classification based on gini coefficient of inequality. In: Proceedings of the Fourth International Workshop on Feature Selection in Data Mining, pp. 76–85 (2010)
16. Salton, G., Wong, A., Yang, C.S.: A vector space model for automatic indexing. ACM Communication 18(11), 613–620 (1975)
17. Sebastiani, F.: Machine learning in automated text categorization. ACM Computing Survey 34(1), 1–47 (2002)
18. Xu, J., Croft, W.B.: Query expansion using local and global document analysis. In: Proceedings of the Nineteenth Annual International ACM SIGIR Conference on Research and Development in Information Retrieval, pp. 4–11 (1996)
19. Xu, J., Croft, W.B.: Improving the effectiveness of information retrieval with local context analysis. ACM Transaction on Information System 18(1), 79–112 (2000)
20. Yang, Y., Pedersen, J.O.: A comparative study on feature selection in text categorization. In: Proceedings of the Fourteenth International Conference on Machine Learning, pp. 412–420. ACM (1997)

21. Cui, H., Wen, J.-R., Nie, J.-Y., Ma, W.-Y.: Probabilistic query expansion using query logs. In: Proceedings of the 11th International Conference on World Wide Web, pp. 325–332. ACM (2002)
22. Hsu, C.-C., Li, Y.-T., Chen, Y.-W., Wu, S.-H.: Query Expansion via Link Analysis of Wikipedia for CLIR. In: Proceedings of NTCIR-7 Workshop Meeting, Tokyo, Japan, December 16-19 (2008)
23. Zhang, J., Deng, B., Li, X.: Concept Based Query Expansion Using WordNet. In: Proceedings of the 2009 International e-Conference on Advanced Science and Technology. IEEE Computer Society, USA (2009)

Appendix

In this section, we discuss the classification framework that we use for labelling visited documents. Like in Section 5.1, we use the same seed based classificationa and WCP feature selector as proposed by [15]. We briefly discussed the framework as follows.

Let F be the set of terms in the subjected document collection. Each group of documents belonging to a class c_i is represented by a term vector $\mathbf{c}_i$ known as seed vector defined over F. The seed vector $\mathbf{c}_i$ is assumed to be the best term vector which can differentiate the documents belonging to c_i from the documents belonging to other categories. Each element in $\mathbf{c}_i$ has a weight defined by $wcp(f, c_i)$. Given a test example d defined over F, d is classified by the following function.

$$\text{classify}(d) = \arg \max_{c_i}\{cosine(\mathbf{d}, \mathbf{c}_i)\} \tag{14}$$

where $cosine(\mathbf{d}, \mathbf{c}_i)$ is the cosine similarity between document d and the seed vector of the class c_i.

Testing and Improving the Performance of SVM Classifier in Intrusion Detection Scenario

Ismail Melih Önem

Department of Computer Science, Middle East Technical University Ankara, Ankara, Turkey
melihonem@hotmail.com

Abstract. Intrusion Detection attempts to detect computer attacks by examining various data records observed in processes on the network. Anomaly discovery has attracted the attention of many researchers to overcome the disadvantage of signature-based IDSs in discovering complex attacks. Although there are some existing mechanisms for Intrusion detection, there is need to improve the performance. Machine Learning techniques are a new approach for Intrusion detection and KDDCUP'99 is the mostly widely used data set for the evaluation of these systems. The goal of this research is using the SVM machine learning model with different kernels and different kernel parameters for classification unwanted behavior on the network with scalable performance. Also elimination of the insignificant and/or useless inputs leads to a simplification of the problem, faster and more accurate detection may result. This work also evaluates the performance of other learning techniques (Filtered J48 clustering, Naïve Bayes) over benchmark intrusion detection dataset for being complementary of SVM. The model generation is computation intensive; hence to reduce the time required for model generation various different algorithms. Various algorithms for cluster to class mapping and instance testing have been proposed to overcome problem of time consuming for real time detection. I show that our proposed variations matured in this paper, contribute significantly in improving the training and classifying process of SVM with high generalization accuracy and outperform the enhanced technique.

Keywords: SVM, SVM kernel and parameter selection, SVM performance, SVM categorizing, Real time network analyses, Intrusion classifier, Intrusion detection, Category discovery.

1   Introduction

With the immense growth of computer network usage and the huge rise in the number of applications running on top of it, network security is becoming more and more arrogant. Therefore, the role of Intrusion Detection System (IDSs), as special-purpose appliances to category anomalies in the network, is becoming further significant. The analysis in the intrusion detection and categorization field has been mostly focused on anomaly-based and misuse-based discovery techniques for a long time. While misuse-based discovery is generally preferred in commercial products due to its predictability and high accuracy, in academic research anomaly classification is typically formulated as a more powerful method due to its theoretical promising for turning to novel attacks.

A. Fred et al. (Eds.): IC3K 2010, CCIS 272, pp. 173–184, 2013.
© Springer-Verlag Berlin Heidelberg 2013

Difficulty is discovering unwanted behavior in network traffic after they have been subject to machine learning methods and processes. There is a great written works on various security methods to defend network objects from unauthorized use or disclosure of their private information and valuable assets. Even so, unconscious or automatic users find a way through much wiser means of get ridding of avoidance methods.

In usual methods located on port numbers and protocols have proven to be ineffective in terms of dynamic port allocation and packet encapsulation. The signature matching methods, on the other hand, require a known signature set and processing of packet payload, can only handle the signatures of a limited number of IP packets in real-time. A machine learning method based on SVM (supporting vector machine) is tendered in this paper for accurate classification and discovery unwanted behavior with scalable performance. The method classifies the Internet traffic into broad application categories according to the network flow parameters obtained from the packet headers. An optimized feature set is acquired via various classifier selection methods.

This paper proposes a new approach for enhancing the training process of SVM when dealing with large training data sets. It is based on the combination of SVM and clustering analysis. Different SVM machine learning models are used for discovering unwanted behavior on the network traffic. LIBSVM and Weka is used in a Java environment for training and testing the learning algorithms. It provides several different SVM implementations along with multiple kernels. I examine three things, the relative importance of features in training the dataset, the choice of kernel algorithm and parameter selection of SVM classifiers. By understanding what features are the most relevant, the dataset can be trimmed to include only the most useful data. The choice of kernel results in different levels of errors when applied to the KDD Cup dataset [1]. Frameworks offer five different kernels: Sigmoid, Linear, Polynomial and RBF. Each kernel offers three parameters for tuning and optimization which values are "gamma, cost and nu".

The performance norm has also been the subject of mine research. Here, the best kernel should maximize a predictive performance criterion as well as a computational performance criterion. That is, I seek the best categorizers that are; good at discover unwanted behavior, are efficient to compute over massive datasets of network traffic. I address the "predictive performance" criterion, what meaning by good, after describing the cost model for this domain.

The extended contributions of this work are as follows: First, to reduce the training time of SVM, we propose a new support vector selection technique using feature and kernel analysis. Here, we combine the analysis and SVM training phases. Second, we show analytically the degree to which our approach is asymptotically quicker than standard SVM, and validate this claim with experimental results. Finally, we compare our approaches with random selection, and other classifiers on a benchmark data set, and demonstrate impressive results in terms of training time, CPU test time and accuracy.

In real time traffic flows, Classifier performance depends greatly on the characteristics of the data to be classified. There is no single classifier that works best on all given problems so I have to test the classifier speed on different algorithms for compare evaluation performance and to find the characteristics of data that determine classifier on the real time network performance. Determining a suitable classifier for a network classification problem is hard work in the domain. J48 and Naïve Bayes classifiers beside SVM are used to evaluate the performance of total classification.

2   Challenges of Improving Categorizing

The approach to this work is done in steps, with supplemental complexity being added to the model at each level. As a prelude to developing any models, the data must first be put into a usable format. I am using the KDDCup 99 dataset, delineated earlier, which includes of features that are either continuous (numerically) valued or discrete. The continuous features in the provided dataset are in the text format (i.e. tcp/udp) and must be transformed.

One of the primary challenges of intrusion discovery is gathering applicable data for training and testing of an algorithm. Lack of the KDD data set is the vast number of redundant records, which causes the learning algorithms to be biased towards the frequent records, and thus prevent them from learning rare records, which are usually more pernicious to networks. In addition, the existence of these repeated records in the test set will cause the evaluation results to be biased by the methods which have better categorizing rates on the frequent records.

The SVM is one of the most successful classification algorithms in the machine learning area, but its long training time limits its use. Many applications, such as Data Mining and Bio-Informatics, require the processing of huge data sets. The training time of SVM is a serious obstacle in the processing of such data sets.

One of the other disadvantages of SVM-based and other supervised machine learning method is the requisite on a large number of labeled training samples [2]. Furthermore, recognizing the traffic after the network flow is collected could be too late should security and interventions become necessary in the early stage of the traffic flow. My intend is using supervised machine learning methods, as well as using feature parameters obtainable in the traffic flow for fast and accurate Network traffic discovery.

Even though, the recommended data set still suffers from some of the problems in complex data set and may not be a perfect stand in of existing real networks, because of the lack of public data sets for network-based IDSs, at the same time it can be applied as an impressive benchmark data set to help researchers compare different machine learning methods.

In real time detection, various techniques have been used for intrusion method for improving the detection rate and performance throughput of intrusion detection. Signature-Matching methods are one of the efficient techniques for this. Machine learning which characterize the datasets into clustering for intrusion detection which removes the some problems based on observation. Most of the detection rate for intrusion detection can be acceptable but no able to solve the times consuming measures are used for deciding the final decision.

3   Problem Definition and Classifiers

Using machine learning techniques to analyze incoming network data, I can decide to determine malicious attacks before they compromise an information system. Research in the field of intrusion detection seems to focus on a variety of support vector machine method, neural networks and cluster algorithms.

In this paper, I tried an effective approach to solve the two mentioned issues, resulting in new train and test sets, which consist of chosen records of the complete KDD data set. The provided data set does not suffer from a large number of tagged training samples. Besides, the numbers of records in the train and test sets are reasonable. This advantage makes it affordable to run the experiments which needed to randomly select a small portion. Inevitably, evaluation results of different research work will be consistent and comparable.

Through the use of correct kernel choice, feature selection and parameter selection, I have shown that it is possible to improve the accuracy and efficiency of a Support Vector Machine applied to an Intrusion Detection Scenario.

Support vector machines (SVMs) are a set of related supervised learning methods used for classification and regression. In simple words, given a set of training examples, each marked as belonging to one of two categories, an SVM training algorithm intensifies a model that predicts whether a new example falls into one category or the other. Additionally, a support vector machine constructs a hyper plane or set of hyper planes in a high or infinite dimensional space, which can be used for discovery of unwanted behavior in network traffic. SVMs use two key concepts to solve this problem: large-margin separation and kernel functions. Classification exercise usually involves separating data into training and testing sets. Each instance in the training set contains one "target value" and "several attributes".

Given a training set of instance-label pairs (x_i, y_i), $i = 1, \ldots, l$ where $x_i \in 2\,R^n$ and $y \in \{1, -1\}^l$ the support vector machines (SVM)[3] require the solution of the following optimization problem:

$$\min_{w, b,} \quad \frac{1}{2} w^T w + C \sum_{i=1}^{l} \varepsilon \tag{1}$$

$$\text{subject to} \quad y_i (w^t \Phi(x_i) + b) \geq 1 - \varepsilon_i$$

$$\varepsilon_i \geq 0$$

Training vectors x_i are mapped into a higher (maybe infinite) dimensional space by the function Φ. SVM finds a linear separating hyper plane with the maximal margin in this higher dimensional space. $C > 0$ is the penalty parameter of the error term. Furthermore, $K(x_i, x_j) \equiv \Phi(x_i)^T \Phi(x_j)$s called the kernel function. γ, r, d are kernel parameters.

Linear Kernel. The simpler kernel achieves to making a classification decision based on the value of a linear combination of the characteristics.

$$K(x_i, x_j) = x_i^T x_j \tag{2}$$

Polynomial Kernel. This kind of kernel represents the inner product of two vector(point) in a feature space of multi-dimension.

$$K(x_i, x_j) = (\gamma x_i^T, x_j + r)^d, \gamma > 0 \tag{3}$$

Radial Basis Function (RBF) Kernel. Nonlinearly maps samples into a higher dimensional space so it, unlike the linear kernel, can handle the case when the relation between class labels and attributes is nonlinear.

$$K(x_i, x_j) = exp\left(-\gamma \|x_i - x_j\|^2\right), \gamma > 0 \tag{4}$$

Sigmoid Kernel. A SVM model using a sigmoid kernel function is equivalent to a two-layer, perceptron neural network.

$$K(x_i, x_j) = tan\left(\gamma x_i^T, x_j + r\right)^d \tag{5}$$

A [4]j48 decision tree is a predictive machine-learning model that decides the target value of a new sample based on various attribute values of the available data. The internal nodes of a decision tree denote the different attributes; the branches between the nodes tell us the possible values that these attributes can have in the observed samples, while the terminal nodes tell us the classification of the dependent variable.

The [4]Naïve Bayes classifier works on a simple, but comparatively intuitive concept. Also, in some cases it is also seen that Naïve Bayes outperforms many other comparatively complex algorithms. It makes use of the variables contained in the data sample, by observing them individually, independent of each other.

4   Approach and Classification

4.1   Data Analysis and Data Partitioning

As I mentioned before, there are some difficulties in the KDD data set, which cause the estimation results on this data set to be deceptive. In this section I perform a set of tests to show the existing deficiencies in KDD.

First steps of executions are on partitioning because of using large volume of data. Before building a model, typically I separate the data using a partition utility. Partitioning produces mutually different datasets of attack types. The five traffic categories are "normal, probe, denial of service (DoS), user-to-root (U2R), remote-to-local (R2L)".

For this purpose, the data set is divided into five segments (*Attacks and Normal Traffic*), where the observations which I perform a set of experiments. That shows the existing deficiencies on the portion of the data set and is then evaluated.

The DoS attack data set is divided into five segments too, where the observations which I perform a set of experiments to show the existing deficiencies on the portion of the data set and is then evaluated.

The consequential deficiency in the KDD data set is the vast number of redundant records, which causes the learning algorithms to be biased towards the frequent records, and thus prevent them from learning unfrequented records which are usually more harmful to networks such as U2R and R2L attacks. In addition, the existence of these repeated records in the test set will cause the evaluation results to be biased by the methods which have better classification rates on the frequent records.

Table 1. Dataset Redundancy

Class	Redundant	Invalids	Unique	Ratio %
Normal	159967	2	812814	83,56
DoS	3636103	0	247267	6,37
U2R	0	0	52	100,00
R2L	127	0	999	88,72
Probe	27242	0	13860	33,72

The typical approach for performing anomaly discovery using the KDD data set is to employ a customized SVM machine learning algorithm to learn the general behavior of the data set in order to be able to differentiate between normal and malicious activities and randomly shuffles the order of all instances passed through it.

However, detailed classification are not of much interest in this paper since most of the anomaly detection systems work with binary labels, i.e., anomalous and normal, rather than identifying the detailed information of the attacks.

Table 2. Dataset subsampling

Class	Inst. Count	Taken	Ratio %
Normal	812814	67943	8,36
U2R	52	52	100,00
R2L	999	999	100,00
Probe	13860	13860	100,00
DoS-Back	968	968	100,00
DoS-Land	19	19	100,00
DoS-Neptune	242149	13626	5,63
DoS-Pod	206	206	100,00
DoS-Smurf	3007	3007	100,00
DoS-TearDrop	918	918	100,00
Total:	1074992	101598	9,45

4.2   Feature Selection

A smaller feature set may result in considerably improved training and classification timing. Hanging on the anomaly discovery application, timing may be critical. Supplementally, some features may not truly relate to the intrusion classification results and should be excluded.

There are also more methodical approaches. From a theoretical perspective, it can be shown that optimal feature selection for supervised learning problems requires an exhaustive search of all possible subsets of features of the chosen cardinality. If large numbers of features are available, this is impractical. For practical supervised learning algorithms, the search is for a satisfactory set of features instead of an optimal set.

Most methods for attribute selection involve searching the space of attributes for the subset that is most likely to predict the class best. For optimal set choosing, I combine three satisfactory methods which normalize the attribute ranking. One way to accelerate the search process is to stop evaluating a subset of attributes as soon as it becomes apparent that it is unlikely to lead to higher accuracy than another candidate subset. This is a job for a paired statistical significance test, performed between the classifier based on this subset and all the other candidate classifiers based on other subsets.

Gain Ratio Attribute, evaluates the worth of an attribute by measuring the gain ratio with respect to the class, ranks attributes by their individual evaluations.

Info Gain Attribute evaluates the worth of an attribute by measuring the information gain with respect to the class, ranks attributes by their individual evaluations.

CFS[5], is a simple filter algorithm that ranks feature subsets according to a correlation based heuristic evaluation function. Irrelevant features should be ignored because they will have low correlation with the class. Redundant features should be screened out as they will be highly correlated with one or more of the remaining features. The acceptance of a feature will depend on the extent to which it predicts classes in areas of the instance space not already predicted by other features. M_S is the heuristic "merit" of a feature subset S containing.

C is the class attributes and the indices j range over all attributes in the set. U is the gain value when gain ratios when selection done.

$$\sum_j U(a_j, C) \Big/ M_s \tag{6}$$

Table 3. Feature Selection Result

| Gain Rat. A. | | Info Gain A. | | Selection Result | | CFS |
Gn.	A.	Gn.	A.	Tot.	Att.	Att. Name
0,69	3	0,26	3	0,95	service	service
0,74	5	0,19	5	0,93	src_bytes	dst_bytes
0,44	12	0,45	12	0,88	logged_in	wrong_fr.
0,62	6	0,23	6	0,84	dst_bytes	logged_in
0,35	30	0,28	30	0,64	diff_srv_rt	srv_ser_rt
0,36	29	0,27	29	0,64	same_s._rt	same_s._rt
0,35	4	0,29	4	0,64	flag	
0,23	26	0,36	26	0,59	srv_ser._rt	
0,26	25	0,31	25	0,57	serror_rate	
0,42	33	0,14	33	0,56	dst_h._s.c.	
0,41	23	0,12	23	0,53	count	

Selected feature set is "service, src_bytes, logged_in, dst_bytes, diff_srv_rate, same_srv_rate, srv_serror_rate"

4.3   Kernel Selection

Another crucial issue for support vector machines is choosing the kernel function. Kernels introduce different nonlinearities into the SVM problem by mapping input data X implicitly into hypothesis space via a function Φ where it may then be hyper plane separable.

However, searching for different kernels either via trial-and-error or other exhaustive means can be a computationally higher one.

Weka provides several different SVM implementations along with multiple kernels via standard parameters. I examine two things, the relative importance of features in training the dataset and the choice of kernel algorithm. By understanding what features are most relevant, the dataset can be trimmed to include only the most useful data.

Random sub-sampling, validation method randomly splits 66% of dataset into training and %33 for validation data. For each such split, the model is fit to the training data, and predictive accuracy is assessed using the validation data.

Table 4. Kernel Accuracies

Kernels	Accuracy (%) with selected attributes	Accuracy(%)
Linear	86.77	73.35
Polynomial	58.57	33.23
RBF	99.67	97.71
Sigmoid	66.76	78.25

4.4   Parameter of Kernel Selection

The most common and reliable approach to parameter selection is to decide on parameter ranges, and to then do an exhaustive grid search over the parameter space to find the best setting. Unfortunately, even moderately high resolution searches can result in a large number of evaluations and unacceptably long run times.

Approach is to start with a very coarse grid covering the whole search space and keeping the number of samples at each iteration constant. I compare the performance of the proposed search method by changing SVM type and gamma value for selected Kernel using LIBSVM and the RBF kernel, both in terms of the quality of the final result and the work required to obtain that result.

By storing the search parameter bounds in a list, the searching itself is independent of the number of parameters in the increasing space. This allows us to re-use the same parameters and code for all four kernels.

For both search methods, the parameter ranges are :

C-SVC
- $\log_y \{0, ..., -8\}$
- $\log_c$ x 10{10, ..., 10}

nu-SVC
- $\log_y \{0, ..., -5\}$
- nu x 100 {10, ..., 70}

Results of search are:

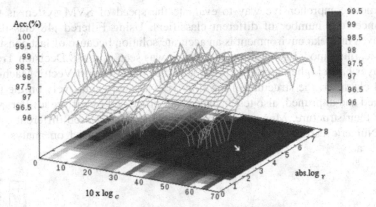

Fig. 1. C-SVC Distribution of Parameter Search

Table 5. nu-SVC Parameter Selection Results

C	Gamma	Accuracy(%)
1000000	5,00E-05	99,867
10000	5,00E-05	99,801
50000	5,00E-05	99,801

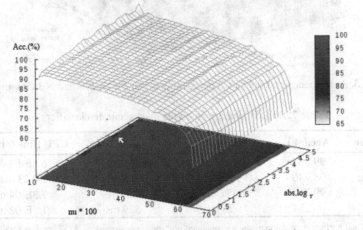

Fig. 2. nu-SVC Distribution of Parameter Search

Table 6. nu-SVC Parameter Selection Results

nu	gamma	Accuracy(%)
0,11	0,001	98,740
0,11	5,00E-04	98,740
0,11	9,00E-04	98,740

4.5   Performance and Speed Comparison

For the most comprehensive way to evaluate the speed of SVM system is test the performance of a number of different classifiers. Using Filtered j48 classifier and Naïve Bayes in Weka environment is an accurate solution because of fast construction speed of learning model generation. In cases, it was seen that J48 Decision Trees and Naïve Bayes had a higher speed but lower accuracy than Support Vector Machines.

In J48 classifier, the structure of the filter is evaluated exclusively on the training data pruned and unpruned, also test instances are processed by different filters without changing their structure. Moreover Naive Bayes classifier is used different estimator classes. Numeric estimator precision values were chosen based on analysis of the training data.

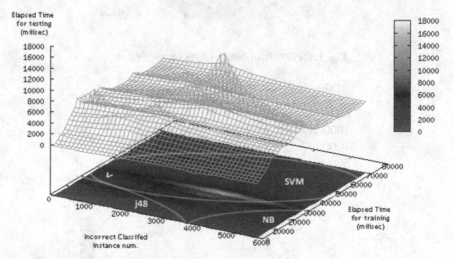

Fig. 3. Performance and Speed Comparison between SVM, J48 and NB classifiers

Table 7. Speed Comparison on the most accurate classifiers

Classifier	Acc.(%)	Learning time	Testing time	CPU time for 1 ins.
SVM-C	99,867	3865 ms	2092 ms	6.8E-02 ms
SVM-nu	98,740	9965 ms	6624 ms	0.23 ms
J48	99.716	172 ms	23 ms	7.8E-04 ms
NB	96.284	26 ms	529 ms	1.9E-02 ms

Test of speed and performance comparison was operated on "*randomly organized 15572 training and 25405 test instances*".

5   Result and Discussion

Firstly, final data set has the following advantages over the original KDD data set. It does not include redundant records in the train set, so the classifiers will not be biased towards more frequent records. There are no duplicate records in the proposed test sets; therefore, the performances of the learners are not biased by the methods which have better classification rates on the frequent records. The numbers of records in the training and test sets are reasonable, which makes it affordable to run the experiments on the complete set without the need to randomly select a small portion. Consequently, evaluation results of different research works will be consistent and comparable.

Table 8. RBF overall accuracy

SVMType	Accuracy (%) with selected params.	Classification
C-SVC	99.66	34594/34711
nu-SVC	99.00	34365/34711

Similarly, the best subset features to be trained on can be successfully identified using the parametric methods described above. By combining the kernel, feature and parameter selection, I arrive at an improved version of the algorithm. This more quickly and more accurately predicts the safety of network traffic.

Combination of clustering and only 3 classifier has been used for model generation and performance test. In all system for real time classification on network traffic one of this clustering and classifiers is used to improve the detection rate of intrusion detection system. Experiments are carried out on 41 class labels datasets and 7 class labels datasets. And also filtered classification is used to improve the accuracy of j48 Decision Tree classifiers. After comparison, I could find the way to do faster detection for network intrusions while protecting a computer network from unauthorized users, including perhaps insiders etc. The learning task about classification is to build a predictive model (a categorizer) capable of distinguishing between *bad* connections, called intrusions or attacks, and *good* normal connections.

6   Conclusions

Through the use of correct kernel choice and feature selection, I think that I have shown that it is possible to improve the accuracy and efficiency of a Support Vector Machine applied to an Intrusion categorizing scenario. The choice of kernel should be made for correct the superior results. Similarly, the best subset features to be trained on can be successfully identified using the parametric methods. By combining the kernel, feature and parameter selection, I arrived at an improved version of the algorithm. This is more quickly and more accurately predict the safety of network traffic.

After comparing the SVM with other classifiers, I believe that better handling of algorithm parameter can give us improved accuracy, while it can be as fast as J48 and Naive Bayes. Moreover, the success of classifier performance suggests that an ensemble approach can give a good result for real time network analyses.

There are some critiques of attack taxonomies and performance measures. However, attack taxonomies are not of much interest in this paper since most of the anomaly categorizing systems work with binary labels, i.e., anomalous and normal, rather than identifying the detailed information of the attacks.

The number of records in the train and test sets is reasonable, which makes it affordable to run the experiments on the complete set. Consequently, evaluation results of different research on different subsets works are consistent and comparable. Based on my approach, I gained performance results that indicate that our approximation of using SVM to represent and detect computer intrusions is workable. Intrusion detection systems have gained not popular acceptance, mainly because of their space requirements and the performance impact that is suffered while running them with regular system activity. I have displayed that it is workable to run an intrusion detection system based on improved SVM method, concurrently with other user activities on multi-user networks, without superfluous degradation in performance.

The proposed methods can be applied to encrypted network traffic, since it does not rely on the application payload for classification. Furthermore, as all the feature parameters are computable without the storage of multiple packets, the method lends itself well for real-time traffic identification.

Currently this work is being extended to work across each of the SVM kernels supported by LIBSVM, but some care must be taken to ensure that the search is robust in the face of infeasible solutions which are more likely with some of the other kernels. In addition, a kernel-searching is being added on top of the per-kernel parameter search so the system can automatically identify the best kernel and its parameter settings.

References

1. McHugh, J.: Testing intrusion detection systems: a critique of the 1998 and 1999 darpa intrusion detection system evaluations as performed by lincoln laboratory. ACM Transactions on Information and System Security 3(4), 262–294 (2000)
2. Yao, J., Zhao, S., Fan, L.: An Enhanced Support Vector Machine Model for Intrusion Detection. In: Wang, G.-Y., Peters, J.F., Skowron, A., Yao, Y. (eds.) RSKT 2006. LNCS (LNAI), vol. 4062, pp. 538–543. Springer, Heidelberg (2006)
3. Mahoney, M.V., Chan, P.K.: An analysis of the 1999 dARPA/Lincoln laboratory evaluation data for network anomaly detection. In: Vigna, G., Krügel, C., Jonsson, E. (eds.) RAID 2003. LNCS, vol. 2820, pp. 220–237. Springer, Heidelberg (2003)
4. Notes about Classification Methods,
   http://www.d.umn.edu/~padhy005/Chapter5.html
5. Boser, E., Guyon, I., Vapnik, V.: A training algorithm for optimal margin classifiers. In: Proceedings of the Fifth Annual Workshop on Computational Learning Theory, pp. 144–152. ACM Press (1992)

Part II

Knowledge Engineering and Ontology Development

A Quantitative Knowledge Measure and Its Applications

Rafik Braham

PRINCE Research Group, College of ICT (ISITCom), University of Sousse, Sousse, Tunisia

Abstract. Several concepts related to knowledge have emerged recently: knowledge management, knowledge society, knowledge engineering, knowledge bases, etc. We are here specifically interested in "scientific knowledge" in the context of student learning assessment. Therefore, we develop a framework within which knowledge is decomposed into grains called knowlets so that it can be quantified. Knowledge becomes then a measurable quantity in very much the same way information is known to be a measurable quantity (in the sense of Shannon's information theory). We then define an appropriate metric that we use in the specific domain of learning assessment. The proposed framework may be utilized for knowledge acquisition in the context of ontology learning and population.

Keywords: Knowledge representation and acquisition, Similarity measure, Ontology, Information theory, Knowledge metrics, Knowlets.

1   Introduction

The nature of "Knowledge" is an old subject that attracted the attention of philosopher for a long time. It's believed that Socrates has said "I only know that I know nothing." In recent years, several concepts related to knowledge have been in use: knowledge management, knowledge society, knowledge engineering, knowledge bases, etc. This is due to the fact that with the fascinating progress in computer science, there is a rush to use computers in the processing of almost everything. Knowledge is no exception. But before we could manipulate knowledge, we need to be able to represent it first.

In many practical situations, there is also the need to quantify and measure knowledge [5]. For example if it is an asset, then we need to know how much this asset is worth.

Our interest in knowledge comes from our study of student learning assessment [8-9]. More precisely, our goal was to design exams that reflect best the material actually covered. We believe that this undertaking is now feasible as a consequence of the development of the field of ontology. For this reason, our work may find several applications related to ontologies.

E-learning has led to renewed focus on student learning assessment. A number of assessment environments and related standards have emerged [10, 12, and 14].

The specific question that we set to answer in our work is this: "how much do students know about the subject of a given course?" But first we define the problem in a general context as a "knowledge acquisition problem."

A. Fred et al. (Eds.): IC3K 2010, CCIS 272, pp. 187–196, 2013.

Consider the following scenario:

1. Take a certain knowledge base (KB) which contains at some point in time an amount of stored Knowledge denoted by K_S (according to some positive metric to be defined later). We may assume $K_S = 0$ initially.

2. Now some knowledge denoted by K_{in} is presented to the system (KB).

3. The system will compare K_{in} to K_S. Only the part of K_{in} that is novel (with respect to K_S) shall be stored. Since the Knowledge increment is greater than or equal to zero, then "new $K_S \geq$ old K_S".

Three basic questions can be raised at this stage:

- What knowledge metric to use?
- How can K_{in} and K_S be compared?
- What use can be made of this metric?

These are the questions we intend to answer in the following sections.

2   Defining Knowledge Levels

The definition of knowledge is an epistemological subject that attracted interest for a long time. Several popular views put knowledge at the intersection between truths and beliefs (see [3] for a recent discussion). The relationship between knowledge and information is also an intriguing question. There is clear evidence that information is somehow transformed and integrated into existing knowledge for the creation of new knowledge. It also clear that learning is a process by which people are making sense of information they receive and discover by thinking and other mental actions how to transform prior knowledge into new knowledge [2]. Fig. 1 below gives a summary of our view on these relationships.

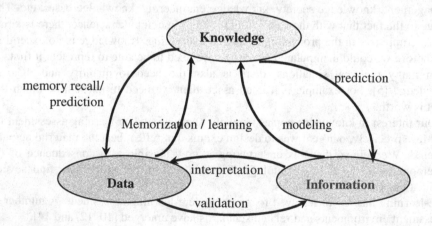

Fig. 1. Triad data-information-knowledge

Once knowledge is defined, researchers and thinkers have attempted to measure it. Two main techniques have been developed: subjective and objective measures. In the subjective measure, a person is asked to assess his/her own knowledge about a given subject. This has been called "perceived knowledge" or awareness [2]. In the latter type, a person is asked to answer certain questions (for example true/false) in very much the same way the knowledge of students is assessed in a final exam (usually from memory and not open book). In this regard, some standards have been created. For example, in the OECD/PISA text reading assessment [4], the following five aspects have been fixed to test understanding of a text:

—forming a broad general understanding;
—retrieving information;
—developing an interpretation;
—reflecting on the content of a text; and
—reflecting on the form of a text.

It can be seen that these aspects assess varying degrees of subject mastering and each level involves a certain type of brain activity. To formalize this idea, Bloom developed a model known as Bloom's taxonomy that is now well accepted and referenced by educators and pedagogy experts. He claims that learning may be categorized into six levels: from the simplest one that he calls curiously "knowledge" to the most elaborate namely "evaluation." At this highest level and according to Bloom, a learner is able to make judgments (see Krathwohl [1] for a thorough discussion).

Inspired by Bloom's six-level knowledge model, several workers tempted to grade exams and course contents using quantitative metrics based on Bloom's taxonomy, for example Oliver et al. [19] and Zheng [26].

Whether it is knowledge in the absolute sense or knowledge as related to learning, KBs etc, a basic element is that of implicit (or tacit) knowledge versus explicit knowledge. This leads us to another aspect: knowledge change. Knight [2] points out the existence of two patterns in the changes of knowledge: the first pattern an additive approach and the second pattern an integrative approach. In the first type, he claims that students have gathered facts, whereas in the second type, they manipulated facts in several ways such as building explanations, synthesizing facts and creating abstract groupings.

In this work, we are interested in explicit knowledge. When students are tested for example, we assume that we are measuring their explicit knowledge. Of course it is related to the quantity of implicit knowledge. This may explain findings by several authors that show differences between perceived (self-reported) and objective knowledge quantities [2].

The ideas we develop here should be regarded within the framework of ontology. In fact we believe that this work is made possible with the emergence of ontology as an important new area in computer science. It is well known that ontology is relative to a domain of study. As explained below, we consider science and engineering domains. This is important for concept definition when we consider specific documents. For example, certain terms such as verbs and nouns may not be important to us and will not be counted as "concepts", whereas they may be capital for someone studying rhetoric or languages.

Any framework of knowledge has to make some assumptions about the levels of granularity because knowledge is necessarily hierarchical. This question may be debated on psychological and cognitive grounds. Although our arguments are purely technical, we reflect first on some epistemological aspects.

We define four Knowledge levels, although it may be argued that more levels may exist. The concept of "Knowledge Level" (KL) in our work should not be confused with the one described by Newell [18], nor with the KLs in the sense of philosopher J. Locke (also called degrees). Our KLs are defined from a logics point of view. They allow us to present corresponding metrics as we shall explain below.

a. Knowledge of Level 1: This is basic knowledge. It describes concepts, items or objects, for example animal, tree, person …

b. Knowledge of Level 2: Here we have properties and relations defined on concepts. Elements of knowledge at this level require two K-elements of Level 1. Examples: a parrot is-a bird; Coca-Cola is-a soft-drink; Mozzarella cheese is-made-in Italy, lions are-faster-than humans. It includes simple relations of the type $5=2+3$ and $5>4$ as well.

c. Knowledge of Level 3: This level incorporates three cases:
- Rules and inferences, for example: hasUncle ← hasParent^hasBrother
- Logical structures of the type IF-THEN
- Equations.

d. Knowledge of Level 4: This is the highest level. It includes logical structures of the form IF-THEN-ELSE such as those encountered in theorems. Algorithms also fall in this category. Algorithms typically consist of one or more IF-THEN-ELSE structures.

To simplify the terminology, we will call elements (grains or items) of knowledge of any level "knowlets," a word inspired from applets and servlets in computer science. Note that this definition is not quite the same as Mons' [17] and knowlets are not just the smallest "piece" of knowledge. They are hierarchical elements of knowledge.

3   Knowledge Entropy

Two Knowledge kinds are of interest to us: "stored Knowledge" (K_S) and "learned Knowledge" (K_L). The latter one is new knowledge actually, i.e. knowledge to be learned and added to K_S. When a person (a learner in our case) is presented with some Knowledge K_{in}, the amount of gained knowledge, denoted by $H(K_L)$ must be computed having the following properties:

1. $H(K_L)$ is positive.
2. $H(K_L) = 0$ if $K_{in} \subset K_S$.

Shannon in his seminal work on information theory [22] was inspired by Hartley and used the well-known logarithmic measure for information. Since then, information theoretic approaches have flourished like those proposed by Smyth and Goodman [24], Lin [15], etc. We employ a logarithmic measure as well.

Using the KLs defined earlier, we have for K_{in} in the general case:

$$K_{in} = K_{in,1} \cup K_{in,2} \cup K_{in,3} \cup K_{in,4}$$

where $K_{in,n}$ is the knowlet of level n contained in K_{in}. In other words, K_{in} must be decomposed according to KLs before proceeding further. Let us assume without any loss of generality, that:

$$K_{in} = K_{in,n} \text{ (just one level), n} = 1, 2, 3 \text{ or } 4.$$

Under these assumptions, we define our knowledge metric with the following fundamental equation:

$$H(K_L) = \alpha_n \log_2(1 + \frac{K_{in}^2 - K_{in} * K_S}{K_{in}^2}) \tag{1}$$

where α_n is the knowledge unit of Level n, $K_{in} * K_S$ is a measure of correlation defined as follows:

$$K_{in} * K_S = \text{Sim}(K_{in}, K_{in} \cap K_S) \tag{2}$$

where "Sim" represents a similarity function that we will discuss in more detail in Section 4 next.

Furthermore, we use the notation:

$$K_{in} * K_{in} = \text{Sim}(K_{in}, K_{in}) = K_{in}^2$$

(this is consistent with notation from the field of signal processing). We give $H(K_L)$ as defined by (1) the name of "knowledge entropy" and we choose the "bit" as a unit of measure in line with information theory since logarithm base 2 is used in (1) and throughout.

When we compare the knowledge levels defined in Section 2 in terms of number of concepts (or ideas) involved, we find appropriate to take $\alpha_n = n\alpha_1$. Furthermore we set $\alpha_1 = \log_2 2^{1/2} = \frac{1}{2}$ (actually $\alpha_n = \log_2 2^{n/2} = n/2$).

Let us go back to Equation (1) and examine its main properties. Two cases are to be considered: $K_{in} \subset K_S$ and $K_{in} \not\subset K_S$.

• $K_{in} \subset K_S$: in this case $K_{in} \cap K_S = K_{in}$ so that $K_{in} * K_S = K_{in}^2$ and thus $H(K_L) = 0$ as precisely desired.

• $K_{in} \not\subset K_S$: if $K_{in} \cap K_S = \emptyset$ then mathematically speaking K_{in} and K_S are orthogonal ($K_{in} \perp K_S$). In this case $H(K_L) = \alpha_n \log_2(2) = \alpha_n$ (its maximum value).

If $K_{in} \cap K_S \neq \emptyset$ then $H(K_L)$ lies anywhere between zero and this maximum value.

4 Knowledge Similarity

At this stage of the discussion, we need to define the similarity employed in Equation (1). Several similarity measures have been proposed in the literature, for example Lin's [15] and Resnik's [21]. More recent similarity metrics have been proposed by d'Amato [6] and Slimani et al. [23]. Mihalcea et al. [16] and Warin et al. [25] give comparative studies of these measures among others. Some of the measures are defined based on information theoretic approaches while others use a logics and/or ontology point-of-view.

The choice of a particular measure depends on the form of the objects to be compared: texts, semantic maps, rules, etc. In our case, we need to compare knowlets (concepts, properties, rules/equations, theorems). We define a similarity measure adapted from Lin's in the following way:

$$\text{Sim}(K_1, K_2) = \frac{K_1 \cap K_2}{K_1 \cup K_2} \tag{3}$$

We could have used cardinals but we prefer to keep the notation simple. Let us illustrate the use of this definition with an example (More on this in Section 7).
Let $K_1 = \{$father $\equiv$ man $\wedge$ parent$\}$ and $K_2 = \{$mother $\equiv$ woman $\wedge$ parent$\}$. Then:

$$\text{Sim}(K_1, K_2) \quad = \frac{\wedge + parent}{man + \wedge + parent + woman} = \frac{2}{4} = 0.5.$$

The obvious cases of $K_1 = K_2$ and $K_1 \perp K_2$ can be easily checked (maximum and minimum similarity values).

In practice and for correct knowledge acquisition, a threshold value μ should be chosen to decide for new versus learned knowledge, for example μ=.25.

5   Knowledge Distance

Instead of similarity, we may use the concept of distance. The Hamming distance (d_H) is perhaps best known in computer science and quite simple as well. Consider the following example. Let $K_{in} = \{$information source$\}$ and $K_S = \{$discrete information source$\}$. K_{in} may be decomposed into three concepts: information, source and information source. K_S contains two additional concepts: discrete and discrete information source (five concepts overall). Let us use as a reference (or base) the ordered list of concepts:

$K_B = \{$discrete information source, discrete, information source, source, information$\}$.

We may then write in vector notation $K_{in} = (0, 0, 1, 1, 1)$ and $K_S = (1, 1, 1, 1, 1)$. The corresponding Hamming distance defined as the number of different positions is then $d_H(K_{in}, K_S) = 2$. To compare two knowledge quantities, we could use similarity or distance with the difference that distance increases when similarity decreases. Several similarity measures are actually defined as distance expressions.

6   Knowlets in Practice

We are interested in this paper in technical documents (course materials, papers, exams) from scientific and engineering fields. These documents comprise four types of knowlets:

(1) Concepts: in the form of one or more words.
(2) Theorems: generally in the form of IF-THEN or IF-THEN-ELSE. This includes algorithms (see Section 2).
(3) Equations: usually definitions or a series of derivations.
(4) Examples: applications of theorems and equations for specific values and conditions.

Transforms such as Fourier, Laplace, Z, may be considered as special cases of equations and transforms come in pairs (analysis and synthesis equations). We may extend this logic to laws of physics and other entities like those suggested by Gruber [11].

The case of examples is less straightforward and requires a more elaborate analysis. Examples may be applied to equations, theorems and so forth. They actually help us understand them. But a fundamental question is the following: how many examples are necessary to fully understand a theorem say? The answer is, in theory, a large number, approaching infinity. Of course all depends on the theorem and the examples themselves. It is however safe to assume that examples may be ranked in a decreasing order of usefulness. We make use of the fact that:

$$\frac{1}{2} + \frac{1}{4} + \frac{1}{8} + \cdots \left(\frac{1}{2}\right)^n + \cdots = 1 \text{ and assume that:}$$

H(n examples) $= \beta \log_2[1 + \frac{1}{2} + \frac{1}{4} + \frac{1}{8} + \cdots \left(\frac{1}{2}\right)^n]$ where $\beta = $ H(theorem) in the case of a theorem for example. Note that $\lim_{n \to \infty}$ H(n examples) $= \beta$.

7   Applications and Case Study

The metrics that we have proposed may find numerous applications such as benchmarking ontologies, concept maps, T-Boxes and A-Boxes. Our own interest lies in the field of e-learning and student learning assessment more specifically. We believe that these metrics can be employed as effective tools to evaluate exams with respect to course contents. Furthermore, they may be quite useful in the automatic generation of assessment items from course material.

We illustrate these ideas with a practical example using a course on information theory (at the senior level) that we have been teaching for a few years now. First of all we need a text reference. We have chosen Shannon's paper [22] as it is known to a wide audience. Furthermore, it may be easily employed as a reference (at least in part) for any course on information theory.

Let us first clarify the use of the two notions of similarity measure defined in Section 4 and correlation metric defined in Section 3 with examples of concepts taken from Shannon's paper which are considered different.

As in Section 5 above, suppose $K_{in} = \{$information source$\}$ and $K_S = \{$discrete information source$\}$. This corresponds to case a of Section 2. K_{in} is composed of three concepts: information, source and information source. K_S contains two additional concepts: discrete and discrete information source (five concepts overall). We have then:

$$\text{Sim}(K_{in}, K_S) = 3/5 = 0.6$$
$$K_{in} * K_S = \text{Sim}(K_{in}, K_{in} \cap K_S) = 1$$
$$H(K_L) = 0.$$

Now let $K_{in} = \{$second-order approximation of English$\}$ and $K_S = \{$first-order approximation of English$\}$. This corresponds to case b of Section 2. Then:

$$\text{Sim}(K_{in}, K_S) = 3/7 = 0.43,$$

$$K_{in} * K_S = \text{Sim}(K_{in}, K_{in} \cap K_S) = 3/5 = 0.6$$

$$H(K_L) = .24 \text{ bit.}$$

The analysis of Shannon's paper (without the appendices) reveals at least 16 concepts, 36 relations/properties (these two numbers can only be more or less subjective), 9 equations, 12 theorems and 17 examples. Out of the 12 theorems, 7 are of the form IF-THEN (equivalent to equations) and the rest 5 are of the form IF-THEN-ELSE. This analysis would have been carried out ideally with automatic techniques. But it was done manually due to the lack of appropriate tools at the present time.

According to our metrics and using the results of the above analysis, we have:

$$H(\text{Sh}1948) \quad = (16+36\cdot2+9\cdot3+7\cdot3+5\cdot4)\cdot\alpha_1 + H(\text{examples}).$$

The examples case is somewhat complex. Seven examples are for concepts (distributed as 1, 1, 1, 1, 1, 2), five for equations (distributed as 1, 1, 3) and five for theorems (one If-Then and 2, 1, 1 If-Then-Else). Therefore:

$$H(\text{examples}) = [5\log_2(1 + \tfrac{1}{2}) + \log_2(1 + \tfrac{1}{2} + \tfrac{1}{4})]\alpha_1 + [2\log_2(1 + \tfrac{1}{2}) + \log_2(1 +$$

$$\tfrac{1}{2} + \tfrac{1}{4} + \tfrac{1}{8})]\cdot3\alpha_1 + \log_2(1 + \tfrac{1}{2})\cdot3\alpha_1 + [2\log_2(1 + \tfrac{1}{2}) + \log_2(1 + \tfrac{1}{2} + \tfrac{1}{4})]\cdot4\alpha_1$$

$H(\text{examples}) = 19.6\alpha_1 = 9.8$ bits.

We have finally: $H(\text{Sh}1948) = 175.6\alpha_1 = 87.8$ bits of Knowledge entropy.

It may be useful to compute the average Knowledge entropy. In this case it is equal to $\frac{87.8}{90} = .975$ b/knowlet.

This figure is relatively low (less than 1), but if we do not take into account the examples, it becomes 1.07 b/knowlet.

We looked at exams for the last three years. We have found an average of 10 concepts and 4 equations per exam. We may conclude that $H(\text{exam}) = 22\cdot\alpha_1 = 11$ bits, i.e. 12% of Shannon's paper. Currently there are no standards in the literature to tell us what value would be acceptable. It may depend on the course and the number of exams (tests, final, etc.).

To have another benchmark, we applied the same principles to another course (at the master's level) on modeling and simulation that meets 1.5 hours a week (the information theory course is 3-hour). We have found $H(\text{course}) = 44\cdot\alpha_1 = 22$ bits and $H(\text{exam}) = 8.75$ bits (almost 40%).

8   Related Work and Discussion

In this paper we gave a foundation for knowledge metrics. Ideally, an automatic generation of knowledge from text or documents should be made. Then documents are compared automatically as well. This undertaking is for that matter impractical at the present time as necessary tools are still under development. Significant progress has been made in the area of ontology learning and population during the last few years. Valuable tools have been proposed in this regard [7, 27], but some time is still needed before they become fully operational.

To the best of the author's knowledge, the only works that grade exams and course contents using quantitative metrics are those based on Bloom's six-level knowledge taxonomy, for example [19, 26]. We therefore believe that we have presented original ideas that would allow us to assess quantitatively our exams with respect to course contents we present to students.

The metrics we defined may be used for comparative purposes but with due precaution. The scientific importance of any specific theorem for example is measured by its impact on the course of science and technology and is by no means an absolute value.

We should note that the knowledge metrics we have defined open a large scope of applications especially those related to ontology development and comparison and not just learning assessment.

Another possible further exploration can be done to assess the validity of these metrics from a cognitive standpoint. Indeed we have not elaborated on how this work would compare against human perception of knowledge. Furthermore, even if we assume that we established ground rules about scientific knowledge measurements, the quantitative evaluation of other forms of knowledge such as ideas, practical skills (student lab work), philosophical essays, etc remains an open question.

Acknowledgements. The author would like to thank ISITCom and Prince Research Group for their financial support.

References

1. Krathwohl, D.R.: A Revision of Bloom's Taxonomy: An Overview. Theory into Practice 41(4) (Autumn 2002)
2. Knight, A.J.: Differential Effects of Perceived and Objective Knowledge Measures on Perceptions of Biotechnology. AgBioForum 8(4), 221–227 (2005)
3. Godfrey-Smith, P.: Knowledge, trade-offs, and tracking truth. Philosophy and Phenomenological Research LXXIX (1), 231–239 (2009)
4. Todd, R.J.: From information to knowledge: charting and measuring changes in students' knowledge of a curriculum topic. Information Research 11(4), paper 264 (2006), http://InformationR.net/ir/11-4/paper264.html
5. Liebowitz, J., Suen, C.Y.: Developing knowledge management metrics for measuring intellectual capital. Journal of Intellectual Capital 1(1), 54–67 (2000)
6. d'Amato, C., et al.: Semantic Similarity Measure for Expressive Description Logics. In: Proc. CILC, Rome, Italy (June 2005)
7. Buitelaar, P., Cimiano, P., Magnini, B.: Ontology Learning from Text: An Overview, pp. 1–9. IOS Press (2003)
8. Cheniti-Belcadhi, L., Braham, R., Henze, N., Nejdl, W.: A Generic Framework for Assessment in Adaptive Educational Hypermedia. In: Proc. IADIS WWW / Internet, Madrid, Spain, pp. 397–404 (2004)
9. Cheniti-Belcadhi, L., Braham, R.: Assessment Personalization on the Semantic Web. Journal of Computational Methods in Sciences and Engineering, Special issue: Intelligent Systems and Knowledge Management 8, 1–20 (2008)
10. Gardner, L., Sheridan, D., White, D.: A web-based learning and assessment system to support flexible education. Journal of Computer Assisted Learning 18(2), 125–136 (2002)
11. Gruber, T.R.: Toward Principles for the Design of Ontologies used for Knowledge Sharing Technical Report KSL 93-04, Knowledge Systems Lab (1993)

12. He, L.: A Novel web-based educational assessment system with Bloom's taxonomy. Current Developments in Technology-Assisted Education, 1861–1865 (2006)
13. University of Victoria: Hot Potatoes (2010), http://hotpot.uvic.ca/
14. IMS Global Learning Consortium: IMS Question and Test Interoperability Overview, Version 2.1 (2006), http://www.imsglobal.org/question/qtiv2p1pd2/imsqti_oviewv2p1pd2.html
15. Lin, D.: An Information Theoretic Definition of Similarity. In: Proc. of the 15th Int. Conference on Machine Learning (1998)
16. Mihalcea, R., et al.: Corpus-based and Knowledge-based Measures of Text Semantic Similarity. In: Proc. of 21st National Conference on AI, pp. 775–780. AAAI (2006)
17. Mons, B.: Calling on a million minds for community annotation. In: WikiProteins (2008), http://genomebiology.com/2008/9/5/R89
18. Newell, A.: The Knowledge Level. AI Magazine, 1–20 (Summer 1981)
19. Oliver, D., Dobele, T., Greber, M., Roberts, T.: This Course Has A Bloom Rating Of 3.9. In: Proc. Sixth Australasian Computing Education Conference (ACE 2004), Conferences in Research and Practice in Information Technology, vol. 30 (2004), http://crpit.com/confpapers/CRPITV30Oliver.pdf
20. Palloff, M., Pratt, K.: How do we know they Know? Student Assessment Online. In: Proc. 22nd Annual Conf. on Distance Teaching and Learning, Wisconsin, pp. 1–5 (2006)
21. Resnik, P.: Semantic Similarity in a Taxonomy: An Information-Based Measure and its Application to Problems of Ambiguity in Natural Language. Journal of Artificial Intelligence Research 11, 95–130 (1999)
22. Shannon, C.: A Mathematical Theory of Communication. The Bell System Technical Journal 27, 379–423/623–656 (1948)
23. Slimani, T., Ben Yaghlane, B., Mellouli, K.: A New Semantic Similarity Measure based on Edge Counting. World Academy of Science, Engineering and Technology 23, 34–38 (2006)
24. Smyth, P., Goodman, R.M.: An Information theoretic Approach to Rule Induction from Databases. IEEE Trans. on Knowledge and Data Engineering 4(4), 301–316 (1992)
25. Warin, M., Oxhammar, H., Volk, M.: Enriching an Ontology with WordNet based on Similarity Measures. In: Proc. of the MEANING Workshop (2005)
26. Zheng, A.Y.: Application of Bloom's Taxonomy Debunks the MCAT Myth. Science 319, 414–415 (2008)
27. Zouaq, A., Nkambou, R.: Building Domain Ontologies from Text for Educational Purposes. IEEE Trans. on Learning Technologies 1(1), 49–62 (2008)

A Model-Driven Development Method for Applying to Management Information Systems

Keinosuke Matsumoto, Tomoki Mizuno, and Naoki Mori

Department of Computer Science and Intelligent Systems, Graduate School of Engineering
Osaka Prefecture University, 1-1 Gakuen-cho, Nakaku, Sakai, Osaka 599-8531, Japan
matsu@cs.osakafu-u.ac.jp

Abstract. Almost every information system is built assuming that it is to be modified during operation. It costs very much to transplant the system to change requirement specifications or implementing technologies. For this purpose, there is a model theory approach that is based on a theory that a system can be modeled by automata and set theory. However, it is very difficult to generate automata of the system to be developed right from the start. In addition, there is a model-driven development method that can flexibly correspond to changes of business logics or implementing technologies. In the model-driven development, a system is modeled using a modeling language. This paper proposes a new development method applying the model-driven development method to a component of the model theory approach. The experiment has shown that a reduced amount of workloads is more than 30% of all the workloads.

Keywords: Model-driven development, Model theory approach, UML, Graphic diagrams, Transaction processing system.

1 Introduction

Almost every information system is built assuming that it is to be modified during operation. However, an end user must have detailed knowledge of the system or prior involvement in its development to modify the system effectively. It costs very much to transplant the system to change requirement specifications or implementing technologies. In order to overcome the problem, various software development methods have been advocated. Traditionally, a Management Information System (MIS) has been developed without using formal methods. By the informal methods, the MIS is developed on its lifecycle without having any models. It causes many problems such as lack of the reliability of system design specifications. In order to overcome the problem, some formal approaches to the MIS development have been developed.

Vienna development method [1] is one of the most popular methods. It designs a system based on a system model, and the model is described in the set theory [2] and the logics. In addition, Takahara et al. proposed a unique systems development method, a model theory approach [3]-[5]. This approach is based on an idea that a system can be modeled by automata and set theory. An automaton consists of two or

A. Fred et al. (Eds.): IC3K 2010, CCIS 272, pp. 197–207, 2013.

more states and a function that defines what processing is performed to an input in each state. However, it is very difficult to generate automata of the system to be developed right from the start. On the other hand, there is a model-driven development method [6]-[9] that can flexibly correspond to changes of business logics or implementing technologies. In the model-driven development, a system is modeled using a modeling language such as UML [10]. Generating source codes automatically reduces developing cost and makes consistency of design and implementing.

It is possible to combine the model-driven development and the model theory approach to bring advantages of the both methods. This paper proposes a new development method applying the model-driven development method to a component of MIS of the model theory approach. This research aims at cutting down the amount of workloads by applying the proposed method.

2   Model Theory Approach

This chapter explains the model theory approach. According to Takahara et al., automata can describe arbitrary information systems. The model theory approach is proposed as a development method of MIS using this theory.

2.1   Management Information Systems

A management information system consists of two components: a Problem Solving System (Solver) and a Transaction Processing System (TPS). The former is a system which offers some supports or answers when, how, and how much to dispatch goods. The latter is a system which deals with daily regular business activities in respect of recording sales or updating goods in stock. They are possible to independently operate on a constructing system, and also can be combined to realize more complicated systems.

This paper targets the TPS of MIS. Fig. 1 shows an outline of structure of the TPS. The TPS is modeled regarding a file system or database as a state, user's operation as an input, and response to a user from the system as an output. Therefore, it is necessary to define states, inputs, outputs, state transition functions, output functions and so on that consist of elements of automata to implement the TPS.

The TPS is implemented as a Web application as shown in a deployment diagram of the TPS shown in Fig. 2. UserModel.p in Fig. 2 is a compiled and executable file of implemented codes of the TPS. Moreover, stdUI.php is a PHP [11] program which is a user interface of the TPS. It changes automatically contents in accordance with UserModel.p. Therefore, it is possible for a developer of the TPS not to edit stdUI.php, but also to edit it to change the user interface arbitrarily. BusinessDataFile.lib is a business data file. The user of the TPS can use the system by accessing Web server, using a browser as a Web client.

A TPS Solver consists of processes that model a problem as automata and solving program goal-seeker. The Solver considers problem specifications as inputs, and solutions as outputs.

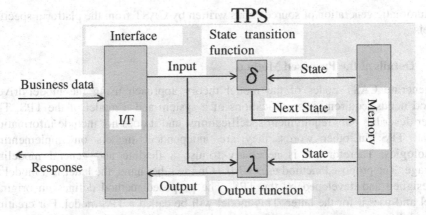

Fig. 1. Structure of the TPS

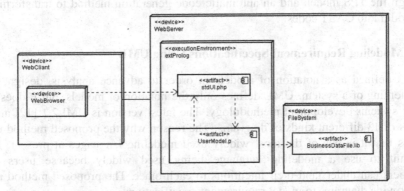

Fig. 2. TSP deployment diagram

2.2   Computer Acceptable Set Theory

The model theory approach uses a description of set theory called CAST (Computer Acceptable Set Theory) as a language to design and implement the TPS and Solver. CAST enables us to express set theory, proposition logic, and predicate calculus, and it can deal with automata. Takahara et al. have improved development environment and execution environment of CAST. All CAST codes are implemented by hand from the automata in the model theory approach.

3   Proposed Method

This chapter describes a proposed method. The method applies the model-driven development method to develop the TPS in the model theory approach, and cuts down the amount of development workloads. A concrete process is as follows:

(1) Design of platform independent models from requirement specifications

(2) Design of platform specific models

(3) Automatic generation of source codes written by CAST from the platform specific models

3.1   Details of the Proposed Method

To generate CAST codes of the model theory approach using the model-driven method needs requirement specifications of a system and a model of the TPS. The former describes only requirement specifications, and it does not include information on the TPS. In other words, they are independent models on implementing technologies. Therefore, it is necessary to use a flexible and general modeling language. The proposed method uses UML. On the other hand, the latter is a model to be designed and developed by the TPS. The proposed method defines an original model and uses it for the latter. This model will be called a TPS model. For creating the model from requirement specifications, the proposed method uses UML diagrams to design the TPS model, and an automatic code generation method to transform the TPS model into CAST codes.

3.2   Modeling Requirement Specifications Using UML

UML is defined as a notation of models in order to advance analysis, design, and implementing of a system. UML defines only the notation of models, but it does not include systems development methodology. The latest version is UML2.1 [12] and it consists of 13 different kinds of diagrams. The reason why the proposed method uses UML is that it is one of the most widely used modeling languages at present. It is important to use a modeling language being used widely because users and developers can understand their intentions to each other. The proposed method uses the following diagrams to model requirement specifications:

(1) Use Case Diagram. A use case diagram expresses functions to be implemented. The functions correspond to interface elements I/F of Fig. 1, and they are defined by a use case diagram. To describe the TPS, you must define functions (main use case) whose granularity are comparatively large, and indivisible functions (sub use case) that are contained in a main use case. This relation of the use cases can be expressed by using inclusion relation of UML2.1. The inclusion relation between a main use case and a sub use case may be in the relation of many-to-many. An inclusion relation is described using include stereotype. Fig. 3 shows this situation.

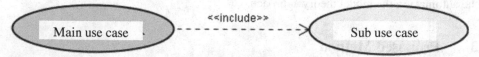

Fig. 3. Inclusion relationship between use cases

(2) Class Diagram. A class diagram defines attributes of a data file for business. The data file is stored in a file system or a database, and corresponds to a memory (DB) in Fig. 1. The usage of a class diagram in the proposed method is closer to an entity in Entity-Relationship diagram than a concept of the class in object-oriented

development. An attribute is defined so that it may be normalized in third normal form in a relational database.

(3) Activity Diagram. An activity diagram expresses a process to realize sub use cases and user defined functions. An activity diagram is created for every sub use case defined by the use case diagram. This corresponds to state transition function δ and output function λ shown in Fig. 1. A user doesn't need to consider that a movement of the system is an automaton, but just to describe them like a flowchart.

A system developer creates these UML diagrams at first, and creates specific models for the TPS. CAST code generation from the model is performed using this TPS model and the activity diagrams.

3.3   Designing TPS Model for Code Generation

A TPS model is an original defined model of the proposed method. We have designed a meta-model which defines a TPS model itself. The structure of the meta-model is shown in Fig. 4. This meta-model contains a structure of the TPS, and it is used as a direct input for code generation. A user designs this TPS model using the above-mentioned UML diagrams. UML diagrams do not depend on the model theory approach, but only depending on specifications of a system. TPS model is designed for implementing the system by the model theory approach.

3.4   Automatic CAST Code Generation

In the model-driven development, models, such as UML diagrams, are stored as a file in XML (eXtensible Markup Language) [13] form. It enables to exchange data of the models between modeling tools for the model-driven development. Therefore, this XML file is analyzed in order to generate CAST codes from the models. Rules that map model elements in the XML file and CAST codes are created. The models are transformed into CAST codes using these rules and JET (Java Emitter Template) [14] technology.

3.5   TPS Development Tool for Realizing the Proposed Method

A TPS development tool is implemented as a plug-in of an integrated development environment Eclipse [15] for realizing the proposed method. The functions of the TPS development tool are listed as follows:

(1) Construction of a Project. This function is to build a specific project for TPS development on a work space of Eclipse. A user can use this function calling a wizard from Eclipse. A specific icon is displayed so that user may be easy to distinguish the project from other projects on the work space. Whenever you build a project, the tool automatically generates necessary folder structure and templates of UML diagrams.

(2) Generation of TPS Model from UML Diagrams. This function is to analyze use case diagrams and class diagrams to generate TPS model. A user calls a wizard and should just input required information.

(3) TPS Model Editor. This is a function to edit TPS model. It can also output TPS model in XML form. It is implemented using EMF (Eclipse Modeling Framework) [16]. A user designs TPS model using this editor.

(4) Code Generation from Models. This function generates codes using JET. That is, JET engine maps TPS model and skeleton codes according to contents of JET templates, and generates concrete codes. Since TPS model is stored in XML form by the TPS model editor, it can be transformed into CAST codes of the TPS using JET templates.

(5) Editing Support Function of CAST Codes. This function supports for editing generated CAST codes.

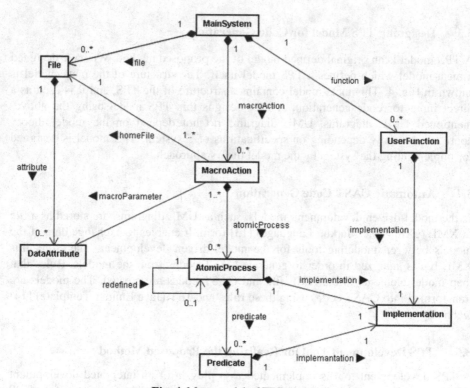

Fig. 4. Meta-model of TPS model

4   Experiment and Results

An experiment has been carried out to show the validity of the proposed method. The proposed method has been applied to one of TPS examples shown in [4] and implemented using the development tool explained in Chapter 3. This chapter shows results and discussions of the application experiment.

4.1   Contents of an Experiment

A bookstore credit sale management system is taken as one of examples of TPS application. The aim of the experiment is to investigate how many workloads for developing TPS can be reduced by using the proposed method. The TPS development tool is used for the proposed method. On the other hand, hand coding of all codes of TPS (we call it a conventional method) is carried out as the comparison of quantity of workloads. We introduce a method [17] how to quantify the amount of development workloads: It supposes that equal amount of workloads to add one node by the model editor and to describe one line of source code. The number of nodes of UML diagrams that are needed for applying the proposed method and the rate of automatic generation of CAST source codes are computed. As a result, the workloads of the proposed method and the conventional method are compared.

4.2   Requirement Specifications

Specifications of the bookstore credit sale management system are as follows:

(1) Customer Management. The system registers a new customer, and update or delete data of an existing customer. In the case of a new customer, the system suggests to register the new customer.

(2) Sales Management. The system adds new sales data and updates the customer's accounts receivable.

(3) Credit Management. Customer's credit data are recorded and its accounts receivable are updated.

(4) Report Generation. The system generates a credit sale balance report.

(5) Bill Generation. The system generates a bill for every customer.

4.3   Model Design

A use case diagram and a class diagram are created from the requirement specifications. The use case diagram and the class diagram are shown in Figs 5 and 6 respectively. But, sub use cases of the report generation function and the bill generation function are omitted in Fig. 5.

TPS model is generated from the completed use case diagram and the class diagram. An activity diagram is created from TPS model. A part of activity diagram corresponding to the customer management use case is shown in Fig. 7. The rectangle nodes described in Fig. 7 are input data; we call them activity parameter nodes. After creating a middle file from the activity diagram, codes are generated using it.

Nodes inputted by hand are 69 in all UML diagrams to model the bookstore credit-sale management system. There are more nodes actually in UML diagrams than this. It is because a part of nodes of UML diagrams are automatically generated as a template at the time of creating a TPS project by a wizard. The nodes generated automatically are two use case nodes in a use case diagram, and one class node in a class diagram. For each activity diagram, one start node, one end node, one activity node, and two activity parameter nodes are added automatically.

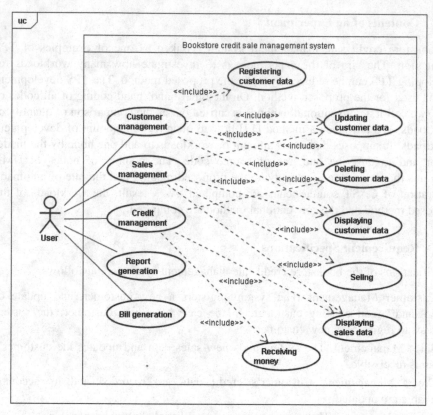

Fig. 5. Use case diagram of bookstore credit sale management system

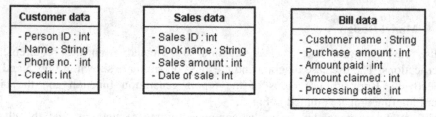

Fig. 6. Class diagram of bookstore credit sale management system

4.4   Code Generation Results of Customer Data

Code generation results are shown in Table 1. Where, logical LOC (Lines Of Code) is a number of lines except for blank lines and comment lines. In addition, "Number of nodes" column shows the number of nodes inputted by hand at modeling. A numerical value of "Total codes" column is a logical LOC of codes shown in the book [5]. A number in Table 2 is a value that the number of nodes generated automatically is subtracted from the number of nodes contained in each model. Therefore, the value of use case column of the customer management function in Table 2 is 3 (=5-2).

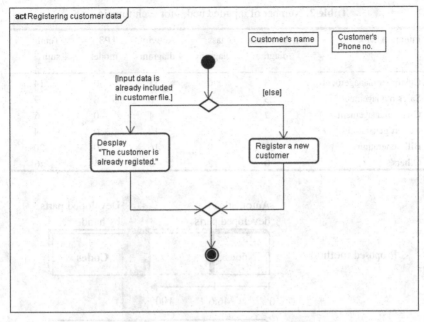

Fig. 7. Activity diagram for registering customer data

Table 1 shows that automatic generation rate is 60.2 (=148×100/246) % of all the completed codes. In addition, the codes generated automatically are the same as that indicated by the book of the model theory approach. There is no necessity for correcting the automatic generated codes in order to complete the codes. In addition, Fig. 8 shows reduced amount of workloads computed from Table 1. Since number of inputted nodes in modeling is 69, the rate of modeling workloads to 148 automatic generated lines of codes is 46.6 (=69×100/148) % because amount of workloads adding one node is supposed to be equal to describing one line of source code. These results proves that the proposed method can cut down 32.1% of all amount of workloads compared to the conventional method of hand coding of all the source codes. Besides reduction of workloads, the models which are not dependent on implementing technologies are acquired as an advantage of the model-driven development.

Table 1. Code generation results for bookstore credit sale management system

Function name	Number of Nodes	Logical LOC		Automatic generation rate (%)
		Total codes	Automatic	
Customer management	14	50	33	66.0
Sales management	9	25	18	72.0
Credit management	6	20	18	90.0
Report generation	14	60	29	48.3
Bill generation	20	79	41	51.9
Others	6	12	9	75.9
Total sum	69	246	148	60.2

Table 2. Number of inputted nodes for each function

Function name	Use case diagram	Class diagram	Activity diagram	TPS model	Total sum
Customer management	3	0	11	0	14
Sales management	3	0	6	0	9
Credit management	2	0	4	0	6
Report generation	4	0	10	0	14
Bill generation	4	0	16	0	20
Others	0	2	0	4	6

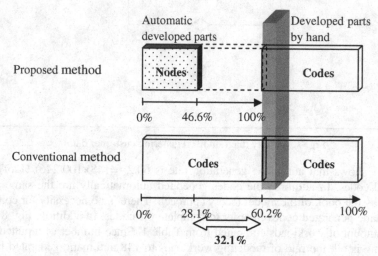

Fig. 8. Workloads ratios of bookstore credit sale management system

5   Conclusions

This paper has proposed a new development method applying the model-driven development method to a component TPS of the model theory approach. We have developed a TPS development tool for the proposed method. To show the validity of the proposed method, an experiment has been carried out using the tool. The experiment of the proposed method has shown good results, and the reduced amount of developing workloads is more than 30% of all the workloads.

As a future subject, improvement of JET templates is necessary to increase the amount of reduced workloads by the proposed method. We also need to establish a modeling method from requirement specifications to UML diagrams, and a better quantification method for workloads.

Acknowledgements. This work was partially supported by JSPS KAKENHI 21560430.

References

1. Fitzgerald, J., Larse, P.G.: Modeling Systems. Cambridge University Press (1998)
2. Cantone, D., Omodeo, E., Policriti, A.: Set Theory for Computing. Springer, Heidelberg (2001)
3. Takahara, Y., et al.: System Development Methodology: Transaction Processing System in MGST Approach. J. of the Japan Society for Management Information 14(1), 1–18 (2005)
4. Takahara, Y., Liu, Y., Chen, X., Yano, Y.: Model Theory Approach to Transaction Processing System Development. Int. J. of General Systems 3(5), 537–557 (2005)
5. Takahara, Y., Liu, Y.: Foundations and Applications of MIS: a Model Theory Approach. Springer, Heidelberg (2006)
6. Kleppe, A., Jos Warmer, J., Bast, W.: MDA Explained, the Model Driven Architecture: Practice and Promise. Addison-Wesley (2003)
7. Mellor, S.J., Clark, A.N., Futagami, T.: Model-Driven Development - Guest Editor's Introduction. J. IEEE Software 20(5), 14–18 (2003)
8. Selic, B.: The Pragmatics of Model-Driven Development. J. IEEE Software 20(5), 19–25 (2003)
9. Völter, M., Stahl, T., Bettin, J., Haase, A., Helsen, S.: Model-Driven Software Development Technology, Engineering, Management. Wiley (2006)
10. UML, http://www.uml.org
11. PHP, http://www.php.net
12. UML2.1 Superstructure Specification, http://www.omg.org/
13. XML, http://www.w3.org/XML
14. Marz, N., Aniszczyk, C.: Create More – Better – Code in Eclipse with JET. IBM Developer Works Article (2006)
15. Eclipse, http://www.eclipse.org
16. EMF, http://www.eclipse.org/modeling/emf
17. Matsumoto, K., Maruo, T., Murakami, M., Mori, N.: A Graphical Development Method for Multiagent Simulators. In: Cakaj, S. (ed.) Modeling, Simulation and Optimization, - Focus on Applications, pp. 147–157. INTECH (2010)

Automated Reasoning Support for Ontology Development

Megan Katsumi and Michael Grüninger

Department of Mechanical and Industrial Engineering,
University of Toronto Toronto, Ontario, Canada M5S 3G8

Abstract. The design and evaluation of ontologies in first-order logic poses many challenges, many of which focus on the specification of the intended models for the ontology's concepts and the relationship between these models and the models of the ontology's axioms. In this paper we present a methodology for the verification of first-order logic ontologies, and provide a lifecycle in which it may be implemented to develop a correct ontology. Automated reasoning plays a critical role in the specification of requirements, design, and verification of the ontology. The application of automated reasoning in the lifecycle is illustrated by examples from the PSL Ontology.

1 Introduction

The design of ontologies is a complicated process, and there has been much work in the literature devoted to the development of methodologies to assist the ontology engineer in this respect. In this paper we present an approach to the semiautomatic verification of first-order logic ontologies, and describe a lifecycle in which it may be implemented to develop a correct ontology. Our objective is to use automated reasoning as the basis for formally defining the steps within a development methodology that focuses on the model-theoretic properties of the ontology. Although the ontology lifecycle described in this paper was developed to address issues that arise as a result of the semidecidable nature of first-order logic, the methodology and lifecycle is applicable to automated reasoning with less expressive languages.

We will discuss some existing methodologies for ontology development, and provide motivation for the methodology presented here. This work is a result of experiences in the development of an extension of the PSL ontology [1,7] that is sufficient to represent flow modelling notations such as UML, BPMN, and IDEF3. The focus of the paper will be a discussion of the role that theorem proving plays in each phase of the lifecycle, with an emphasis on pragmatic guidance for the semiautomatic verification of ontologies.

2 Ontology Development in Literature

High-level methodologies for ontology development tend to cover the breadth of the development process, however they do not provide techniques at the more detailed level.

A. Fred et al. (Eds.): IC3K 2010, CCIS 272, pp. 208–225, 2013.

The On-To-Knowledge Management (OTKM) [11] methodology covers the full lifecycle of ontology development. It provides useful insights into the steps required both pre- and post- application, but it does not provide exact details on how activities like testing should be performed, or what is to be done if an ontology fails to satisfy a requirement. METHONTOLOGY [2] covers areas of the lifecycle similar to what is presented in the OTKM methodology, with a focus on the explanation of the concepts at each stage in development. DILIGENT [10] provides guidance to support the distributed (geographically or organizationally) development of ontologies; in particular, the authors present a method to enable the adaptation of an ontology from different parties.

Low-level ontology design methodologies provide detailed instruction, concentrating on means to accomplish some necessary step in ontology development. The methodology that was used in the design of the TOVE Ontology for enterprise modelling [5] introduced the notion of competency questions to define the requirements of an ontology and to guide the formal definition of a set of axioms for the ontology. The Enterprise Ontology [13], also arising from work with enterprise ontologies, presents an approach for ontology capture that emphasizes the notion of "middle-out" design, giving equal weight to top-down issues (such as requirements) and bottom-up issues (such as reuse).

The evaluation of an ontology is generally accepted as a key process in its development. The main efforts in this area focus on taxonomy evaluation [3]; to the best of our knowledge, there have been no efforts towards evaluation methodologies that are semantically deeper than this. This is possibly due to the issue of semidecidability of first-order logic and the inherent intractability of theorem provers that presents a challenge for the evaluation of test results. High level methodologies typically do not specify the requirements in a verifiable form. In any case, the existing methodologies do not sufficiently address the issue of ontology verification, and our goal is to address this hole, in particular, the challenges encountered in first-order logic development.

3   The Ontology Lifecycle

The lifecycle presented in Figure 1 is intended to serve as a structured framework that provides guidance during the ontology development process. This lifecycle addresses the limitations of existing verification techniques while providing a structure within which existing techniques for ontology design and requirements identification may be applied. The feedback and interactions between the various phases illustrates the tight integration between the design and the verification of the ontology. Throughout this section, we illustrate the efficacy of this lifecycle framework using examples from the design and maintenance of the PSL Ontology [1].

---

[1] The Process Specification Language (PSL) [1,7] has been designed to facilitate correct and complete exchange of process information. PSL has been published as an International Standard (ISO 18629) within the International Organisation of Standardisation and the full set of axioms (which we call T_{psl}) in the Common Logic Interchange Format is available at http://www.mel.nist.gov/psl/ontology.html. This paper only uses the PSL Ontology as the context of a case study in ontology development; the focus is not on the content of the PSL Ontology, but rather on how various axioms were proposed and revised at different stages of the ontology lifecycle.

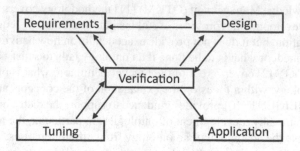

Fig. 1. The Ontology Lifecycle

We define five phases in the ontology lifecycle:

- The Requirements Phase produces a specification of the intended models of the ontology.
- The Design Phase produces an ontology to axiomatize the class of models that captures the requirements. The feedback loop that is shown in Figure 1 between the Design Phase and the Requirements Phase occurs as the ontology develops. In the Requirements Phase the intended models must initially be specified informally, until the design of the ontology has matured such that its requirements may be specified using its vocabulary.
- The Verification Phase guarantees that the intended models of the ontology which are specified in the Requirements Phase are equivalent to the models of the axioms which are produced in the Design Phase.
- The Tuning Phase addresses the pragmatic issue of dealing with cases in which the theorem provers used in the Verification Phase fail to return a definitive answer.
- The Application Phase covers the different ways in which the ontology is used, such as decision support systems, semantic integration, and search.

Each phase in the ontology lifecycle is associated with a set of reasoning problems that are defined with respect to the axioms of the ontology T_{onto}, a domain theory Σ_{domain} that uses the ontology, and a query Φ that formalizes the competency questions, and/or intended models produced in the Requirements Phase. Since these reasoning problems are entailment problems of the form:

$$T_{onto} \cup \Sigma_{domain} \models \Phi$$

we can utilize theorem provers to verify that the axiomatization of the ontology satisfies its requirements. The relationships between the stages of the ontology lifecycle and the different aspects of the reasoning problems are shown in Figure 2.

The Design Phase provides the axioms $T_{onto} \cup \Sigma_{domain}$ that form the antecedent of the reasoning problem. In the Verification Phase, we use theorem provers to determine whether or not the sentences that capture the requirements are indeed entailed by the axioms of the ontology.

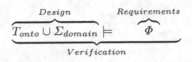

Fig. 2. Reasoning problems and stages of the ontology lifecycle

3.1 Requirements

Requirements for an ontology are specified with respect to the intended semantics of the terminology, and the challenge of the initial phase of the ontology lifecycle is to cast these requirements in a testable form. From a mathematical perspective the requirements may be characterized by their intended models; to allow for semiautomatic verification we can specify these semantic requirements as entailment problems with the use of competency questions. From a reasoning problem perspective the output of the Requirements Phase is a set of consequents for the entailment problems described above.

For more mature ontologies, we may be able to use the representation theorem, to completely characterize the requirements with the use of a relative interpretation. In this case, the output of the Requirements Phase is the set of resulting entailment problems stating the necessary and sufficient conditions for the relative interpretation; these entailment problems differ slightly from Figure 2 and will be presented later in more detail.

Intended Models. In current ontology research, the languages for formal ontologies (such as RDFS, OWL, and Common Logic) are closely related to mathematical logic, in which the semantics are based on the notion of an interpretation. If a sentence is true in the interpretation, we say that the sentence is satisfied by the interpretation. If every axiom in the ontology is satisfied by the interpretation, then the interpretation is called a model of the ontology. With a formal ontology, the content of the ontology is specified as a theory, so that a sentence is consistent with that theory if there exists a model of the theory that satisfies the sentence; a sentence can be deduced if it is satisfied by all models of the theory. Therefore, the semantics of the ontology's terminology can be characterized by this implicit set of models, which we refer to as the set of intended models.

We aim to define the requirements for a *semantically correct* ontology with the relationship between the intended models for the ontology, and the actual models of its axiomatization, (see Figure 3). An axiomatization is semantically correct if and only if it does not include any unintended models, *and* it does not omit any intended models. Formally, we define these potential *semantic errors* as follows:

Definition 1. *An error of* unintended models *is present in the ontology if and only if there exists any model $\mathcal{M}$ such that*

$$\mathcal{M} \in Mod(T_{onto}) \Rightarrow \mathcal{M} \notin \mathfrak{M}^{intended}$$

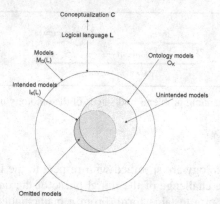

Fig. 3. The relationship between intended models for an ontology and the models of the ontology's axioms (from [9]). In this paper we refer to the models of the ontology O_k as $Mod(T_{onto})$, and we refer to the intended models $I_k(L)$ as $\mathfrak{M}^{intended}$.

Definition 2. *An error of* model omission *is present in the ontology if and only if there exists any model $\mathcal{M}$ such that*

$$\mathcal{M} \in \mathfrak{M}^{intended} \Rightarrow \mathcal{M} \notin Mod(T_{onto})$$

Intended models are specified with respect to some well-understood class of mathematical structures (such as partial orderings, graph theory, and geometry). The extensions of the relations in the model are then specified with respect to properties of these mathematical structures.

For example, the intended models for the subactivity occurrence ordering extension to the PSL Ontology are intuitively specified by two properties:

1. the partial ordering over the subactivity occurrences;
2. the mapping that embeds the partial ordering into the activity tree.

Formally, the intended models for the subactivity occurrence ordering extension are defined by

Definition 3. *Let $\mathfrak{M}^{soo}$ be the following class of structures such that for any $\mathcal{M} \in \mathfrak{M}^{soo}$,*

1. *$\mathcal{M}$ is an extension of a model of T_{psl} (i.e. the PSL Ontology);*
2. *for each activity tree τ_i, there exists a unique partial ordering $\varrho_i = (P_i, \prec)$ and a mapping*
   $\theta : \tau_i \to \varrho_i$ such that
   (a) $\langle s_1, s_2, a \rangle \in \mathbf{min\_precedes} \Rightarrow \theta(s_2) \not\prec \theta(s_1)$
   (b) $\langle \theta(s), s, a \rangle \in \mathbf{mono}$;
   (c) comparable elements in ϱ are the image of comparable elements in τ.
3. *$\langle s, a \rangle \in \mathbf{soo}$ iff $s \in P_i$;*
4. *$\langle s_1, s_2, a \rangle \in \mathbf{soo\_precedes}$ iff $s_1 \prec s_2$.*

One can take this approach to explicitly specify the intended models as a class of mathematical structures- however, early in development we often lack the knowledge to do so. Therefore, in practice initially the specification of the intended models is based on use cases for the application of the ontology. The two primary application areas for the PSL Ontology have been in semantic integration and decision support systems. In this approach, the intended models are defined implicitly with respect either to the set of sentences that are satisfied by all of the intended models or to sets of sentences that should be satisfied by some model. If we recall that sentences satisfied by all models are entailed by the axiomatization, then this leads us to the idea of implicitly defining the intended models of the ontology with respect to reasoning problems. The reasoning problems that are associated with the Requirements Phase are competency questions (which arise primarily from decision support use cases), and relative interpretations (derived from the specification of intended models).

Representation Theorems. We use representation theorems to formalize the definition of semantic correctness for our ontology. They are proven in two parts – we first prove every structure in the class is a model of the ontology and then prove that every model of the ontology is elementary equivalent to some structure in the class.

For the new extension T_{soo} to the PSL Ontology, the representation theorem is stated as follows:

Theorem 1. *Any structure $M \in \mathfrak{M}^{soo}$ is isomorphic to a model of $T_{soo} \cup T_{psl}$.*
Any model of $T_{soo} \cup T_{psl}$ is isomorphic to a structure in $\mathfrak{M}^{soo}$.

The characterization up to isomorphism of the models of an ontology through a representation theorem has several distinct advantages. First, unintended models are more easily identified, since the representation theorems characterize *all* models of the ontology. We also gain insight into any implicit assumptions within the axiomatization which may actually eliminate models that were intended. Second, any decidability and complexity results that have been established for the classes of mathematical structures in the representation theorems can be extended to the ontology itself. Finally, the characterization of models supports the specification of semantic mappings to other ontologies, since such mappings between ontologies preserve substructures of their models.

Representation theorems are distinct from the notion of the completeness of an ontology. A logical theory T is complete if and only if for any sentence Φ, either $T \models \Phi$ or $T \not\models \Phi$. The ontologies that we develop are almost never complete in this sense. Nevertheless, we can consider representation theorems to be demonstration that the ontology T_{onto} is complete with respect to its requirements (i.e. set of intended models $\mathfrak{M}^{onto}$). This allows us to say that $T_{onto} \models \Phi$ if and only if $\mathfrak{M}^{onto} \models \Phi$ (that is, the ontology entails a sentence if and only if the class of intended models entails the sentence).

The typical way to prove the Representation Theorem for an ontology is to explicitly construct the models of the ontology in the metatheory and then show that these models are equivalent to the specification of the intended models of the ontology using classes of mathematical structures. An alternative to the mathematical specification of the representation theorem is to employ a theorem prover to verify that the models of the ontology are equivalent to the intended models. Depending on the maturity of the ontology, this can be accomplished in two ways - with the use of competency questions or a relative interpretation.

Competency Questions. Following [5,8,12], competency questions are queries that impose demands on the expressiveness of the underlying ontology. Intuitively, the ontology must be able to represent these questions and characterize the answers using the terminology. Examples of competency questions for the subactivity occurrence ordering extension include the following:

Which subactivities can possibly occur next after an occurrence of the activity a_1?

$$(\forall o, s_1) \, occurrence(s_1, a_1) \wedge occurrence(o, a) \quad (1)$$
$$\wedge \, subactivity\_occurrence(s_1, o)$$
$$\supset (\exists a_2, s_2) \, occurrence(s_2, a_2) \wedge next\_subocc(s_1, s_2, a))$$

Does there exist a point in an activity tree for a after which the same subactivities occur?

$$(\exists a, a_1, s_1) \, subactivity(a_1, a) \wedge occurrence\_of(s_1, a_1)$$
$$\wedge \, ((\forall o_1, o_2) \, occurrence\_of(o_1, a) \wedge occurrence\_of(o_2, a)$$
$$\wedge \, subactivity\_occurrence(s_1, o_1)$$
$$\wedge \, subactivity\_occurrence(s_1, o_2)$$
$$\wedge \, min\_precedes(s_1, s_2, a)$$
$$\supset (\exists s_3) \, subactivity\_occurrence(s_3, o_2)$$
$$\wedge \, min\_precedes(s_1, s_3, a) \wedge mono(s_2, s_3, a) \quad (2)$$

Recall that sentences such as these constitute the consequent Φ of a reasoning problem (see Figure 2), and that they are supposed to be entailed by the axioms of the ontology T_{onto} together with a domain theory T_{domain} (which in this case is a process description that formalizes a specific UML activity diagram). It is in this sense that competency questions are requirements for the ontology – there must be sufficient axioms in $T_{onto} \cup T_{domain}$ to entail the sentences that formalize the competency questions.

In the area of decision support, the verification of an ontology allows us to make the claim that any inferences drawn by a reasoning engine using the ontology are actually entailed by the ontology's intended models. If an ontology's axiomatization has unintended models, then it is possible to find sentences that are entailed by the intended models, but which are not provable from the axioms of the ontology.

To specify requirements, competency questions can be developed from use cases (as above), or with the goal of approximating the Representation Theorem. The relationship between competency questions and the requirements is that the associated query must be provable from the axioms of the ontology alone. Since a sentence is provable if and only if it is satisfied by all models, competency questions implicitly specify the intended models of the ontology. Rather than proving the Representation Theorem directly, we can utilize competency questions for the specification of the requirements for the ontology. We can do this by proving that the extensions of the relations in the models of the ontology have properties that are equivalent to those satisfied by the extensions of the relations in the intended models. For example, the subactivity occurrence ordering ϱ introduced in the definition of the intended models $\mathfrak{M}^{soo}$ is a partial ordering, and this corresponds to the competency questions which asserts that the *soo_precedes* relation is also a partial ordering, and hence is a transitive relation:

$$(\forall s_1, s_2, s_3, a)\ soo\_precedes(s_1, s_2, a) \tag{3}$$
$$\wedge\ soo\_precedes(s_2, s_3, a) \supset soo\_precedes(s_1, s_3, a)$$

This illustrates the technique of identifying competency questions directly from the mathematical definition of the intended models.

Relative Interpretation. With more mature ontologies, we may have a better understanding of both the ontology, and its intended models. If we are able to identify and axiomatize the class of intended models, a theorem about the relationship between the class of the ontology's models and the class of intended models can be replaced by a theorem about the relationship between the ontology (a theory) and the theory axiomatizing the intended models (assuming that such axiomatization is known). We can use automated reasoners to prove the latter relationship and thus verify an ontology in a (semi-)automated way. The relationship between two theories, T_A and T_B, is the notion of interpretation, which is a mapping from the language of T_A to the language of T_B that preserves the theorems of T_A. We will say that two theories T_A and T_B are definably equivalent iff they are mutually interpretable, i.e. T_A is interpretable in T_B and T_B is interpretable in T_A. The key to representing the requirements for a relative interpretation as entailment problems is the following theorem of reducibility from [6]:

Theorem 2. *A theory T is definably equivalent with a set of theories $T_1, ..., T_n$ iff the class of models $Mod(T)$ can be represented by $Mod(T_1) \cup ... \cup Mod(T_n)$.*

The necessary direction of a representation theorem (i.e. if a structure is intended, then it is a model of the ontology's axiomatization) can be stated as

$$\mathcal{M} \in \mathfrak{M}^{intended} \Rightarrow \mathcal{M} \in Mod(T_{onto})$$

If we suppose that the theory that axiomatizes $\mathfrak{M}^{intended}$ is the union of some previously known theories $T_1, ..., T_n$, then by Theorem 2 we need to show that T_{onto} interprets $T_1 \cup ... \cup T_n$. If Δ is the set of translation definitions for this relative interpretation, then the necessary direction of the representation theorem is equivalent to the following reasoning task:

$$T_{onto} \cup \Delta \models T_1 \cup ... \cup T_n \tag{Rep-1}$$

The sufficient direction of a representation theorem (any model of the ontology's axiomatization is also an intended model) can be stated as

$$\mathcal{M} \in Mod(T_{onto}) \Rightarrow \mathcal{M} \in \mathfrak{M}^{intended}$$

In this case, we need to show that $T_1 \cup ... \cup T_n$ interprets T_{onto}. If Π is the set of translation definitions for this relative interpretation, the sufficient direction of the representation theorem is equivalent to the following reasoning task:

$$T_1 \cup ... \cup T_n \cup \Pi \models T_{onto} \tag{Rep-2}$$

By Theorem 2, $Mod(T_{onto})$ is representable by $\mathfrak{M}^{intended}$ iff $T_1 \cup \cdots \cup T_n$ is definably equivalent to T_{onto}, which we can show by proving both of the above reasoning tasks. Note that the requirement **Rep-2** varies from Figure 2 in that the axioms being designed for the ontology form the consequent of the reasoning problem. Also both **Rep-1** and **Rep-2** differ in that they include a set of translation definitions as opposed (or in addition) to a domain theory. Regardless, all requirements specified as entailment problems are verified in the same way with the use of an automated theorem prover.

3.2   Design

Recall that the Requirements Phase specified intended models either implicitly or explicitly. The task of the Design Phase is to produce a set of axioms that will entail exactly these intended models. The reasoning problems associated with the Design Phase focus on the notion of the design rationale for axioms. The task in this case is to characterize why specific axioms are required, tracing each axiom back to the original set of requirements for the ontology. While we do not prescribe a specific methodology for the design of the axioms, a key task to assist the process of axiom design in early stages of development is model exploration.

In early phases of the ontology lifecycle, we often lack a complete understanding of the models of the ontology; in fact, we may not even possess sufficient knowledge to formally express our requirements. In model exploration, we can use an automated model builder to generate models of the axioms in an effort to develop an understanding of both the intended structures and the models of the ontology. We attempt to generate models with specific sizes and/or properties. These specific properties are specified by asserting the existence of elements satisfying certain relations in the ontology.

For example, if T_{onto} is a satisfiable axiomatization of an ontology with some n-ary predicate $P(a_1, a_2, \ldots, a_n)$ with finite $n \geq 1$ in the language of T_{onto}. Then, we might generate a model for $T_\exists$, where:

$$T_\exists = T \cup \{\exists x_1, x_2, \ldots, x_n [\text{all } x_k \text{ distinct} \wedge P(x_1, x_2, \ldots, x_n)]$$

Models generated in this way serve to illustrate the semantics of the ontology's axioms.

3.3   Verification

As we saw earlier, existing approaches to ontology verification focus on taxonomic relationships and obvious logical criteria for ontology evaluation such as consistency. Strictly speaking, we only need to show that a model exists in order to demonstrate that a first-order theory is consistent. Constructing a single model, however, runs the risk of having demonstrated satisfiability for a restricted case; for example, one could construct a model of a process ontology in which no processes occur, without realizing that the axiomatization might mistakenly be inconsistent with any theory in which processes do occur. We therefore need a complete characterization of all models of the ontology up to isomorphism.

In general, verification is concerned with the relationship between the intended models of an ontology and the models of the axiomatization of the ontology. From a mathematical perspective this is formalized by the notion of representation theorems. From a reasoning problem perspective this amounts to evaluating the entailment problems, and we are able to perform this verification semiautomatically with the use of a theorem prover.

Outcomes of Verification. As we have just seen, the reasoning problems associated with Verification support the proofs of the Representation Theorem. When verifying any semantic requirement there are three possible outcomes, as illustrated in Figure 4. Each outcome is discussed in more detail below.

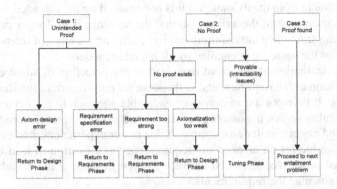

Fig. 4. Outcomes of Semantic Requirement Verification

Case 1: Unintended Proof Found. In this case, a proof was found of a sentence that contradicts the intended semantics of the ontology. This is often encountered when the theorem prover finds a proof without using all, or any clauses from the proposition or query. Given this possibility, a thorough inspection of all proofs found must be performed; if this case is detected, it is indicative of at least one of two possible errors with the ontology:

1. An examination of the proof may lead to the identification of some axiom in the ontology which is not entailed by the intended models; in this case we must return to the Design Phase.

   One such result with the design of the subactivity occurrence ordering extension T_{soo} to the PSL ontology arose when testing its consistency with the first addition of T_{soo}. A proof was found, and normally this would indicate an error in the set of definitions that was being tested. However, upon examination we realized that the definition of the *same_tree* relation made it inconsistent for any model of the ontology to have an occurrence in the same activity tree as the root of the tree:

$$T_{soo} \cup T_{psl} \models (\forall s_1, a) \, \text{root}(s_1, a) \supset \neg(\exists s_2) \, \text{same\_tree}(s_1, s_2, a)$$

This sentence should not be entailed by the intended models of the PSL Ontology itself, yet it was only identified when using the axioms of T_{soo}; it was a hidden consequence of the PSL Ontology. In particular, it was the axiom below from T_{soo} that played the critical role in deriving the unintended sentence:

$$(\forall s_1, a)\mathbf{root}(\mathbf{s_1}, \mathbf{a}) \supset (\exists \mathbf{s_2})\,\mathbf{same\_tree}(\mathbf{s_1}, \mathbf{s_2}, \mathbf{a})$$
$$\wedge\ mono(s_1, s_2, a) \wedge soo(s_2, a) \tag{4}$$

As a result of this discovery, the axiom for *same_tree* was modified.

It is interesting to see the relationship between this case and the failure of a potential representation theorem for the ontology. Part of the representation theorem shows that a sentence that is entailed by axioms is also entailed by the set of intended models. Hidden consequences such as we have just considered are counterexamples to this part of the representation, since it is a sentence that is provable from the axioms, yet it is not entailed by the intended models. One can either weaken the axioms (so that the sentence is no longer provable), or one can strengthen the requirements by restricting the class of intended models (so that the sentence is entailed by all intended models).

2. An examination of the proof may lead to the detection of some error in the definition of the requirements; in this case we must return to the Requirements Phase. It is important to devote considerable attention to the detection of this possibility so that the ontology is not revised to satisfy incorrect requirements. We did not encounter an example of this in our experiences, however it would be possible for an error in the requirements specification to lead to an unintended proof. As discussed above, this type of error could be addressed by strengthening the requirements.

Case 2: No Proof Found. As a result of the semi-decidability of first-order logic and the inherent intractability of automated theorem proving, if no proof is found when testing for a particular requirement then a heuristic decision regarding the state of the lifecycle must be made. It could be the case that no proof is found because the sentence really is not provable; there may be a mistake in the definition of the requirement that we are testing, or there may be an error in the axiomatization of the ontology, (i.e. we cannot prove that the requirement is met because it is not met). However, it could be the case that due to the intractability of automated theorem provers, a proof exists but the theorem prover is unable to find it (at least within some time limit). To avoid unnecessary work, some effort must be made to ensure that we are as informed as possible when making this decision; in particular, previously encountered errors and the nature of the requirement that we are attempting to prove is satisfied must be taken into account. In some cases, we may be able to use a model building tool to search for a counter-example, potentially resolving the uncertainty in this case. We discuss this possibility further in the following section; regardless, if no proof is found we must choose between the following options:

1. If we believe there may be some error in the requirements or the design of the ontology, then we must revisit the Requirements Phase or the Design Phase, respectively. It is recommended that the Requirements Phase is revisited first, as generally a much smaller investment is required to investigate the correctness of a requirement, rather than that of the ontology.

In the course of proving the representation theorem for T_{soo}, the theorem prover failed to entail a property regarding the *mono* relation, namely, that the *mono* relation should only hold between different occurrences of the same subactivity. This requirement was initially expressed as the following proposition:

$$(\forall s_1, s_2, o, a) \; mono(s_1, s_2, a) \land occurrence\_of(o, a)$$
$$\land \; subactivity\_occurrence(s_1, o)$$
$$\supset \neg subactivity\_occurrence(s_2, o) \tag{5}$$

It seemed clear that the axiomatizations of the *mono* relation should have restricted all satisfying models to instances where s_1 and s_2 were not in the same activity tree. Returning to the Requirements Phase, an examination of the above sentence led to the realization that the axiomatization of the proposition had been incorrect. We had neglected to specify the condition that s_1 was not the same occurrence as s_2, (in which case the *subactivity_occurrence* relation clearly holds for s_2 if it holds for s_1). Once this issue was addressed, we continued to the Verification Phase (no revisions were required to the ontology's design) and the theorem prover was able to show that the ontology entailed the corrected property, shown below.

$$(\forall s_1, s_2, o, a) \; mono(s_1, s_2, a) \land occurrence\_of(o, a)$$
$$\land \; subactivity\_occurrence(s_1, o) \land (\mathbf{s_1} \neq \mathbf{s_2})$$
$$\supset \neg subactivity\_occurrence(s_2, o) \tag{6}$$

The previous example was a case where the error that was corrected resulted from a misrepresentation of the intended semantics of the requirements. Another interesting situation was encountered where the requirements' semantics were redefined following a series of inconclusive test results. Originally, T_{soo} was to be a conservative extension of PSL[2]. In part, this meant that the axiomatization of T_{soo} had to account for all of the kinds of activity trees that were represented in PSL. One particular class of activity trees was represented by the zigzag relation, defined below.

$$(\forall s_1, s_3, a) \; zigzag(s_1, s_3, a) \tag{7}$$
$$\equiv (\exists s_2) preserve(s_1, s_2, a)$$
$$\land \; preserve(s_2, s_3, a) \land \neg preserve(s_1, s_3, a)$$

With the inclusion of the zigzag class in T_{soo} we were unable to entail the transitivity of the preserve relation. Review of the inconclusive test results led to the belief that with the inclusion of the zigzag class of activity trees there was in fact, no proof of the transitivity of the preserve relation:

$$T_{soo} \cup T_{psl} \not\vdash (\forall s_1, s_2, s_3, a) \; preserve(s_1, s_2, a)$$
$$\land \; preserve(s_2, s_3, a) \supset preserve(s_1, s_3, a) \tag{8}$$

Careful consideration of the situation led to the decision that the ability of T_{soo} to entail the transitivity of the preserve relation was more important than developing it as a non-conservative extension of PSL. This decision resulted in a

---

[2] A detailed discussion of conservative extensions and the role that they play in ontology development can be found in [4].

change in the axiomatization of T_{soo}, however this change represented a change in the requirements. We were no longer considering the zigzag class, because we had revised the requirements such that T_{soo} did not have to be a conservative extension of PSL. After this change was implemented, we were able to successfully prove that the axioms of the ontology entailed the transitivity of the preserve relation.

2. If we are strongly confident about the correctness of both the requirements and the design of the ontology, then we consider the possibility that it is the intractable nature of the theorem prover that is preventing a proof from being found. In this case, we proceed to the Tuning Phase and attempt to adapt the ontology so that the theorem prover performance is improved to a level where the requirements can be verified.

In another case, we were unsuccessful in proving a particular property about the *min_precedes* and the *preserve* relations. Based on the intended semantics of the two relations, it would appear straightforward that in any model where *min_precedes* holds for two occurrences, *preserve* must hold as well. We attempted to verify this by proving that we could entail the following proposition from the ontology with the theorem prover, however the results were inconclusive:

$$(\forall s_1, s_2, a)min\_precedes(s_1, s_2, a)$$
$$\supset preserve(s_1, s_2, a) \tag{9}$$

Being fairly certain about the correctness of the ontology and specification of the property, we moved to the Tuning Phase. In consideration of the definition of the *preserve* relation and the proposition we were attempting to verify, the reflexivity of the *preserve* relation was an intuitively important property. Two lemmas regarding the reflexivity of the *mono* relation that had already been shown to satisfy the models of the ontology were:

$$(\forall s_1, s_2, a)min\_precedes(s_1, s_2, a)$$
$$\supset mono(s_2, s_2, a) \tag{10}$$

and

$$(\forall s, o, a)subactivity\_occurrence(s, o) \wedge legal(s)$$
$$\wedge occurrence\_of(o, a) \supset mono(s, s, a) \tag{11}$$

The addition of these lemmas to the original reasoning problem aided the theorem prover sufficiently so that the property was proved and we could continue testing the other requirements.

This example illustrates a phenomenon that distinguishes theorem proving with ontologies from more traditional theorem proving – we are not certain that a particular sentence is actually provable. Effectively, every theorem proving task with an ontology is an open problem.

Case 3: All Requirements Met. If we obtain a proof that a requirement is satisfied, and it is consistent with the intended semantics of the ontology, then we may proceed with testing the remaining requirements. Once we obtain such proofs for each requirement, the ontology we have developed satisfies our requirements and we can proceed to the Application Phase.

The results of the Verification Phase may lead to revisions of either the design or the requirements of the ontology. The Design Phase is revisited in the case that an error is detected or suspected in the Verification Phase. If an error in the design is identified, it is noted that the developer must be cautious and consider the entire design when making the correction so that further errors are not created. Additionally, when a correction is made to the design, all previous test results should be reviewed, and any tests (proofs) that were related to the error should be rerun; an error in the design has the potential to positively or negatively impact the results of any tests run prior to its identification and correction. The Requirements Phase is revisited in the case that an error in the requirements is found or suspected during the Verification Phase. It may also be revisited if revisions to the requirements are necessary because of corrections to the ontology that were implemented in the Design Phase.

Note that due to the structure of the entailment problem Rep-2, verification results may require a slightly different interpretation. When attempting to diagnose an error, we should take into account that we are likely reasoning with a well-established axiomatization of some mathematical theory.

Generation of Counter-Examples. Although the theorem prover's results are inconclusive in Case 2, we have the potential to use an automated model building tool to determine whether or not a proof exists. If an ontology's axiomatization has unintended models, then it is possible to find sentences φ that are entailed by the intended structures, but which are not provable from the axioms of the ontology. Therefore we can address the ambiguity of Case 2 by evaluating the satisfiability[3], of

$$T_{onto} \cup \neg\varphi$$

where φ is the requirement we are attempting to prove. In other words, if we are unable to find a proof for a particular requirement, we search for a model that is a counter-example of the requirement. The (non-)existence of such a model will tell us the cause of Case 2 with certainty. If a counter-example is found, it can then be examined (and evaluated) to determine the course of corrective action required. In other words if the counter-example is a desired model then the requirements are too strong should be corrected accordingly, otherwise this indicates that the axiomatization is too weak and must be corrected to exclude the counter-example. In the latter case, we may attempt to generate multiple counter-examples to improve our understanding of the error. In this way, model generation may be used not only to resolve the uncertainty of Case 2, but to assist in the decision-making process required if a proof does not exist.

---

[3] We would be evaluating the satisfiability of $T_1 \cup ...T_n \cup \neg\varphi$ if the requirement in question was from Rep-2.

3.4 Tuning

The Tuning Phase focuses on the mitigation of theorem prover intractability. Similar to the Design Phase, the Tuning Phase develops a set of axioms, however the input and the function of each phase makes the two distinct. In contrast to the Design Phase, the input of the Tuning Phase is a version of the ontology that we believe to be correct, but have not been able to conclusively test. Automated theorem provers sometimes have difficulty finding a proof, though one exists; the aim of this phase is to apply techniques to streamline the ontology in an effort to mitigate theorem prover intractability. This can be accomplished with the development of subsets and the use of lemmas, discussed in further detail below.

Subsets. To develop a subset requires that we remove some of the axioms of the ontology that are not relevant (in the axiomatization of the ontology) to the particular reasoning problem we are considering. The idea is that by excluding these axioms from the reasoning problem, we can increase the efficiency of the theorem prover, as it will not be considering axioms that would not be used for the proof. By reducing the number of clauses that must be considered by the theorem prover, there is the potential to reduce the time required to find a solution, if one exists. The selection of a subset could be, but is not necessarily based on the modules of an ontology. We have currently not explored all aspects of this technique in depth, however it appears that subset selection introduces a new complication that must be considered. When selecting a subset of the ontology, we must ensure that all of the necessary axioms have been included. If any related axiom is excluded, this could allow for an inconclusive test result (we might not prove a sentence because some of the axioms required to prove it have been excluded). At this point, we do not provide any methodology to address this challenge; a clear and complete understanding of the ontology is required to be certain of what axioms must be selected for a subset for a particular reasoning problem.

For example, when working with the PSL ontology, we could exclude axioms from PSL-Core that had to do with timepoints when testing reasoning problems that were related to the composition of activities. This was possible because of our understanding of the concepts of the ontology. The size of the PSL ontology makes the use of subsets necessary for most reasoning problems, however we have observed that these subsets are often successfully reused for other related reasoning problems. While testing the ontology, we were able to use the same subset[4] to successfully entail 14 of 16 propositions related to the ordering of subactivity occurrences.

Lemmas. We use the term lemma in its traditional, mathematical sense. Their use to improve theorem prover performance is a commonly accepted technique. In the Tuning Phase, we can apply this technique to assist in obtaining conclusive test results. Lemmas may be used (in conjunction or independent of subsets) to improve performance as a means of reducing the number of steps required to obtain a proof. By adding a lemma, we hope to provide the theorem prover with a sentence that will be used in the proof; in this case the number of steps required to obtain the proof is reduced (since the theorem

---

[4] This subset can be found at http://stl.mie.utoronto.ca/colore/process/psl-subset-519.clif.

prover no longer needs to prove the lemma before using it). We also speculate that the addition of such a lemma provides a sort of guidance for the search, in the direction of the desired proof. Lemmas should be developed intelligently, with some idea of how the addition of the sentence to the ontology will assist in finding a proof. Another point to consider is that of reusability. Some effort should be made to design a lemma that is general enough to be applied for other reasoning problems. In the event that we have already developed a lemma for one reasoning problem, we should consider its potential use in the Tuning Phase for a related reasoning problem.

An example of the use of a lemma while testing the PSL ontology is discussed in Case 2 in the previous section.

4 Discussion

The description of the Verification Phase highlighted the heuristic decisions in the methodology that result from the semidecidability of first-order logic, and the intractability of theorem proving in this case. As discussed, if we do not obtain a proof when testing a requirement, then we cannot always be certain if this is because a proof does not exist (if this is the case, then we know that our current ontology does not satisfy the requirement, unless it was incorrectly specified) or if a proof does exist but the theorem prover reaches a specified time limit before it is able to find it. We suspect that it is this issue of uncertain paths resulting from Case 2 that will stimulate some criticism of our verification methodology and the lifecycle as it is proposed here. We address this concern with the following remarks:

- In two of the three possible cases that we have identified, we can be certain of the direction we must proceed in (the cases when a proof is found).
- In Case 2, when a proof is not found and there is uncertainty about the cause, we can be certain that the requirements for the ontology's application have not been met. Applications of the ontology that utilize a theorem prover must be able to answer a query (competency questions) or entail a proposition (infer a property). In other words a theorem prover must be able to find a proof of the requirements in some reasonable amount of time. Therefore, we can say that in all cases the verification methodology is capable of testing if the requirements are met; the uncertainty exists in how to proceed in development when a requirement is not met.
- The uncertainty of which path should be followed when a requirement is not met may be mitigated. If thorough documentation practices are followed through development, we may seek out trends to indicate the most likely source of the error. We also presented an application of model generation to search for counter-examples; this has the potential to completely resolve any uncertainty for this case.
- Furthermore, we have demonstrated the feasibility and effectiveness of our methodology in practice with examples of each possible case in the Verification Phase.

5 Future Work

One direction of future work that arose from the development of the subactivity occurrence ordering extension of PSL is in the area of development environments.

Specifically, an environment capable of capturing the development history of an ontology - including test results, changes made to axioms and requirements, and different versions of the ontology developed in the Tuning Phase. This would be valuable not only from a documentation perspective, but as a potential aid to heuristic decisions. A summary of test results could indicate trends (design errors, or lemmas required) that would assist the decision process in the case that no proof is found. Also, if a particular test remains inconclusive after considerable effort, an environment capable of tracking test details could allow the developer to temporarily leave the test unresolved and return to it after additional tests have been performed. With the necessary test history information available, the developer could easily and efficiently apply any corrections, subsets, or lemmas discovered in subsequent tests to the "problem" test. An environment dedicated to this type of documentation has the potential to be theorem prover independent. This would be beneficial since different theorem provers may be more useful for different applications or with different ontologies, so committing to one theorem prover restricts the potential application of the methodology presented here.

Additional areas for future work that we have identified are briefly described below:

- Perform experiments to identify the possibility of problem-general heuristics or techniques to reduce uncertainty in the heuristic decision following a Case 2 of the Verification Phase.
- We acknowledge that techniques involving model generation are not possible with all ontologies, i.e. when all models are infinite. An interesting question to investigate in this case is if it is possible to identify a "finite version" of the ontology that is able to preserve the semantics, without forcing infinite models.
- Include maintenance considerations in the lifecycle phases. Different types of maintenance activities (bug fixes, changes in the ontology's domain) should be performed differently within the lifecycle.
- Investigate the role of non-functional requirements in the ontology lifecycle. In keeping with the analogy typically drawn between software and ontology development, we identify the specification of intended models as the functional requirements of an ontology. This leaves the design and evaluation of non-functional requirements to be explored: how can these qualities be identified and measured? and how can they be integrated in the development lifecycle of ontologies?

6 Conclusions

Existing ontology development methodologies do not provide an account of ontology evaluation that is adequate for verifying ontologies with respect to their model-theoretic properties. In this paper, we have provided an approach to the ontology lifecycle that focuses on support for semiautomatic verification of ontologies, including a methodology that takes into account the pragmatic issues of semi-decidability in first-order logic. The effective use of such a methodology addresses the challenges posed by ontologies that use more expressive languages, such as first-order logic.

Our presentation of the ontology lifecycle rests on the connection between the mathematical definition of the intended models of an ontology and the reasoning problems that are equivalent to the verification of these intended models with respect to the ax-iomatization of the ontology. It is this connection which allows theorem provers to play a pivotal role in ontology design, analysis, and evaluation.

Nothing in the methodology presented here is specific to the PSL ontology, or to first-order logic. The lifecycle accounts for the difficulties of development with first-order logic; however, since the semiautomatic verification of requirements satisfaction could be beneficial in any application of ontology development, we close with an invitation for the techniques presented here to be applied with other ontologies.

References

1. Bock, C., Grüninger, M.: PSL: A semantic domain for flow models. Software and Systems Modeling 4, 209–231 (2004)
2. Fernández, M., Gómez-Pérez, A., Juristo, N.: Methontology from ontological art towards ontological engineering. In: Symposium on Ontological Engineering of AAAI (1997)
3. Gómez-Pérez, A., Corcho, O., Fernández-Lopez, M.: Ontological Engineering: with examples from the areas of Knowledge Management. In: e-Commerce and the Semantic Web, 1st edn. Advanced Information and Knowledge Processing. Springer, Heidelberg (2004), http://www.worldcat.org/isbn/1852335513
4. Cuenca Grau, B., Parsia, B., Sirin, E.: Ontology Integration using ϵ-Connections. In: Stuckenschmidt, H., Parent, C., Spaccapietra, S. (eds.) Modular Ontologies. LNCS, vol. 5445, pp. 293–320. Springer, Heidelberg (2009)
5. Grüninger, M., Fox, M.S.: Methodology for the design and evaluation of ontologies. In: International Joint Conference on Artificial Inteligence (IJCAI 1995), Workshop on Basic Ontological Issues in Knowledge Sharing (1995)
6. Grüninger, M., Hahmann, T., Hashemi, A., Ong, D.: Ontology verification with repositories. In: Conference on Formal Ontology in Information Systems (FOIS 2010), pp. 317–333. IOS Press (2010)
7. Grüninger, M.: Using the PSL ontology. In: Handbook of Ontologies, pp. 419–431. Springer, Berlin (2009)
8. Gruninger, M., Fox, M.S.: The role of competency questions in enterprise engineering. In: Proceedings of the IFIP WG5.7 Workshop on Benchmarking – Theory and Practice (1994)
9. Guarino, N., Oberle, D., Staab, S.: What is an ontology? In: Handbook of Ontologies, pp. 1–17. Springer, Berlin (2009)
10. Pinto, H.S., Tempich, C., Staab, S.: Ontology engineering and evolution in a distributed world using diligent. In: Handbook on Ontologies, International Handbooks on Information Systems, pp. 153–176. Springer, Heidelberg (2003)
11. Sure, Y., Staab, S., Studer, R., Gmbh, O.: On-to-knowledge methodology (otkm). In: Handbook on Ontologies, International Handbooks on Information Systems, pp. 117–132. Springer, Heidelberg (2003)
12. Uschold, M., Grüninger, M.: Ontologies: Principles, methods and applications. Knowledge Engineering Review 11, 93–136 (1996)
13. Uschold, M., King, M.: Towards a methodology for building ontologies. In: Workshop on Basic Ontological Issues in Knowledge Sharing, held in conjunction with IJCAI 1995 (1995)

Modeling the International Classification of Diseases (ICD-10) in OWL

Manuel Möller, Daniel Sonntag, and Patrick Ernst

German Research Center for AI (DFKI), Stuhlsatzenhausweg 3, 66123 Saarbrücken, Germany
manuelm@manuelm.org, {sonntag,patrick.ernst}@dfki.de

Abstract. Current efforts in healthcare focus on establishing interoperability and data integration of medical resources for better collaboration between medical personal and doctors, especially in the patient treatment process. In covering human diseases, one of the major international standards in clinical practice is the International Classification for Diseases (ICD), maintained by the World Health Organization (WHO). Several country- and language-specific adaptations exist which share the general structure of the WHO version but differ in certain details. This complicates the exchange of patient records and hampers data integration across language borders. We present our approach for modeling the hierarchy of the ICD-10 using the Web Ontology Language (OWL). OWL, which we will introduce shortly, should provide a formal ontological basis for ICD-10 with enough expressivity to model interoperability and data integration of several medical resources such as ICD. Our resulting model captures the hierarchical information of the ICD-10 as well as comprehensive class labels for English and German. Specialities such as "Exclusion" statements, which make statements about the disjointness of certain ICD-10 categories, are modeled in a formal way. For properties which exceed the expressivity of OWL-DL, we provide a separate OWL-Full component which allows us to use the hierarchical knowledge and class labels with existing OWL-DL reasoners and capture the additional information in a Semantic Web format.

1 Introduction

Over the last decades healthcare has changed from isolated treatments towards a distributed treatment process. This process depends greatly on the cooperation of specialized medical disciplines. Moreover, medicine questions require to take an enormous amount of expert knowledge into account before decisions are made. To facilitate information exchange and knowledge sharing in medical domains, standardization is playing an important role. The goal is to increase the interoperability within all domains of the healthcare industry so that the interchange of documents can be simplified and medical workflows can be improved.

The difficulty in the area of medical knowledge management is the high diversity of knowledge about single entities. Let us consider a patient in a clinical environment. Even if he does not have a complex disease, he would have to undergo a high number of examinations in different clinical departments. In each step of this treatment process, huge amounts of metadata are created and stored, based on single specific models every

A. Fred et al. (Eds.): IC3K 2010, CCIS 272, pp. 226–240, 2013.
© Springer-Verlag Berlin Heidelberg 2013

time. The challenge is to integrate these *islands of information* [10], so that an overall knowledge base can emerge.

A second problem of this integration process is *semantic heterogeneity*, which means, that there are disagreements about the semantics and names of concepts between the terminologies. Additionally, medical knowledge is very complex and evolves continuously over time. Therefore, new architectures and standards are needed that deal with these problems [18], [17].

Standardized terminologies have a long history in medicine. For human diseases, the first approaches date back to the 18th century. The roots of the modern International Classification of Diseases (ICD) can be traced back to the Bertillon Classification of Causes of Death. The ICD was introduced in 1893 at the International Statistical Institute in Chicago. Five years later, the American Public Health Association (APHA) recommended that Canada, Mexico, and the United States should also adopt it. Many other countries joined subsequently. It was revised several times over the last 100 years. The sixth revision included morbidity and mortality conditions and was renamed the "Manual of International Statistical Classification of Diseases, Injuries and Causes of Death (ICD)." Since 1948 the World Health Organization assumed the responsibility for maintaining and publishing revised versions of the ICD. The current version is ICD-10 from 2006.

Although the overall structure of the ICD-10 was accepted by numerous countries, different versions exist which are maintained by national institutions. For instance, the German version of the ICD-10, maintained by the DIMDI[1], is under the authority of the German Federal Ministry of Health. While major parts of the ICD-10 hierarchy are equal both in the DIMDI version and the WHO version, we found out that the structure and content of certain parts of the ICD-10 varies. Section 3.8 provides details of these differences.

The aim of the work presented here is to generate an ontology covering the domain of human diseases based on the classifications of the two country-specific ICD-10 versions described above. The ultimate goal is to leverage technologies from the Semantic Web to ease the work of medical experts (1) by supporting them in making medical image data as well as patient records available for semantic search, and (2) by providing intelligent annotation suggestions based on rich formal models for medical domain knowledge. This research was triggered by the broader effort within the research projects MEDICO and RadSpeech (http://www.dfki.de/RadSpeech/). From our discussions with clinicians we learned that a representation of the ICD-10 is an absolute necessity for the efficient semantic radiological image annotation in the everyday practice of the university hospitals participating in MEDICO [12].

2   OWL for Medical Ontologies and Related Work

OWL is based on First Order Logic and is currently the most commonly used language to represent formal ontologies in the Semantic Web. It is similar to XML Schema [7] in so far as it allows for a specification of constraints for the structure of XML documents.

---

[1] Deutsches Institut für Medizinische Dokumentation und Information.

Additional to that, OWL provides information about the interpretation of RDF statements and constructs for representing classes, subclasses as well as typed properties and sub-properties. Domains and ranges (restrictions regarding the subjects and objects of properties) can be specified and cardinality constraints can be formulated. Concrete objects are instances belonging to a certain class. In a simple example ontology, the actual liver of the author and the liver of the reader are both *instances* of the *class* liver. These instances might differ in size, weight, etc., and are composed of different physical matter, of course. But they share certain features like function, being part of the respective human body, and so on to qualify them as belonging to the same class liver. A class in an ontology can have any number of instances.

Typical questions that an OWL reasoner can answer can be roughly divided into two groups: (1) questions that are limited to the class level model and (2) questions also involving instances. Typical reasoners for OWL-DL are Pellet [16] and KAON2 [9]. A typical task for reasoners is to check for the global consistency of a (medical) ontology. Consistency means that there are no logical contradictions in the ontology. This is usually used to detect modeling mistakes. An examples for the application of such tests for finding incorrect modeling in the medical terminology SNOMED[5] (Systematized Nomenclature of Human and Veterinary Medicine) can be found in [4].

The initial idea for generating an OWL version of ICD-10 from data available on the web dates back to a similar approach for generating an OWL model for ICD-9 as presented in [11]. Biomedical ontologies and terminologies received high attention in the last decade and provide promising technologies for data integration. Bodenreider et al. evaluated popular large scale ontologies such as SNOMED, FMA, and Gene Ontology and stated that "ontologies play an important role in biomedical research through a variety of applications" [2]. In this context, a number of semi-structured medical terminologies and classification systems have been converted to formally structured formats recently. For SNOMED, an OWL ontology was created and used to detect weaknesses in the original modeling [14,15]. Noy and Rubin have presented an approach for translating the Foundational Model of Anatomy ontology (FMA) to OWL [13]. From their approach we adopted the idea to split the generated ontology into an OWL-DL and an OWL-Full component. Cardillo et al. presented an approach for a formal representation of mappings between ICD-10 and the International Classification of Primary Care version 2 (ICPC-2) [3]. However, their focus was on the formal representation of mappings between ICD-10 and ICPC-2. The work presented in this paper tries to complement their efforts by providing a formal model of additional relations within the ICD-10.

3 Modeling Approach

This section describes our general approach for the generation of the ICD-10 in OWL. Figure 1 shows the data flow during the ontology generation process. Following the elements in this diagram, the subsequent sections will discuss the different processing steps and give details about the applied techniques and algorithms.

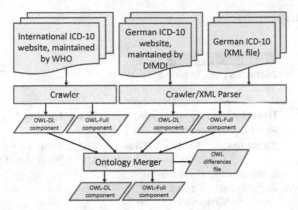

Fig. 1. Data flow of the ontology generation process

3.1 Data Sources

The OWL ontology which we generated is based on data available via the websites of the organizations responsible for maintaining the respective ICD versions. As we will show, the data which is publicly available on the Internet is well suited to generate a rich formal model of the classification of human diseases. The websites are highly structured and contain enough information to fit the use case in the MEDICO project. Another advantage is that we can reflect updates of the ICD published on the websites by re-running our crawlers.

For the English version of the ICD we used the official WHO website.[2] The website only partly reflects the original hierarchical structure of the ICD-10. As an additional source we used the ICD-10 manual [21]. Figure 2 (a) shows a screenshot covering the first of "Nutritional anaemias (D50-D53)."

From the different German sources available we chose to use the current ICD-10-GM, "GM" being the "German Modification" (see Section 3.8). Figure 2 (b) shows the same fragment of the ICD-10 as the previous screenshot, but this time in German. Our starting points for the German ICD-10 is the respective website and a publicly available XML file which is structured using the Classification Markup Language (ClaML). As the name suggests, ClaML is special language designed to represent classification hierarchies [8]. It provides special notations to state super- and subclass relations, declare attributes, and to specify metadata elements, among other things. To interpret the notation correctly, we use the WHO manual [21] and a supplementary documentation for attributes exclusively stated in the DIMDI version [6] using the the DIMDI website.[3]

3.2 OWL Model Generation

The general structure of the ICD-10 is as follows. It consists of "Chapters" using Roman numerals from I to XXI. The chapters again contain "Blocks of categories" (e. g.,

---

[2] http://apps.who.int/classifications/apps/icd/icd10online/
[3] http://www.dimdi.de/static/de/klassi/diagnosen/icd10/htmlgm2009/index.htm

Nutritional anaemias
(D50-D53)

D50	Iron deficiency anaemia
	Includes: anaemia:
	· asiderotic
	· hypochromic
D50.0	**Iron deficiency anaemia secondary to blood loss (chronic)**
	Posthaemorrhagic anaemia (chronic)
	Excludes: acute posthaemorrhagic anaemia (<u>D62</u>)
	congenital anaemia from fetal blood loss (<u>P61.3</u>)
D50.1	**Sideropenic dysphagia**
	Kelly-Paterson syndrome
	Plummer-Vinson syndrome
D50.8	**Other iron deficiency anaemias**
D50.9	**Iron deficiency anaemia, unspecified**

(a) Example for an English entry from the WHO ICD-10 website

Alimentäre Anämien
(D50-D53)

D50	Eisenmangelanämie
	Inkl.: Anämie:
	· hypochrom
	· sideropenisch
D50.0	**Eisenmangelanämie nach Blutverlust (chronisch)**
	Posthämorrhagische Anämie (chronisch)
	Exkl.: Akute Blutungsanämie (<u>D62</u>)
	Angeborene Anämie durch fetalen Blutverlust (<u>P61.3</u>)
D50.1	**Sideropenische Dysphagie**
	Kelly-Paterson-Syndrom
	Plummer-Vinson-Syndrom
D50.8	**Sonstige Eisenmangelanämien**
D50.9	**Eisenmangelanämie, nicht näher bezeichnet**

(b) Respective entry from the German DIMDI ICD-10 website

Fig. 2. Language-specific ICD-10 online versions: *Nutritional anaemias* in English and *Alimentäre Anämien* in German

Chapter III: "Diseases of the blood and blood-forming organs and certain disorders involving the immune mechanism") which specify a range of categories of a particular aspect (e. g., D50-D89). These blocks then contain "Categories," denoted by an ICD-10 code, a capital letter, and Arabic numbers (e. g., D50-D53: Nutritional anaemias; D55-D59: Haemolytic anaemias; etc.). They are further subdivided into "Subcategories." The subcategories are coded by attaching an additional digit after the decimal point (e. g., D50.1: Sideropenic dysphagia) . These codes differentiate from the specific language or writing system of the different ICD-10 versions.

Contrary to the WHO website, the DIMDI XML file constrains the subcategories by using an additional decimal number, which is appended to the codes of their parent categories. These subcategories are also defined using "Modifiers" which specify a set of "ModifierClasses" as subclasses. Each of these classes possesses a number, a label, and additional information such as "Exclusions" or "Inclusions." If a category contains a "Modifier," it will be specialized by generating new categories with each particular "ModifierClass." These relations are represented in OWL by creating a new OWL class for each "Modifier," which is a subclass of `icd10:Modifier`, and defining the appropriate ModifierClasses as subclasses. The combination is denoted using an `owl:unionOf` of the specific category and each ModifierClass. This form of specification is used extensively. Our analysis has shown that 4488 of the 16214 classes are specified in this way by DIMDI.

ICD-10 is not only a classification of diseases, but the terminology also includes links to other related aspects, such as symptoms, signs and consequences of other external causes. Therefore, the manual describes an additional level of order which groups certain chapters according to their particular aspects. For example, "Chapters I to XVII relate to diseases and other morbid conditions." It is worth mentioning that this systematic level is not available from the website but only from the manual.

Using this information about chapters and groups of chapters we modeled the first two hierarchy levels by hand. The OWL class `icd10:Entry` is the super class of the bilingual ICD-10 hierarchy. As mentioned before, differences between the German and English ICD-10 exist. Our analysis shows that there are ICD-10 categories which are present in the German ICD-10 but not in the English ICD-10 and vice versa. Section 3.8 gives details about these differences.

In addition, the origins of the concepts are also encoded in our OWL model. The first approach was to add a super class for each class to denote its origin. But this proved to be incorrect. Let us consider the block R00-R99, which is present in both terminologies and thus gets both super classes. The symptom R65, which has the super class R00-R99, is only stated by the DIMDI. Consequently, a reasoner would infer that R65 is in the DIMDI and WHO version, because it would build the transitive closure and R65 would get both super classes. For that reason, we decided to denote the provenance using the two boolean OWL-Full properties `icd10:isDIMDIEntry` and `icd10:isWHOEntry`.

Figure 3 shows an (abbreviated) example of the generated class hierarchy. We use two HTTP crawlers, implemented in Java, to generate OWL models for each of the two input sources. OWL classes and axioms are generated using the Jena Ontology API.[4] Other libraries—such as the OWL API [1]—were not able to handle the OWL-Full expressivity of our modeling.

The generated OWL model consists of two components. The OWL-DL part contains the hierarchy of the ICD-10 according to the hierarchical structure described above. All ICD-10 categories and subcategories are reflected by OWL classes. The hierarchical information is reflected by `owl:subClassOf` axioms. For a discussion of the contents of the OWL-Full part see further below.

---

[4] `http://jena.sourceforge.net/ontology/`

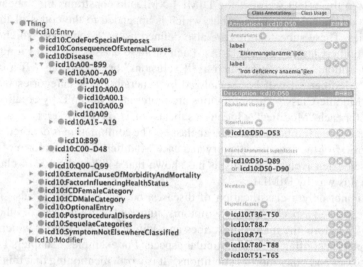

Fig. 3. (Left) General class hierarchy of the OWL model; (Right) Screenshot of the OWL-DL version of the ICD-10 in the Ontology Editor Protégé

We will explain the next steps by giving an example. Figure 2 shows the first part of the WHO ICD-10 website about "Nutritional anaemias (D50-D53)." We will focus on the entry "D50.0 Iron deficiency anaemia secondary to blood loss (chronic)" as a guiding example throughout this paper.

Each class is identified by an URL, which consists of a specific ICD-10 name space and the special term as the URL anchor. The terms for the categories are simply the particular ICD-10 codes. For blocks and chapters, a range pattern is used which covers their content, e. g., D50-D53. From this we create an OWL class with the local name "D50.0." The ".0" indicates that this is a sub-category of "D50 Iron deficiency anaemia." Thus, we add an `owl:subClassOf` axiom which represents this relationship. The bold-faced name of the sub-category becomes the English `rdfs:label` of this class. Later, by merging with the OWL model of the German ICD-10, we can also add the German labels "Eisenmangelanämie nach Blutverlust (chronisch)." For some concepts, the DIMDI specifies up to three labels, which differ in their length and detail. The smaller labels are, thereby, necessary because some print formats require them. In our case we can neglect this limitation and use always the most detailed label available. We use the standard XML language tags to differentiate between these languages.

3.3   ICD-10 Characteristics

Despite a specialization hierarchy, multiple characteristics are stated in the WHO manual [21] and the DIMDI supplement [6], and they can be shared by different classes. These are:

Dagger and Asterisk Categories.   Statements containing information about an underlying disease with a particular additional manifestation can be expressed thanks to asterisk and dagger codes. Underlying diseases are marked with a dagger and are the primary criterion. Therefore, they have to appear in the diagnostic statement, whereas the manifestation marked with an asterisk is only additional. These circumstances are represented in OWL using two properties, namely `icd10:hasAdditionalManifestation` with its inverse `icd10:hasUnderlyingDisease`. The first one's domain is all classes which represent a dagger category and have a range of all asterisk categories. The restriction that an additional manifestation needs at least one underlying disease is expressed by the property's cardinality, which is at least one.

Optional Concepts. The DIMDI defined a supplemental characteristic and this is only applied in their version of the terminology. Optional concepts are similar to dagger and asterisk categories. If marked as optional, a concept will be mandatory for some diagnoses but only supplemental for other ones.

Categories Limited to One Gender. The ICD-10 contains several categories which are only applicable to either males or females. Consider, for instance, diseases of the genitals, like "D40 Neoplasm of uncertain or unknown behavior of male genital organs" or conditions which occur during the pregnancy of women, e. g., "O00 Ectopic pregnancy." The facts are represented by one super class for each gender.

Sequelae Categories. Sequelae categories are used for mortality cause encoding. They indicate that the death is not caused by the main effect of a given disease. Instead it is caused by residual effects.

Postprocedural Disorders. Categories which fall under this characteristic point out conditions and complications which occur after treatment, e. g., surgical wound infections or shock.

Contrary to dagger and asterisk categories, each characteristic is represented using a specific super class. All classes which share the characteristic are subclasses of this class.

3.4   Handling ICD-10 Exclusions

For some ICD-10 categories so-called "Exclusions" also exist. According to the ICD-10 manual [21], they exclude certain conditions that, "although the rubric title might suggest that they were to be classified there, are in fact classified elsewhere." The example in Figure 2 (a) lists two such excludes for D50.0: "acute posthaemorrhagic anaemia" with a link to ICD-10 category D62 and "congenital anaemia from fetal blood loss" with a link to category P61.3. We capture this information by adding `owl:disjointWith` axioms between D50.0 and D62 as well as between D50.0 and P61.3. This can be expressed using the expressivity of OWL-DL (see Figure 4).

However, by relying exclusively on `owl:disjointWith` axioms, we would lose important information. As the ICD-10 manual states, exclusions can be extended using additional strings and constructions of braces. They indicate that neither the words that precede them nor the words after them are proper terms. Thus, a more precise qualification has to be applied [21]. If we compare the brace constructs with the encoding in the XML file, it becomes clear that they are used to provide a more comprehensive structuring of the data. The XML file reflects them by splitting up the "Exclusions" and

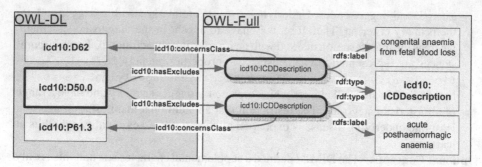

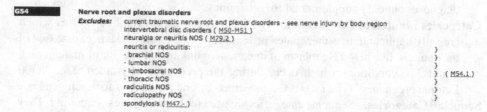

Fig. 4. Structure and relationship of OWL-DL and OWL-Full component by example

Fig. 5. Example for ICD-10 category with "Exclusions" that do not point to other ICD-10 categories

adding a new fragment for each brace element to the "Exclusions." Figure 5 depicts "Inclusions" for concept "O71.6: Obstetric damage to pelvic joints and ligaments" using braces and figure 6 shows them using the German XML encoding. As we see, without proper post-processing we are not able to relate that the "Exclusion" of concept M54.1 is shared by multiple Exclusions.

These qualifications are covered in additional OWL individuals of the upper level class icd10:Description. The individuals can have several properties of type icd10:concernsClasses. Because this property concerns other ICD-10 categories, it needs to have a class-valued range. Thus, the individuals require OWL-Full expressivity. To encode the string information of an "Exclusion," we use rdfs:label. For each excluded statement, we generate one individual. We also encode the information contained in the brace constructs, which appear in the WHO version. Therefore, we create an individual for the information which occurs after the braces. They are related to each Exclusion using an OWL-Full property icd10:qualifiedBy. Extracting this information for the DIMDI data is among our next steps.

Additionally, a closer look at the ICD-10 revealed that for numerous categories, the "Exclusions" do not point to other ICD-10 categories but to arbitrary descriptions of certain medical symptoms. Figure 5 gives an example showing ICD-10 subcategory "O71.6: Obstetric damage to pelvic joints and ligaments." As we cannot generate proper disjointness axioms for these exclude expressions we decided to store them using the exclude individuals described above without pointing to another ICD-10 category.

```
<Rubric kind="exclusion">
    <Label xml:lang="de" xml:space="default">
        <Fragment type="list">Neuritis oder Radikulitis:</Fragment>
        <Fragment type="list">brachial o.n.A.<Reference class="in brackets" code="M54.1">M54.1-</Reference></Fragment>
    </Label>
</Rubric>
<Rubric kind="exclusion">
    <Label xml:lang="de" xml:space="default">
        <Fragment type="list">Neuritis oder Radikulitis:</Fragment>
        <Fragment type="list">lumbal o.n.A.<Reference class="in brackets" code="M54.1">M54.1-</Reference></Fragment>
    </Label>
</Rubric>
```

Fig. 6. Example for ICD-10-GM braces constructs within "Exclusions"

3.5 Handling ICD-10 Inclusions and Notes

Similar to the "Exclusions" two other properties for categories are part of the ICD-10. "Inclusions" are additions to the rubric in which they occur. The ICD-10 manual describes them as a guide and provides examples to formulate diagnostic statements. "Inclusions" are represented using the individuals in OWL-Full in the same way as "Exclusions."

In addition, it is possible that a note is provided for an ICD-10 element. These notes give hints how to use the particular category, block, or chapter. For example, a physician who is writing a medical report sees from these hints that the category "G09 Sequelae of inflammatory diseases of central nervous system" is to be used to indicate conditions whose primary classification is G00-G08 (i. e., excluding those marked with an asterisk) as the cause of sequelae, themselves classifiable elsewhere. The only purpose of the notes is to support human beings in interpreting the ICD-10. Also, they are not interpretable for reasoners because they only contain continuous text. For that reason, we do not relate the notes individuals to classes with OWL-Full properties, instead we use owl:AnnotationProperties.

3.6 Merging the English and German OWL Models

To merge the two ICD-10 variants, we have to distinguish between the OWL-DL and Full parts. The two OWL-Full parts are merged by just importing them into a new ontology. This is possible because they only contain properties of the classes defined in the DL versions and the class definitions were not altered during the merging process.

The merging of the DL ontologies can be divided into two phases. First, an automatic integration is performed, which is then refined by a manual step. The automatic merging process starts with the WHO ontology. As stated in the last paragraph of section 3.2, all classes are identified using their particular ICD-10 code. In most of the cases, these codes are the same in the DIMDI and WHO version. Therefore, we begin the integration by checking if each DIMDI class is present in the WHO version. If so, we add the label and owl:subClassOf properties of the class. Adding owl:subClassOf axioms is necessary because only some of them exist in the DIMDI version. For example, if a block only appears in the DIMDI version, all owl:subClassOf relations concerning this block only occur in the DIMDI ontology. If the class is not contained, it will be created and all properties will be copied.

In addition, a few classes exist which are semantically very similar, but differ by their ICD code. For example, the DIMDI chapter D50-D90 only varies in the range and the existence of the class D90 from the WHO chapter D50-D89, but both concern the same diseases. To merge these classes, we first manually determine all possible pairs $(classA, classB)$ which differ in this sense. After that, we define a new super class $unionAB$ for each pair. This is the `owl:unionOf` of the pair's classes and gets the a concatenation of the local names of both. To determine the location of the new class, we search the first super class which $classA$ and $classB$ have in common. $unionAB$ is then added as a subclass of this class. This traversing is necessary, because there can be super classes not covering the entire range of $unionAB$. For example, we merge the blocks D80-D89 and D80-D90 and the direct super classes are D50-89 and D50-D90 respectively, which are only present either in the WHO or DIMDI version. Therefore, we have to traverse the hierarchy one step further and find the appropriate super class, which is `owl:unionOf` of the classes D50-D89 and D50-D90.

Besides the new integrated ontology, a difference ontology is generated during the merging process, which distinguishes between WHO and DIMDI classes only occurring in one version. We know that this knowledge is already contained in the ontology by the label and `icd10:isDIMDIEntry` or `icd10:isWHOEntry`. However, we decided to produce an explicit representation of the differences, because it makes the merging process more transparent and the differences are easier to examine. For these reasons, we denote the differences in both ontologies using OWL-Full properties in the difference ontology. We are using one property for every ICD source to denote the exclusiveness and one property to denote the classes later manually added by the merging process. Figure 3 shows a screenshot of the generated OWL class for ICD-10 category D50.0 in Protégé[5].

3.7   Results and Discussion

By the definition given in the ICD-10 manual, the ICD is a classification system with "a hierarchical structure with subdivisions." And further: "A statistical classification of diseases should retain the ability both to identify specific disease entities and to allow statistical presentation of data for broader groups, to enable useful and understandable information to be obtained." From this we concluded that the ICD is based on a hierarchical system of classes. The relations between these classes are proper subset relations in the sense of set theory. Thus, we decided to represent the relations of the ICD using OWL and its `subClassOf` relations. Table 1 lists some general metrics for the generated ontology. It also lists differences between the German and the English ICD-10 versions in terms of number of classes. The majority of all ICD-10 categories, i.e., about 60%, share the same ICD-10 code (for details see Section 3.2) and thus could be mapped using this as an identifier. However, there were some discrepancies between the WHO and German versions. One reason were different modeling granularities producing more categories in some branches. These differences are discussed subsequently.

We decided to split our OWL model into two components similar to the approach of Noy and Rubin in [13] for translating the Foundational Model of Anatomy ontology to OWL. The OWL-DL component allows to perform DL-reasoning using standard OWL-DL reasoners like Pellet [16]. Information from the ICD which requires modeling in

---

[5] http://protege.stanford.edu

Table 1. Metrics for the generated ICD-10 ontology

	WHO ICD-10	German ICD-10
OWL classes	11,308	16,214
disjointness axioms (see Section 3.4)	13,094	27,899
excludes pointing to another category	5,150	4,417
excludes without a proper link to other categories	35	73

OWL-Full is still available in the OWL-Full component. To use the complete model, the OWL-Full component can be loaded. This variant imports the OWL-DL model.

3.8 Differences between WHO and DIMDI Versions of ICD-10

In both terminologies we located classes that either only occur in the DIMDI or WHO version. These differences can be classified into two categories: (1) classes which only appear in one ICD-10 variant and have no particular counterpart in the other; and (2) classes which have a slightly different identification, but can be merged manually.

The first category is the most extensive. We identified 1,145 classes which are exclusively part of the WHO version and 5,707 classes exclusively part of the DIMDI version. It is important to note that we are only regarding the classes of the actual ICD-10 entries and not the classes for constructs like modifiers; these will be discussed later. Furthermore, there are blocks which differ in both versions. This means that one version specifies some parts of its terminology with more granularity than the other or that some concepts were simply left out. For example, the WHO subdivides the block "V01-X59 Accidents" into 27 sub-blocks using two hierarchy levels. In contrast, the DIMDI version does not make any further subdivisions here. Moreover, the DIMDI describes the block "U60-U61 Stadieneinteilung der HIV-Infektion" ("Staging of HIV-Infection"), which is not present in the WHO version at all.

We derived the second category during a manual examination of all differences. Hereby, we identified five blocks, which vary in their ICD-10 code. Table 2 lists them and opposes the WHO blocks with the details which the DIMDI contains. In all cases, there are differences in the range of the blocks. This is interesting, as even though a block can specify a broader range, it can be semantically more restricted. We will illustrate this by an example. The WHO version has the block "U80-U89 Bacterial agents resistant to antibiotics." The German version "U80-U85 Infektionserreger mit Resistenzen gegen bestimmte Antibiotika oder Chemotherapeutika" has almost the same range ("U80-U89" vs. "U80-85"). It could be assumed that the block with the smaller range is also more specific. But in this example the opposite is true since the German term "Infektionserreger" ("infectious agent") includes diseases caused by both bacteria and viruses. In contrast, the WHO block only covers diseases caused by bacteria. The generated OWL ontology is used in the research project MEDICO to allow for semantic annotation and retrieval across medical documents and images annotated with ICD-10 terms both in English and German.

Table 2. ICD-10 blocks, which can be merged together, although they exclusively appear in the DIMDI or WHO version

WHO ICD-10	DIMDI ICD-10
D50-D89 Diseases of the blood and blood-forming organs and certain disorders involving the immune mechanism	D50-D90 Krankheiten des Blutes und der blutbildenden Organe sowie bestimmte Störungen mit Beteiligung des Immunsystems
D80-D89 Certain disorders involving the immune mechanism	D80-D90 Bestimmte Störungen mit Beteiligung des Immunsystems
V01-Y98 External causes of morbidity and mortality	V01-Y84 Äußere Ursachen von Morbidität und Mortalität
O80-O84 Delivery	O80-O82 Entbindung
U80-U89 Bacterial agents resistant to antibiotics	U80-U85 Infektionserreger mit Resistenzen gegen bestimmte Antibiotika oder Chemotherapeutika

4   Conclusions and Future Work

We presented an OWL model of ICD-10. Our model captures hierarchical information of ICD-10 as well as comprehensive class labels for both English and German. Peculiarities such as "Exclusions" statements, which make statements about the disjointness of certain ICD-10 categories, are provided in a separate OWL-Full component. This component allows us to use hierarchical knowledge and class labels with existing OWL-DL reasoners. Our automatic generation and merging method also revealed systematic differences between the German DIMDI and the English WHO version. The goal of this approach was to reduce *Semantic Heterogeneity* in healthcare by integrating two semiformal terminologies. The current ontology represents the main and most important parts of both ICD-10 variants.

We plan to combine the results with additional conceptualizations, so that ontology networks can be created and interconnected. For example, Cardillo et al. describe an approach to map the ICD with the International Classification of Primary Care Version 2 (ICPC-2) [3]. This would foster the creation of expressive medical knowledge bases which would improve the retrieval and reuse of knowledge by reducing ambiguity. The XML file states additional information to be integrated in the future. For example, the dagger and asterisks are distinguished depending on the treatment. Different asterisk and dagger terms have to be used to formulate diagnoses for clinical treatments if diagnosis reports are created for accounting documents. During our examination of the generated OWL files and the respective ICD-10 websites, we recognized that certain verbal structures occur very often, e. g., "Injury of X," where X is an anatomical designation, like arm or leg. Moreover, the manuals describe predefined terms which are used very often, for example, the two acronyms NOS, meaning "not otherwise specified," and NEC, standing for "not elsewhere classified." Linguistic analysis can exploit this information to extract relations, e. g., to concepts represented in anatomical ontologies. This approach has already been applied successfully to other corpora within the

MEDICO project [20,19] and we plan to extend it to the ICD-10 as well. Another possibility to enhance our OWL models is to include more languages. This would generate an international representation of the ICD-10. For example, it would be possible to include the French version[6] of the ICD-10. The website is structured like the German DIMDI version. Consequently, either our HTML crawler could be used to extract the necessary information or, if a XML encoding is available, this could be parsed directly. However, one can see at first sight that the French version also differs from the other two version, e. g., it covers only 21 chapters and omits the chapter about "Codes for special purposes." Therefore, an examination of the differences would be necessary.

Acknowledgements. This research has been supported in part by the THESEUS Program in the MEDICO Project, which is funded by the German Federal Ministry of Economics and Technology under grant number 01MQ07016. The responsibility for this publication lies with the authors.

References

1. Bechhofer, S., Volz, R., Lord, P.: Cooking the Semantic Web with the OWL API. In: Fensel, D., Sycara, K., Mylopoulos, J. (eds.) ISWC 2003. LNCS, vol. 2870, pp. 659–675. Springer, Heidelberg (2003)
2. Bodenreider, O.: The Unified Medical Language System (UMLS): Integrating biomedical terminology. Nucleic Acids Research 32(Database Issue), D267–D270 (2004), http://nar.oxfordjournals.org/cgi/content/abstract/32/suppl_1/D267
3. Cardillo, E., Eccher, C., Serafini, L., Tamilin, A.: Logical Analysis of Mappings Between Medical Classification Systems. In: Dochev, D., Pistore, M., Traverso, P. (eds.) AIMSA 2008. LNCS (LNAI), vol. 5253, pp. 311–321. Springer, Heidelberg (2008), http://dblp.uni-trier.de/db/conf/aimsa/aimsa2008.html#CardilloEST08
4. Ceusters, W., Smith, B., Kumar, A., Dhaen, C.: Mistakes in medical ontologies: where do they come from and how can they be detected? Stud. Health Technol. Inform. 102, 145–163 (2004), http://view.ncbi.nlm.nih.gov/pubmed/15853269
5. Cote, R., Rothwell, D., Palotay, J., Beckett, R., Brochu, L.: The systematized nomenclature of human and veterinary medicine. Tech. rep., SNOMED International. College of American Pathologists, Northfield, IL (1993)
6. Deutsche Krankenhausgesellschaft: Deutsche kodierrichtlinien - allgemeine und spezielle kodierrichtlinien für die verschlüsselung von krankheiten und prozeduren. Tech. rep., Institut für das Entgeltsystem im Krankenhaus, InEK GmbH (2009)
7. Fallside, D.C., Walmsley, P.: XML Schema Part 0: Primer Second Edition. W3C Recommendation, October 28 (2004), http://www.w3.org/TR/xmlschema-0/
8. Hoelzer, S., Schweiger, R.K., Liu, R., Rudolf, D., Rieger, J., Dudeck, J.: Xml representation of hierarchical classification systems: from conceptual models to real applications. In: Proc. AMIA Symp., pp. 330–334 (2002)

---

[6] http://www.dimdi.de/dynamic/en/klassi/diagnosen/icd10/htmlfren/fr-icd.htm

9. Hustadt, U., Motik, B., Sattler, U.: Reducing SHIQ- Description Logic to Disjunctive Datalog Programs. In: Proc. of the 9th International Conference on Knowledge Representation and Reasoning (KR 2004), Whistler, Canada, pp. 152–162 (June 2004)

10. Lenz, R.: Information management in distributed healthcare networks. Data Management In a Connected World 3551, 315–334 (2005)

11. Möller, M., Mukherjee, S.: Context-Driven Ontological Annotations in DICOM Images – Towards Semantic PACS. In: Azevedo, L., Londral, A.R. (eds.) Proceedings of the Second International Conference on Health Informatics, HEALTHINF, pp. 294–299. INSTICC Press (2009)

12. Möller, M., Sintek, M., Buitelaar, P., Mukherjee, S., Zhou, X.S., Freund, J.: Medical image understanding through the integration of cross-modal object recognition with formal domain knowledge. In: Proc. of HEALTHINF 2008, Funchal, Madeira, Portugal, vol. 1, pp. 134–141 (2008)

13. Noy, N.F., Rubin, D.L.: Translating the Foundational Model of Anatomy into OWL. Web Semantics: Science, Services and Agents on the World Wide Web 6(2), 133–136 (2008)

14. Schulz, S., Suntisrivaraporn, B., Baader, F.: SNOMED CT's problem list: Ontologists' and logicians' therapy suggestions. In: Proc. of The Medinfo 2007 Congress. Studies in Health Technology and Informatics (SHTI-series). IOS Press (2007),
http://lat.inf.tu-dresden.de/research/papers/2007/
SchSunBaa-Medinfo-07.pdf

15. Schulz, S., Suntisrivaraporn, B., Baader, F., Boeker, M.: SNOMED reaching its adolescence: Ontologists' and logicians' health check. International Journal of Medical Informatics 78(Supplement 1), S86–S94 (2009)

16. Sirin, E., Parsia, B., Grau, B., Kalyanpur, A., Katz, Y.: Pellet: A practical owl-dl reasoner. Web Semantics: Science, Services and Agents on the World Wide Web 5(2), 51–53 (2007),
http://dx.doi.org/10.1016/j.websem.2007.03.004

17. Sonntag, D.: Ontologies and Adaptivity in Dialogue for Question Answering. AKA and IOS Press, Heidelberg (2010)

18. Sonntag, D., Wennerberg, P., Buitelaar, P., Zillner, S.: Pillars of ontology treatment in the medical domain. Journal of Cases on Information Technology (JCIT) 11(4), 47–73 (2009)

19. Wennerberg, P.: Aligning medical domain ontologies for clinical query extraction. In: Proc. of the 12th Conference of the European Chapter of the Association for Computational Linguistics: Student Research Workshop (EACL 2009), pp. 79–87. Association for Computational Linguistics, Morristown (2009)

20. Wennerberg, P., Möller, M., Zillner, S.: A linguistic approach to aligning representations of human anatomy and radiology. In: Proc. of the International Conference on Biomedical Ontologies, ICBO 2009 (July 2009),
http://precedings.nature.com/documents/3521/version/2

21. WHO: International statistical classification of diseases and related health problems. Tech. rep., World Health Organization (2004),
http://www.who.int/classifications

A Systemic Approach to Multi-Party Relationship Modeling

Anshuman B. Saxena[1,2] and Alain Wegmann[1]

[1] Systemic Modeling Laboratory, EPFL Station 14, 1015 Lausanne, Switzerland
[2] TATA Consultancy Services Innovation Labs, EPIP 96 Whitefield, 560 066 Bangalore, India

anshuman.saxena@epfl.ch, anshuman.saxena@tcs.com,
alain.wegmann@epfl.ch

Abstract. Socio-economic systems exist in a wide variety of activity domains and are composed of multiple stakeholder groups. These groups pursue objectives which are often entirely motivated from within their local context. Domain specificities in the form of institutional design, for example the de-regulation of Public Utility systems, can further fragment this context. Nevertheless, for these systems to be viable, a management subsystem that maintains a holistic view of the system is required. From a Systems perspective, this highlights the need to invest in methods that capture the interactions between the different stakeholders of the system. It is the understanding of the individual interactions that can help piece together a holistic view of the system thereby enabling system level discourse. In this paper we present a modeling technique that models industry interactions as a multi-party value realization process and takes a Systems approach in analyzing them. Every interaction is analyzed both from outside – system as a black box and from within – system as a white box. The design patterns that emerge from this whole/composite view of value realization provide the necessary foundation to analyze the working of multi-stakeholder systems. An explicit specification of these concepts is presented as Regulation Enabling Ontology, REGENT. As an example, we instantiate REGENT for the urban residential electricity market and demonstrate its effectiveness in identifying the requirements for time-based electricity supply systems.

Keywords: Value modeling, Ontology design, Industry de-regulation.

1   Introduction

Socio-economic systems are composed of multiple stakeholder groups. These groups can be viewed as bound together in one system by virtue of their membership to a shared relationship - the interactions between individual groups being different instances of this relation. Nevertheless, the diversity and plurality of stakeholder beliefs encourage individual stakeholder groups to pursue objectives entirely motivated from within their local context. In addition, domain specificities in the form of institutional design, for example the unbundling of large, vertically integrated Public Utilities into lean, efficient and more focused entities can further fragment this context [3]. Public Utilities [19], such as electricity, telecommunication, transportation, posts, gas and water supply, are most representative of such

A. Fred et al. (Eds.): IC3K 2010, CCIS 272, pp. 241–257, 2013.
© Springer-Verlag Berlin Heidelberg 2013

restructuring. From a management perspective, such unbundling results in the dissolution of the high level management structures which, in the pre-deregulated era, were responsible for the end-to-end delivery process. A deregulated industry is, instead, composed of multiple smaller management structures - each restricted in scope to some specific aspect of the industry. The absence of a holistic industry wide management structure makes deregulated industries vulnerable to systemic failure. For these unbundled entities to work as a viable whole [5], a management subsystem that maintains a holistic view of the constituent sub-systems is required. In the Public Utility domain, such a system wide management structure is usually available in the form of Industry Regulator. Nevertheless, modern regulatory systems need to go beyond the usual concerns of price, quality, output and access, and invest in schemes that capture the interactions among the stakeholders of the industry. Understanding the individual interactions help piece together a holistic view of the industry, thereby allowing the regulator to devise well informed interventions aimed at strengthening the four major quality systems constituting sustainable development – people, economic development, environment and availability of resources [27].

To address these challenges, we invoke the notion of value and model every relationship in an industry as a set of value realization processes. Value is a qualitative concept and, thus, well suited for an interdisciplinary discourse. Taking a Systems perspective, we analyze the value realization process both at the industry level and at the level of individual stakeholders within the industry. Two important design patterns emerge from this whole/composite view of value exchange: any value created in an industry has an associated supplier and adopter; a supplier of one set of value is an adopter of some other set of value. These design patterns form the basis for formalizing the concepts required to explain multi-party relationships in an industry.

This paper is an attempt to provide an explicit specification of these concepts as ontology. The ontology will provide regulators with a standard representational vocabulary with which they can document the material and information interplay between the different stakeholders of an industry. It is the abstraction of industry specific configuration details as shared pan-industry concepts that will facilitate the knowledge-level communication among the community of regulators, thereby enabling more effective and speedy sharing of regulatory best practices. Section 2 provides a brief overview of Systems thinking approach and presents a Systems perspective of the de-regulated electricity supply industry. Section 3 explores the notion of value in greater detail and introduces the concepts of resource and feature as building blocks of the value realization process. Section 4 describes the Regulation Enabling Ontology, REGENT, in detail, highlighting the different design choices that were made during the development of REGENT. Section 5 instantiates REGENT for the urban residential electricity market and, as an example, demonstrates its effectiveness in establishing regulatory oversight. Section 6 presents some related work in this field. The paper concludes with future work directions in Section 7.

2   A Systems Perspective of Industry

A Systems approach to understanding the relationship between the stakeholders of an industry allows taking a holistic view of the industry and analyzing how these relationships influence one another in the context of the overall well being of the industry. This is particularly useful for deregulated industries where management structures only exhibit knowledge about local relationships and the relevance of these relationships to the entire system remains largely unexplored. For a regulator to act as a true custodian of the industry, it is important that it has the complete knowledge about the different interactions that occur in an industry and the bearing these relationships may have on the overall working of the industry. To further illustrate the affect of deregulation on the overall management of the industry, we use the visual semantics of SEAM to analyze the evolution of Electricity Supply Industry (ESI).

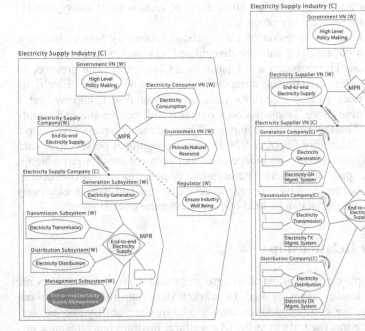

Fig. 1. Pre-deregulated ESI   **Fig. 2.** Deregulated ESI

SEAM is a set of Systemic Enterprise Architecture Methods [25] that exploit the principles of General Systems Thinking (GST) [26]. GST advocates that the component parts of a system can be best understood in the context of relationships with each other and with other systems, rather than in isolation. An important way to fully analyze a system is to understand the part in relation to the whole. SEAM represents any perceived reality as a hierarchy of systems. Each system can be analyzed as a whole [W] - showing its externally visible characteristics or as a composite [C] – showing its' constituents as a set of interrelated parts. When applying SEAM to an industry, two main aspects are analyzed: (1) How different stakeholders cooperate together to achieve some common objective; these groups of stakeholders are referred to as value network, VN. (2) How these value networks interact within an industry; these interactions are referred to as Multi-Party Relationship, MPR. The visual syntax of SEAM includes block arrows for systems, annotated ovals for externally visible properties, diamonds for relations, simple lines for active participation to a relation, dashed lines for pseudo participation to a relation and rounded end-point lines for emphasizing the identical nature of modeling elements.

Figure 1 presents a SEAM depiction of a pre-deregulated Electricity Supply industry. The four prominent entities that engage in the activities of this industry are the Electricity Supply Company (ESC), Electricity Consumer VN, Government VN and the Environment VN. When viewed as a whole, the ESC [W] exhibits the overall responsibility of maintaining an end-to-end supply of electricity – from generation to distribution. When viewed as a composite, the ESC [C] reveals its' constituent subsystems. ESCs can have different architectures. Nevertheless, for these subsystems to work as a viable whole, each ESC has some form of management subsystem [5] that oversees the end-to-end delivery process.

Figure 2 presents a SEAM depiction of a deregulated Electricity Supply Industry. The vertically integrated ESC of the pre-deregulated era stands unbundled into independent Generation, Transmission and Distribution Companies. The presence of multiple such companies constitutes competition, and provides the Electricity Consumer VN the choice to buy electricity from one Generation Company, get it transmitted through some other Transmission Company and receive the end supply service from yet another Distribution Company. These three companies when put together represent the Electricity Supplier VN. From a management perspective, each of these companies is controlled by an independent management sub-system which is strictly limited to its' part of industry operations, e.g. generation, transmission or distribution. Unlike the pre-deregulated era, there exists no end-to-end electricity supply management system that can be held responsible for the overall supply.

3   Multi-party Relationship

An industry is a complex composition of diverse stakeholder groups. Suppliers are primarily concerned about issues related to market share, profit and return-on-investment; consumers are concerned about cost, availability, reliability and ease-of-use; governments are concerned about collective welfare, institutional relevance and political indispensability; and the issues of interest from an environment point of view include habitat and climate related ecological concerns. To realize the benefits of Systems approach in analyzing the different facets of an industry, it is important to first identify a unifying concept that can act as a generic platform for the interdisciplinary discourse required in an industry. In this paper we exploit the notion of *value* as the unifying concept and treat the above mentioned stakeholder concerns as context specific manifestations of the value concept.

Based on the analysis presented in [19], we define value as the tangible or intangible effect accrued by a stakeholder through the consumption or trade of a service or good. The notion of value is at the heart of MPR modeling. Stakeholders aspiring for a common set of value are grouped together as a VN. MPR models industry interactions as a value realization process between VNs. VNs exchange resources, material and information. Any resource addition to the VN affects the stakeholders of the VN either in a favorable way, realizing positive value, or in an unfavorable way, realizing negative value. Figure 3 depicts MPR as a bi-directional value realization process between the different VNs in an Electricity Supply Industry.

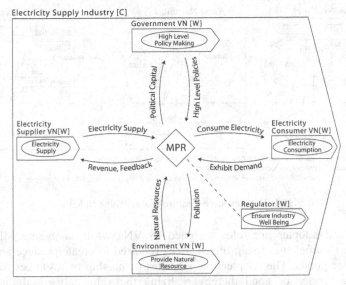

Fig. 3. Bi-directional value realization in MPR

Value is a subjective notion, dependent exclusively on stakeholder perceptions. An effect welcome by some stakeholders may be completely rejected by others. For example, time based electricity pricing schemes where a consumer can pay less for off peak electricity usage is perceived by many as a positive value as it provides an opportunity to reduce electricity bills by shifting workloads to low cost off peak durations. For others this may not be a welcome change as it results in increased night time activity in the neighbourhood. As a result it is desirable to explicitly specify the context in which a value is created, delivered or consumed. We accomplish this by introducing the concepts of resource and feature.

We follow the definition given in [4], where *resources* are defined as "... assets, capabilities, processes, and information" in control of the stakeholder. Thus resource can be considered as the contribution an individual stakeholder can bring to a VN. *Feature* on the other hand is a composite attribute which exists only at the VN level. Based on the resources available with the different stakeholders of a VN, the VN may exhibit different properties. These properties emerge from the different combinations between these resources, and are known as the features of the VN. For a given industry, an MPR identifies the different resources available with each VN, the set of possible features that may emerge from them and the value these features may bring to the other VNs. The same is presented in Figure 4. The use of the term enterprise in the figure is a more formal way of referring to stakeholders constituting a VN. The resource, feature and value concepts coupled with the GST inspired whole-composite view of value exchange guides our ontology design activity. Two important design patters emerge from this combination.

D1. For every value created in an industry there exists a supplier VN and an adopter VN

D2. Each VN in an industry acts as a supplier of one set of value and an adopter of another set of value

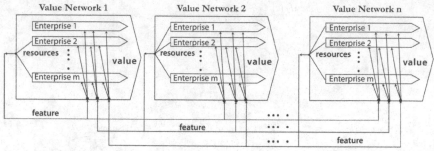

Resource : assests, capabilites, processes, information owned by stakeholders
Feature   : properties exhibited by Value Networks
Value   : tangible and intangible benefits received through the consumption of service or good

Fig. 4. The Resource-Feature-Value triune in MPR

Supplier and adopter are roles assigned to VNs while analyzing MPRs. The supplier role signifies ownership of resources required to create/produce and deliver the services or goods. The adopter role signifies ownership of resources required to consume the service or good thereby realizing the value advertised through the features of the service or good.

Design Patterns have their genesis in the field of architecture where they were first proposed as an architectural concept by Christopher Alexander [1]. These were later adopted in software engineering, and are defined as an artifact in the form of a construct, a model, a method or an instantiation, which is general enough to be reusable in solving commonly occurring problems [8]. In this paper we use these two design patterns as the basic constructs for formally specifying the knowledge required to formulate an overall understanding of any industry.

4   REGENT: A Regulation Enabling Ontology

As defined in [12], ontology is an explicit specification of a shared conceptualization. It is aimed at formalizing a specific view point that enables/enriches the discourse on some aspect of interest in the real world. The purpose of REGENT is to enable the discourse on industry regulation. Formalization of the concepts that constitute an industry and the relationships that hold among these concepts provides a common vocabulary with which regulators can represent their understanding of the industry. Such a standardized way of documenting information is particularly useful in promoting knowledge-level communication between the different industry regulators.

Various ontology languages exist to represent these concepts and relationships. The most prominent of these is OWL [24]. It is developed by the World Wide Web Consortium and consists of individuals, properties, and classes. Individuals represent the objects in the domain of interest, properties are binary relations on these individuals, and classes are interpreted as sets that contain these individuals. Our reference to concept and relationship maps to the notion of class and property in OWL. Individuals are instantiation of concept. OWL has three sub-languages: OWL-Lite, OWL-DL and OWL-Full. The expressiveness of OWL-DL falls between that of OWL-Lite and OWL-Full. It is based on Description Logics [2] which are a decidable fragment of First Order Logic and are thus conducive for automated reasoning. For this purpose we use OWL-DL as the language for specifying REGENT. The development of REGENT was done using the ontology development tool, Protégé [22]. The visualizations presented in this paper have been created using the OntoViz graphical plug-in in Protégé. In the following, we present our design choices for REGENT.

REGENT has two top level classes: `IndustryConcept` class and `ConceptSpacePartition` class. `IndustryConcept` is the foundational class for all the concepts in an industry. It is based on the Resource-Feature-Value triune detailed in sub-section 2.3. `ConceptSpacePartition` is the class which subsumes the different viewpoints that can be useful in analyzing the set of concepts detailed in the `IndustryConcept` class.

4.1   The IndustryConcept Class

The `IndustryConcept` class formalizes the concepts of resource, feature and value. Figure 5 presents the taxonomy of the `Resource` class. The `Resource` class has two subclasses: `Commercial` and `Operational`. This refinement of the `Resource` class is a manifestation of the design pattern D2. As depicted in Figure 3, every value realization is a bi-directional process. We exploit the dual nature of VN, i.e. the simultaneous role of a supplier of one value and an adopter of some other value, to classify the resources available with a VN. From an industry perspective, a product or service creation process has two parts – the operational process of bringing the service or good into existence and the commercial process of making it tradable [21]. The operational process is related to the supplier role of VN; the supplier has complete control over this process. On the other hand, the commercial process is related to the adopter role of VN. It is aimed at making the service or good conducive for consumption and, thus, requires taking an adopter perspective. Accordingly, the set of resources in an industry can be divided into two – the ones required to realize the operational process, the `RS_Operational` class, and the others required to realize the commercial process, defined as the `RS_Commercial` class.

We can further refine this classification by exploiting the insights of the supplier and adopter process. At the supplier end, bringing a service or good into existence entails two aspects – production and delivery. For instance, in the Electricity Supply Industry it is not sufficient for the electricity to be generated at the generation units, it is equally important that it is available at the prospective location of consumption. Operational resources that contribute towards the production of the industry offering are categorized as the `RS_OP_Production` class while the ones that contribute towards the delivery of the industry offering are categorized as the `RS_OP_Delivery` class. At the adopter end, realizing the benefits of the offering entails two aspects – reception and consumption. For instance, the complementary nature of electricity requires the availability of electrical appliances to consume electricity. Commercial resources that contribute towards the consumption of the industry offering are categorized as the `RS_CM_Consumption` class while the ones that contribute towards the reception of the industry offering are categorized as the `RS_CM_Reception` class. Finally, based on their cognitive orientation a resource can be further classified as tangible and intangible. The leaf nodes of the taxonomy presented in Figure 6 refine the higher level `RS_CM_*` and `RS_OP_*` classes as `RS_*_*_Tangible` and `RS_*_*_Intangible` subclasses.

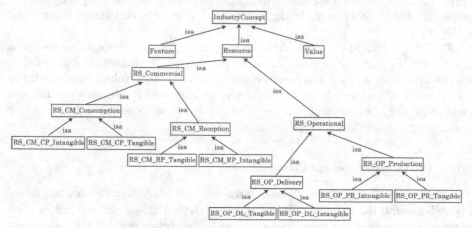

Fig. 5. Taxonomy of the Resource class

Figure 6 presents the taxonomy of the `Feature` class. The `Feature` class is a manifestation of the design pattern D1. As argued in [19], we do not treat value as an intrinsic characteristic of a product or service, and hence do not subscribe to the value chain metaphor [17] which is often interpreted to suggest that a value can be moved from the supplier to the adopter. The notion of supplier and adopter in D1 is to highlight the role of VNs in supplying resources that lead to the realization of some value at the adopter VN. Nevertheless, connecting resources directly to value will bypass an intermediate composition level where resources from different enterprises within a VN come together to define artifacts with some potential value content. This concept of composition is concretized in the `Feature` class. Features can, thus, be viewed as the potential value of a combination of one or more resources of a supplier VN. This potential value gets transformed into realized value when the adopter VN consumes the underlying artifact i.e. the industry offering. Thus feature and value differ only in the context of the observer. Feature expresses the view of the supplier of his product or service and value is the view of the adopter of the consumed product or service. This difference is captured as property constraints and is further detailed in Section 4.3.

From a taxonomy point of view, interpretation of features as potential value results in similar refinements of the `Feature` and `Value` classes. The taxonomy of the `Feature` class is presented in Figure 6. We posit that the `Value` class has a similar taxonomy tree hence do not present it separately. The following discussion on the specificities of feature refinement applies equally to the value concept.

The `Feature` class has two subclasses: `FT_Utility` and `FT_Warranty`. Utility and warranty are two concepts publicized as part of the Information Technology Infrastructure Library (ITIL) [16], developed by the UK's Office of Government Commerce (OGC) for Information Technology Services Management. Utility captures the functionality offered by a product or service and is informally interpreted as 'what the industry offering does'. On the other hand, warranty is the promise that a product or service will meet its' agreed requirements, informally interpreted as 'how the industry offering is done'. In the Requirements Engineering field, these are often termed as the function and non-functional requirements [9].

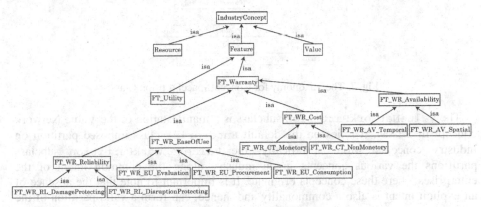

Fig. 6. Taxonomy of the Feature class

The utility of a service or good is usually well understood. It is the warranty aspect that is open to interpretation and is hence further refined. A warranty can be related to the availability, reliability, ease of use and cost of the service or good. The FT_WR_Availability class represents the attributes that capture the readiness of the service or good to be consumed by the adopter. The readiness can be both temporal, FT_WR_AV_Temporal class, and spatial, FT_WR_AV_Spatial class. The presence of electricity supply at the time and place of consumption will constitute the temporal and spatial availability of the service provided by the ECN. The objects of the FT_WR_Reliability class represent the appropriateness of the service or good for consumption. Appropriateness can be achieved by ensuring safeguards against disruptive failures, the FT_WR_RL_DisruptionProtecting class, and damaging failures, the FT_WR_RL_DamageProtecting class. For instance, the use of surge protector equipment can protect against slight variations in electricity supply but a line breaker would be required to stop the supply in the event of very high variations in supply. The FT_WR_EaseOfUse class represents the (in) convenience of evaluating – FT_WR_EU_Evaluation, procuring - FT_WR_EU_Procurement, and consuming - FT_WR_EU_Consumption, a product or service. The FT_WR_Cost class captures the attributes that define the cost of the service or good. The cost can be interpreted both in monetary, FT_WR_CT_Monetary, and in non-monetary terms, FT_WR_CT_NonMonetary.

4.2   The ConceptSpacePartition Class

The taxonomy of the ConceptSpacePartition class is presented in Figure 7. As the name suggests, this class creates a partition on the set of concepts represented in the IndustryConcept Class. A partition imposes a certain view of the industry. The Enterprise subclass partitions the various concepts in an Industry along the well established boundaries of legal ownership and undertaking. For instance every resource in an industry is owned by some enterprise. `Enterprise` subclass is the default partition of the objects represented by `IndustryConcept` class.

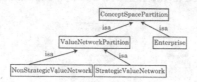

Fig. 7. The Taxonomy for ConceptSpacePartition Class

The `ValueNetworkPartition` subclass is a manifestation of the Value Network concept in SEAM. It relies on the default `Enterprise` class imposed partition on industry concepts. More specifically, the `ValueNetworkPartition` subclass partitions the various concepts in an industry along the common intent of the enterprises where these concepts originate. It is important to note that the absence of an explicit intent is also a commonality and, hence, can form a valid partition of the Industry concepts. As a result, the `ValueNetworkPartition` class is further sub-divided into `VNP_Strategic` and `VNP_NonStrategic`. The strategic subclass refers to a partition that is based on some maximizing something – profit, welfare, power, etc. By contrast, the non-strategic subclass is blind and has no objective, no preferences, and no foresight, for instance the Environment [6].

4.3   Property Constraints

The properties that bind the different concepts in REGENT are depicted in Figure 8. Properties in OWL are binary relations constraining the interaction between any two classes. For any property connecting an object o1 to object o2 an inverse property can also be specified which connects object o2 with o1. In the following, we discuss these properties on a class by class basis. For the sake of clarity, words starting with upper case alphabet are class names and the same when written in lowercase represent objects of that class.

The objects in the Resource class are constrained through two properties. 1) The *hasOwner* property mandates that each resource is connected to some enterprise. To ensure the uniqueness of this relation we limit the property to have a single value i.e. each resource has only one owner. In OWL this is accomplished by setting the property characteristics as functional. The corresponding inverse property that connects an enterprise to its resources is the *isOwnerOf* property. The one-to-many nature of this relation is visually represented with an asterisk (*). An enterprise can own more than one resource. 2) The *isProviderOf* property links a resource to the feature it contributes. The corresponding inverse property that connects a feature to its constituent resources is the *hasProvider* property. Both of these properties represent a one-to-many relation – a resource can enable more than one feature and a feature can be enabled by more than one resource.

The objects in the Feature class are constrained through four properties. 1) The *hasTransformationTo* relation specifies the values that are realization of the features.

The corresponding inverse property *isTransformationFrom* specifies the features that constitute the value. Both of these relations exhibit multiplicity – multiple features can aid a value creation and multiple values can be enabled by a feature. 2) The *hasSupplier* relation specifies the supplier value network for a feature. This is a single value relation which restricts each feature to have a unique supplier. The same is imposed by setting the functional characteristic of this property. The corresponding inverse property, *isSupplierOf*, is a multi-valued relation. A value network can be a supplier of more than one feature. 3) The *hasProvider* relation is already discussed above. 4) The *hasAdopter* relation specifies the adopter value network for a feature. The corresponding inverse property, *isAdopterOf*, specifies the set of features that a value network adopts. Both of these are multi-valued properties – a value network can adopt multiple features and a feature can be adopted by multiple VNs.

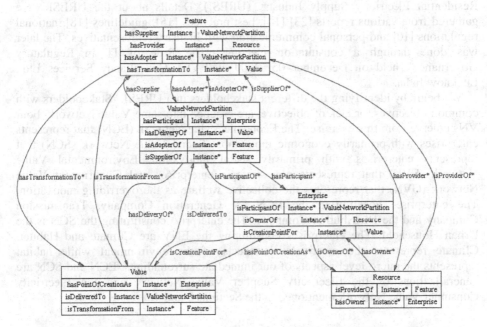

Fig. 8. A visual representation of properties constraining REGENT concepts

The objects in the Value class are constrained through three properties. 1) The *isDeliveredTo* property specifies the value network where a value is realized. This is a single value property; a value is closely associated to the perception of the consumer and, is hence, unique to the value network. We do this by setting the functional characteristic of the property. The corresponding inverse property, *hasDeliveryOf*, specifies the value that a value network consumes. 2) The *hasPointOfCreationAs* property specifies the precise enterprise which consumes this value. Again, consumption is unique to an enterprise; hence, this property is a single-valued function. The corresponding inverse property, *isCreationPointFor*, identifies all the values that are consumed by an enterprise. This is a multi-valued property. 3) The *isTransformationFrom* property has been detailed earlier.

In addition to the properties exhibited by the Feature, Resource and Value class. There exists an additional relation between the objects of the Enterprise class and the objects of the ValueNetworkPartition class. The property *isParticipantOf* identifies the value network to which the enterprise participates. To highlight the fact that an enterprise when part of two value networks does so in different roles, we model this relation as a single-value property – setting its functional characteristic. The corresponding inverse property, *hasParticipant*, is a multi-valued property and identifies all the enterprises that are members of a VN.

5   The Case of Urban Residential Electricity Supply

In this section, we use REGENT to provide a systematic view of the Urban Residential Electricity Supply Industry (URESI). Details about the URESI were gathered from various reports [23] [14], best practices [15], guidelines [18], national regulations [10] and personal communication with Industry representatives. The later was done through a consultation meeting, 'The Role of IT in Regulatory Governance', held on December 05, 2009 at TATA Consultancy Services Ltd., Lucknow India.

We begin by identifying the different stakeholders in a URESI. Stakeholders with common objectives, or lack of objective, are grouped into same Value Network. Four VNs emerge from this exercise: The Economic Value Network (ECN) that represents enterprises with primarily economic motivation, Social Value Network (SCN) that represents enterprises with primarily social motivation, Environmental Value Network (EVN) that represents non strategic enterprises and Government Value Network (GVN) that represents the collective welfare as the overriding motivation. The enterprises constituting the ECN are Generation Company, Transmission Company and the Distribution Company. The enterprise constituting the SCN is the Urban Household. The enterprises constituting the ENV are Climate and Habitat. Climate represents the macro level aspects of the environment while habitat represents the micro level aspects of our immediate surroundings. ECN and SCN are generalizations of the Electricity Supplier Value Network and the Electricity Consumer Value Network mentioned in the Sections 2 and 4.

5.1   Resource Identification

For each of these VN, we take a commercial and operational view of the value exchange and identify the tangible/intangible resources that aid the production/delivery of the VN offering and the reception/consumption of the counter offering from other VNs. These resources along with the related Enterprise and Value Network are listed in Table 1.

5.2   Feature Identification

Every VN in an industry contributes some service or good to other VNs in the industry. As described in Section 2.4.1, a VN offering can be detailed along the utility and warranty dimensions. Table 2 lists the utility and warranty details of the VN offerings in the Urban Residential Electricity Industry.

5.3   Value Identification

Every VN in an industry receives some value in return to his contribution to the Industry. Value can either be positive or negative, solicited in the case of strategic VNs or unsolicited in the case of non-strategic players. Table 3 lists the utility and warranty of the different value created in the Urban Residential Electricity Supply Industry, the VNs that adopt these value and the enterprises in the adopter VN where these value are realized.

Table 1. List of Resource identified in URESI

TAXONOMY	ID	VN	ENTERPRISE	RESOURCE LIST
Commercial / Consumption / Intangible	r1	ECN	Generation Co	MarginalCostBasedGXStrategy
	r2	ECN	Generation Co	MarginalEmissionBasedGXStrategy
	r3	EVN	Habitat	AuditoryHumanSense
	r4	EVN	Habitat	OlfactoryHumanSense
	r5	EVN	Habitat	TactileHumanSense
	r6	EVN	Habitat	VisualHumanSense
	r7	EVN	Habitat	GustatoryHumanSense
	r8	GVN	Ministry of Power	AuthorityReinforcement
	r9	GVN	Ministry of Power	InformationReinforcement
	r10	GVN	Ministry of Power	OrganizationReinforcement
	r11	SCN	Household	UsageBehavior
Tangible	r12	GVN	Ministry of Power	TreasureReinforcement
	r13	SCN	Household	AssistedLivingElectricalAppliance
	r14	SCN	Household	BulkChargingElectricalAppliance
	r15	SCN	Household	ClimateControlElectricalAppliance
	r16	SCN	Household	HeatingCoolingElectricalAppliance
	r17	SCN	Household	HomeOfficeElectricalAppliance
	r18	SCN	Household	LightingElectricalAppliance
	r19	SCN	Household	MotorDrivenMiscElectricalAppliance
	r20	SCN	Household	PersonalUseElectricalAppliance
Reception / Intangible	r21	ECN	Distribution Co	BillCollectionCapability
	r22	ECN	Distribution Co	BillGenerationCapability
	r23	ECN	Distribution Co	BillTransmissionCapability
	r24	ECN	Distribution Co	SupplyRepairCapability
	r25	ECN	Distribution Co	SupplySupportCapability
	r26	GVN	Ministry of Power	PublicOpinion
Tangible	r27	ECN	Distribution Co	Cash
	r28	ECN	Distribution Co	Credit
	r29	ECN	Distribution Co	ExternalCounter
	r30	ECN	Distribution Co	InternalCounter
	r31	ECN	Distribution Co	LightMaintainenceEquipment
	r32	ECN	Distribution Co	HeavyMaintainenceEquipment
	r33	ECN	Distribution Co	InternetAsCommChannel
	r34	ECN	Distribution Co	Phone
	r35	ECN	Distribution Co	SnailMail
	r36	EVN	Climate	Air
	r37	EVN	Climate	Land
	r38	EVN	Climate	Water
	r39	GVN	Ministry of Power	Election
	r40	GVN	Ministry of Power	Nomination
	r41	SCN	Household	Identity
Operational / Delivery / Intangible	r42	EVN	Climate	ProcurementFeasability
	r43	GVN	Ministry of Power	Campaign
	r44	GVN	Ministry of Power	MoralSuasion
	r45	GVN	Ministry of Power	Propaganda
Tangible	r46	ECN	Distribution Co	AutomaticSwitch
	r47	ECN	Distribution Co	MannualSwitch
	r48	ECN	Distribution Co	ConventionalMeter
	r49	ECN	Distribution Co	SmartMeter
	r50	ECN	Distribution Co	SinglePhaseLoad

TAXONOMY	ID	VN	ENTERPRISE	RESOURCE LIST
Delivery / Tangible	r51	ECN	Distribution Co	ThreePhaseLoad
	r52	ECN	Distribution Co	UndergroundCable
	r53	ECN	Distribution Co	OverheadCable
	r54	ECN	Distribution Co	OilTransformer
	r55	ECN	Distribution Co	FerroTransformer
	r56	ECN	Transmission Co	HighVoltagePowerLine
	r57	ECN	Transmission Co	LowVoltagePowerLine
	r58	ECN	Transmission Co	VeryHighVoltagePowerLine
Operational / Intangible	r59	GVN	Ministry of Power	Grant
	r60	GVN	Ministry of Power	Loan
	r61	GVN	Ministry of Power	Tax
	r62	GVN	Ministry of Power	Rebate
	r63	GVN	Ministry of Power	Authority
	r64	GVN	Ministry of Power	Information
	r65	GVN	Ministry of Power	Organisation
	r66	SCN	Household	SpendingStrategy
	r67	SCN	Household	WorkloadCharacteristics
Production / Tangible	r68	ECN	Generation Co	BioWastePlant
	r69	ECN	Generation Co	CentralisedPlant
	r70	ECN	Generation Co	CoalPlant
	r71	ECN	Generation Co	DistributedGenerationPlant
	r72	ECN	Generation Co	GasPlant
	r73	ECN	Generation Co	HydroPlant
	r74	ECN	Generation Co	NuclearPlant
	r75	ECN	Generation Co	PetroleumPlant
	r76	ECN	Generation Co	SolarFarm
	r77	ECN	Generation Co	TidalUnit
	r78	ECN	Generation Co	WindFarm
	r79	ECN	Generation Co	VariableOutputPlant
	r80	ECN	Generation Co	FixedOutputPlant
	r81	ECN	Generation Co	LargeCapacityPlant
	r82	ECN	Generation Co	SmallCapacityPlant
	r83	EVN	Habitat	BioWaste
	r84	EVN	Habitat	Coal
	r85	EVN	Habitat	Gas
	r86	EVN	Habitat	Hydro
	r87	EVN	Habitat	Nuclear
	r88	EVN	Habitat	Petroleum
	r89	EVN	Habitat	Solar
	r90	EVN	Habitat	Tidal
	r91	EVN	Habitat	Wind
	r92	GVN	Ministry of Power	Treasure
	r93	SCN	Household	MonthlyLoad
	r94	SCN	Household	MonthlyBudget

5.4   Establishing Regulatory Oversight

Balancing the supply and demand for electricity is central to the proper functioning of an electricity grid. The demand, however, tends to exhibit time sensitivities with more electricity required during specific times of the day or year, for example increased lighting requirements during the night and higher climate control needs during peak winter/summer season. In the absence of efficient large scale electricity storage techniques such variability in demand can only be met through flexible generation capabilities. Not all generation units support variable output. For example, nuclear power plants must be run at close to-full capacity at all times whereas production from other sources such as wind and solar, though inherently variable in nature, remains hard to predict. Further, the cost of electricity production varies from one type of generation unit to another. Generation Company operates these units in an increasing order of marginal costs (r1). Thus increased generation required to meet higher demands (peak hours) results in a higher per-unit cost of electricity. Similarly, during periods of low demand (off-peak hours) generation units with high marginal costs are cycled down resulting in a lower per-unit cost of electricity. Installation of smart meters (r49) allows the Distribution Co. to extend its billing capability (r22) and help the ECN introduce time of use (ToU) electricity pricing tariffs (f44). ToU presents economic incentives to enterprises in ECN and SCN alike. Electricity suppliers can increase profits by charging a higher per-unit cost during peak hours and consumers can minimize their bill (f48) by moving their time insensitive workloads (f35) to off-peak hours when the per-unit cost is low. The sensitivity of households to electricity bill is a function of their monthly budget (r94) and spending strategy (r66). Any attempt by households to move electricity workloads to off-peak hours is limited to the rescheduling of time insensitive workloads (f35) which in turn depends on the availability of requisite electrical appliances (r14, 19) and batch oriented workload characteristics (r67).

Table 2. List of Feature identified in URESI

TAXONOMY			ID	VN	FEATURE LIST
Warranty	Utility		f1	ECN	ElectricitySupply
			f2	EVN	ElectricityFuel
			f3	GVN	HighLevelPolicy
			f4	SCN	ElectricityDemand
	Reliability	Disruption	f5	ECN	TimeToRepair
			f6	ECN	GenerationFromRenewable
			f7	GVN	NoticePeriodForNewPolicyAdoption
			f8	SCN	BackupSupport
			f9	SCN	IncomeStability
		Dmg	f10	ECN	ApplianceProtectionInsurance
			f11	GVN	PolicyReviewOption
	Availability	Spatial	f12	ECN	ConnectionTransfer
			f13	SCN	ResidentialStability
			f14	GVN	UniformityInPolicyAcrossSupplyRegion
		Temporal	f15	ECN	FrequencyOfInterruption
			f16	ECN	DurationOfInterruption
			f17	EVN	RenewableFuelSource
			f18	EVN	NonRenewableFuelSource
			f19	GVN	FrequencyOfPolicyChange
			f20	SCN	PaymentTimeliness
	EaseOfUse	Evaluation	f21	GVN	Command&Control
			f22	GVN	Reward&Penalty
			f23	SCN	LoadVerifiability
			f24	SCN	IncomeVerifiability

TAXONOMY			ID	VN	FEATURE LIST
Warranty	EaseOfUse	Procurement	f25	ECN	DistanceFromGrid
			f26	ECN	DistanceFromSource
			f27	ECN	InitialCostOfConnection
			f28	ECN	InitialTimeToConnection
			f29	ECN	IndividualContract
			f30	ECN	CommunityContract
			f31	ECN	EaseOfBillPayment
			f32	ECN	EaseOfServiceSupport
			f33	ECN	VariableLoadSupport
		Consumption	f34	GVN	UniquenessOfInterpretation
			f35	SCN	TimeOfDayInSensitiveConsumption
			f36	SCN	TimeOfDaySensitiveConsumption
			f37	SCN	TimeOfWeekInSensitiveConsumption
			f38	SCN	TimeOfWeekSensitiveConsumption
			f39	SCN	TimeOfYearInSensitiveConsumption
			f40	SCN	TimeOfYearSensitiveConsumption
			f41	SCN	LoadVariance
Cost	Monetory		f42	ECN	FixedSupplyTariff
			f43	ECN	QuantityBasedSupplyTariff
			f44	ECN	TimeofUseSupplyTariff
			f45	ECN	FrequentTarrifVariationUnknownApriori
			f46	ECN	OccassionalTarrifVariationKnownApriori
			f47	GVN	ComplianceCost
			f48	SCN	CostSensitivityOfWorkload
	NonMonetory		f49	ECN	ToxicWasteOutput(emission)
			f50	ECN	WasteDisposal
			f51	EVN	CarbonIntenseNaturalResource
			f52	EVN	CarbonNeutralNaturalResource
			f53	GVN	ComplainceOutcomeOnTrade
			f54	SCN	QualitySensitivityOfWorkload

Table 3. List of Value identified in URESI

TAXONOMY		ID	VN	ENTERPRISE	VALUE LIST
Warranty	Utility	v1	ECN	DistributionCo.	Profit
		v2	ECN	TransmissionCo.	Profit
		v3	ECN	GenerationCo.	Profit
		v4	EVN	Habitat	HabitatPollution
		v5	EVN	Climate	ClimatePollution
		v6	GVN	MinistryOfPower	PoliticalCapital
		v7	SCN	Household	Comfortable Living
	Reliability / Damage / Disrupt	v8	ECN	DistributionCo.	Demand Foresight
		v9	GVN	MinistryOfPower	ElectricitySupplyIndependence
		v10	SCN	Household	RestrictionOnTypeOfAppliance
		v11	ECN	DistributionCo.	PlanningOpportunity
		v12	ECN	DistributionCo.	OpinionSharing
		v13	SCN	Household	SafeOperationOfAppliance
	Availability / Spatial / Temporal	v14	ECN	TransmissionCo.	SpatialDiversityOfDemand
		v15	GVN	MinistryOfPower	CapitalInvestmentInTx.
		v16	SCN	Household	EaseOfResidenceChange
		v17	ECN	GenerationCo.	ContinuedAccessToFuel
		v18	ECN	DistributionCo.	ContinuedDemandForElectricity
		v19	ECN	DistributionCo.	ContinuedRevenueInflow
		v20	GVN	MinistryOfPower	CapitalInvestmentInGeneration
		v21	SCN	Household	ContinuedOperationOfAppliance

TAXONOMY		ID	VN	ENTERPRISE	VALUE LIST
Warranty	EaseOfUse / Consum / Proc Evalu	v22	SCN	Household	RestrictedChoice
		v23	SCN	Household	ChoiceOfElectricitySource
		v24	GVN	MinistryOfPower	UniversalAvailabilityOfSupply
		v25	SCN	Household	InitialPayment
		v26	GVN	MinistryOfPower	CapitalInvestmentInDistribution
		v27	GVN	MinistryOfPower	MinimumQualityofSupply
		v28	SCN	Household	CommercialConvenience
Cost	Monetary	v29	ECN	GenerationCo.	TransactionCost - emission, distributed gx.
		v30	ECN	TransmissionCo.	TransactionCost - infrastructure expansion
		v31	ECN	DistributionCo.	TransactionCost - service delivery
		v32	GVN	MinistryOfPower	FairPrice
		v33	SCN	Household	OpportunityToReduceMonthlyBill
	NonMonetary	v34	EVN	Habitat	AuditoryDispleasure
		v35	EVN	Habitat	GustatoryDispleasure
		v36	EVN	Habitat	OlfactoryDispleasure
		v37	EVN	Habitat	TactitoryDispleasure
		v38	EVN	Habitat	VisibleDispleasure
		v39	EVN	Climate	AirPollution
		v40	EVN	Climate	LandPollution
		v41	EVN	Climate	WaterPollution
		v42	GVN	MinistryOfPower	EnvironmentalStandardComplaince
		v43	SCN	Household	InconvenienceOfReschedulingHouseholdWork

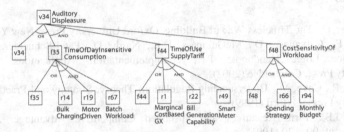

Fig. 9. An Example: Monitoring Auditory Displeasure

The temptation to move workloads to hours of low overall activity, e.g. night time, may result in increased noise levels during odd hours leading to the realization of a negative value of auditory displeasure (v34) to surrounding neighborhoods, the habitat. Use of REGENT to formally represent the value realization process exposes the industry concepts that enable it and the relationship these concepts have with the real world. Industry regulators can use this knowledge, for instance, to clearly identify the different industry elements that need to be monitored so as to track the realization of a given value of interest. An AND/OR graph depicting the value realization process for auditory displeasure (v34) is depicted in Figure 9.

6   Related Work

The role of ontology in formalizing the concepts in a knowledge system is well established. In the context of industry, ontology development has primarily focused on formalizing the domain specificities. The concepts and relationships that occur between entities from different domains have not attracted much ontological attention. E3 value [11] is one of the few attempts to study the value exchange between the stakeholders in an industry. It is, however, restricted to analyzing the economic exchange between companies active in an e-commerce business. Some ontology development has also been recently noticed in understanding regulation, for example

IPROnto [7] which presents a formalization of the concepts in digital rights management. In the Electricity industry power quality measurement related ontology has been presented in PQONT [13].

7   Conclusions

REGENT enables an explicit specification of multi-party relationships in an industry by formalizing the concepts that influence the realization of stakeholder value. A systematic representation of industry knowledge will expose any deficiencies in regulators' understanding of the industry, thereby assisting the regulator in developing a holistic view of the industry. REGENT is an important first step in our larger effort of developing a knowledge system for the regulation of utilities.

References

1. Alexander, C.: The Timeless Way of Building. Oxford University Press, New York (1979)
2. Baader, F., Calvanese, D., McGuinness, D.L., Nardi, D., Patel-Schneider, P.F.: The Description Logic Handbook: Theory, Implementation, Applications. Cambridge University Press, Cambridge (2003)
3. Baldwin, R., Cave, M.: Understanding Regulation: Theory, Strategy and Practice. Oxford University Press, Oxford (1999)
4. Barney, J.B.: Firm Resources and Sustained Competitive Advantage. Journal of Management, 99–120 (1991)
5. Beer, S.: Diagnosing the System for Organizations. John Wiley, London (1985)
6. Birchler, U., Bütler, M.: Information Economics. Routledge, London (2007)
7. Delgado, J., Gallego, I., Llorente, S., García, R.: On The Move to Meaningful Internet Systems. In: Workshop on Regulatory Ontologies and the Modelling of Complaint Regulations, Catania, Italy (2003)
8. Gamma, E., Helm, R., Johnson, R., Vlissides, J.: Design Patterns: Elements of Reusable Object-Oriented Software. Addison-Wesley, Holland (1995)
9. Gause, D., Weinberg, G.: Exploring Requirements: Quality Before Design. Dorset House Publishing, New York (1989)
10. Planning Commission (Power & Energy Division): Annual Report on The Working of State Electricity Boards & Electricity Departments, Government of India (2002)
11. Gordijn, J., Akkermans, J.: Value-based requirements engineering: exploring innovative e-commerce ideas. Requirement Engineering Journal, 114–134 (2003)
12. Gruber, T.: Toward Principles for the Design of Ontologies Used for Knowledge Sharing. International Journal Human-Computer Studies, 907–928 (2003)
13. Küçük, D., Salor, O., Inan, T., Çadırcı, I., Ermis, M.: PQONT: A domain ontology for electrical power quality. Advanced Engineering Informatics, 84–95 (2010)
14. Malaman, R.: Quality of Electricity Supply: Initial Benchmarking on Actual Levels, Standards and Regulatory Strategies. CERC, Milano (2001)
15. OECD: Regulatory Impact Analysis: Best Practice in OECD Countries, Paris (1997)
16. OGC: ITIL, Version 3. The Stationery Office of Government Commerce, London (2007)
17. Porter, M.E.: Competitive Advantage: Creating and Sustaining Superior Performance. Free Press, New York (1985)

18. Queensland Competition Authority: Electricity Distribution: Service Quality Guidelines. Queensland Competition Authority, Australia (2001)
19. Ramsay, J.: The Real Meaning of Value in Trading Relationships. International Journal of Operations & Production Management, 549–565 (2005)
20. Shy, O.: The Economics of Network Industries. Cambridge University Press, UK (2004)
21. Smith, A.: Wealth of Nations. Methuen, London (1904)
22. Ontology Editor and Knowledge Acquisition System, http://protege.stanford.edu/
23. US Aid: Measuring the Quality of Electricity Supply in India. US Aid, India (2007)
24. OWL Web Ontology Language, http://www.w3.org/TR/owl-guide/
25. Wegmann, A., Julia, P., Regev, G., Rychkova, I.: Early Requirements and Business-IT Alignment with SEAM for Business. In: 14th IEEE International Requirements Engineering Conference, Delhi, India, pp. 111–114 (2007)
26. Weinberg, G.: An Introduction to General System Thinking. Wiley & Sons, NJ (1975)
27. World Commission on Environment and Development: Our Common Future. Oxford University Press, Oxford (1987)
28. Zaccour, G.: Deregulation of electric utilities. Kluwer Academic, Boston (1998)

Evaluating Dynamic Ontologies

Jaimie Murdock, Cameron Buckner, and Colin Allen

Indiana University, Bloomington, IN, U.S.A.
{jammurdo,cbuckner,colallen}@indiana.edu
https://inpho.cogs.indiana.edu/

Abstract. Ontology evaluation poses a number of difficult challenges requiring different evaluation methodologies, particularly for a "dynamic ontology" generated by a combination of automatic and semi-automatic methods. We review evaluation methods that focus solely on syntactic (formal) correctness, on the preservation of semantic structure, or on pragmatic utility. We propose two novel methods for dynamic ontology evaluation and describe the use of these methods for evaluating the different taxonomic representations that are generated at different times or with different amounts of expert feedback. These methods are then applied to the Indiana Philosophy Ontology (InPhO), and used to guide the ontology enrichment process.

Keywords: ontology evaluation, taxonomy, ontology population, dynamic ontology.

1   Introduction

The evaluation of domain ontologies that are generated by automated and semi-automated methods presents an enduring challenge. A wide variety of evaluation methods have been proposed; but it should not be assumed that one or even a handful of evaluation methods will cover the needs of all applications. Ontology evaluation is as multifaceted as the domains that ontology designers aspire to model. Projects differ in the resources available for validation, such as a "gold standard" ontology, measures of user satisfaction, explicitly stated assumptions about the logical or semantic structure of the domain's conceptualization, or a textual corpus or dictionary whose fit to the ontology can be measured. They will also differ in the goals of the evaluation – for instance, whether they aim to use evaluation to select amongst a set of available ontologies or to tune their methods of ontology design. Further, the methods will differ in the assumptions they make about their subject domains – for no evaluation method is possible without substantive normative assumptions as to the nature of the "best" ontology.

At the Indiana Philosophy Ontology (InPhO) project, we are developing techniques to evaluate the taxonomic structures generated by machine reasoning on expert feedback about automatically extracted statistical relationships from our starting corpus, the Stanford Encyclopedia of Philosophy (SEP). InPhO does not assume that a single, correct view of the discipline is possible, but rather takes the pragmatic approach that some representation is better than no representation at all [1]. Evaluation allows us to quantify our model and makes explicit the specific biases and assumptions underlying each candidate taxonomy.

A. Fred et al. (Eds.): IC3K 2010, CCIS 272, pp. 258–275, 2013.

In this paper, we describe a pair of evaluation metrics we have found useful for evaluating ontologies and our methods of ontology design. The volatility score (section 4.1) measures the structural stability over the course of ontology extension and evolution. The violation score (section 4.2) measures the semantic fit between an ontology's taxonomic structure and the distribution of terms in an underlying text corpus.

Before diving into these methodologies, we will first situate them within the broader evaluation literature (section 2). Then we will describe the InPhO in further detail, along with the raw materials we will be evaluating (section 3). After this, we explore each of the two new measures, labeling their assumptions and demonstrating their capacity to guide the process of ontology design.

2   State of the Art

Approaches to ontology evaluation are heavily dependent on the positions taken towards ontology structure and design. Different assumptions underlying these positions are often left implicit and this has led to a tangled web of conflicting opinions in the literature. However, Gangemi et al. [2] provide an excellent conceptual scaffolding for use in detangling the web by establishing three categories of evaluation techniques:

- **Structural evaluation** inspects the logical rigor and consistency of an ontology's encoding scheme, typically as a directed graph (digraph) of taxonomic and non-taxonomic relations. Structural evaluations are a measure of syntactic correctness. A few examples of structural evaluation include the OntoClean system [3] and Gómez-Pérez's paradigm of correctness, consistency and completeness [4], which was extended by Fahad & Qadir [5]. Our proposed *volatility score* (Section 4.1) is a structural evaluation of consistency in an ontology.
- **Functional evaluation** measures the suitability of the ontology as a representation of the target domain. Many functional evaluations follow a "gold standard" approach, in which the candidate ontology is compared to another work deemed a good representation of the target domain (e.g. Dellschaft & Staab and Maedche & Staab [6,7]). Another approach is to compare the candidate ontology to a corpus from which terms and relations are extracted [8]. Our proposed violation score (Section 4.2) is a corpus-based functional evaluation of semantic ontological fit.
- **Usability evaluation** examines the pragmatics of an ontology's metadata and annotation by focusing on recognition, efficiency (computational and/or economic), and interfacing. The recognition level emerges from complete documentation and effective access schemes. The efficency level deals with proper division of ontological resources, and proper annotation for each. The interfacing level is limited by Gangemi et al. [2] to the examination of inline annotations for interface design, but these are not essential properties.

  One chief measure of usability is compliance to standards such as OWL and RDFa. Several frameworks for social usability evaluation have been proposed by Supekar and Gómez-Pérez [9,10]. ONTOMETRIC is an attempt to codify the various factors in usability evaluation by detailing 160 characteristics of an ontology and then weighting these factors using semi-automatic decision-making procedures [11].

These three paradigms of evaluation are realized in different evaluation contexts, as identified by Brank, Mladenic and Grobelnik [12]:

- **Applied.** For functional and usability evaluation, using the ontology to power an experimental task can provide valuable feedback about suitability and interoperability. Applied approaches require access to experts trained in the target domain and/or ontology design. Velardi, Navigli, Cucchiarelli, and Neri's OntoLearn system [13] utilizes this type of applied evaluation metric. Porzel and Malaka [14] also use this approach within speech recognition classification.
- **Social.** Methods for usability evaluation proposed by Lozano-Tello and Gómez-Pérez, Supekar and Noy [11,9,10] for networks of peer-reviewed ontologies, in a similar manner to online shopping reviews. Most social evaluation revolves around the ontology selection task. These evaluations involve a purely qualitative assessment and may be prone to wide variation.
- **Gold Standard.** As mentioned above, the gold standard approach compares the candidate ontologies to a fixed representation judged to be a good representation [7,6]. These approaches draw strength from the trainability of the automatic methods against a static target, but the possibility of over-training of automated and semi-automated methods for ontology population means that the methods may not generalize well.
- **Corpus-based.** Approaches such as those used by Brewster, Alani, Dasmahapatra and Wilks [8] calculate the "ontological fit" by identifying the proportion of terms that overlap between the ontology and the corpus. This is a particularly well-suited measure for evaluating ontology learning algorithms. Our methods expand this measurement approach to cover term relations through both the violation and volatility measures.

This collection of evaluation paradigms and contextual backdrops allows us finally to consider the type of information content being evaluated. A "computational ontology", such as the InPhO, is a formally-encoded specification of the concepts and a collection of directed taxonomic and non-taxonomic relations between them [1,15,16]. When evaluating information content, we must be careful to delineate those which are node-centric (focusing on concepts) from those which are edge-centric (focusing on relations). Many authors [7,3,8,4,13] focus upon node-centric techniques, asking "Are the terms specified representative of the domain?" These investigate the lexical content of an ontology. However, the semantic content of an ontology is not defined solely by the collection of terms within it, but rather by the relations of these terms. Maedche & Staab [7] take this initial lexical evaluation and extend it to an edge-based approach which measures the number of shared edges in two taxonomies. The proposed violation and volatility scores (Section 4) are novel edge-based measures which address the semantic content of an ontology by comparing them to statistics derived from a relevant corpus as a proxy for domain knowledge. Additionally, these scores can provide insight to the ontology design process by highlighting particular changes in domain content and measuring convergence towards a relatively stable representation.

3   Our Dynamic Ontology

A wide variety of projects can benefit from the development of a computational ontology of some subject domain. Ontology science has evolved in large part to suit the needs of large projects in medicine, business, and the natural sciences. These domains share a cluster of features: the underlying structures of these domains have a relatively stable consensus, projects are amply funded, and a primary goal is often to render interoperable large bodies of data. In these projects, the best practices often require hiring so-called "double experts" – knowledge modelers highly trained in both ontology design and the subject domains – to produce a representation in the early stages of a project which is optimally comprehensive and technically precise.

There is another cluster of applications, however, for which these practices are not ideal. These involve projects with principles of open-access and domains without the ample funding of the natural sciences. Additionally, ontologies for domains in which our structural understanding is controversial or constantly evolving and projects which utilize computational ontologies to enhance search or navigation through asynchronously updated digital resources must account for the dynamic nature of their resources – whether it is in the underlying corpus or in the judgments of the experts providing feedback on domain structure. On the positive side, these areas often have more opportunities to collect feedback from users who are domain experts (but who lack expertise in ontology design).

For the latter type of project we have recommended an approach to design which we call *dynamic ontology*. While a project in the former group properly focuses the bulk of its design effort on the production of a single, optimally correct domain representation, the latter cluster is better served by treating the domain representation as tentative and disposable, and directing its design efforts towards automating as much of the design process as possible. Dynamic ontology, broadly speaking, tries to take advantage of many data sources to iteratively derive the most useful domain representation obtainable at the current time. Two primary sources of data are domain experts and text corpora. Domain experts provide abstract information about presently-held assumptions and emergent trends within a field from a source, namely their own ideas, that is hard to examine directly. Text corpora make it possible to quantify what is meant by "domain" by providing a concrete encoding of the semantic space that is available for empirical analysis, in contrast to the ill-defined abstraction of "the domain is what the experts conceive of it as". From both kinds of sources many types of data may be gathered: statistical relationships among terms, feedback from domain experts, user search and navigation traces, existing metadata relationships (e.g. cross-references or citations), and so on. As more data become available and our understanding of the subject domain continues to evolve, the domain representation will be be dynamically extended, edited, and improved.

In dynamic ontology, problems of validation loom especially large due to the combination of heterogenous data sources. Each step in the design process presents modelers with a panoply of choices for inconsistency mitigation – e.g., which sources of data to favor over others (e.g. statistical vs. user feedbck), how to leverage user expertise to resolve disagreements, which reasoning methods to use for population, and how much feedback to solicit. The automation of ontology design is a field in its infancy, and very

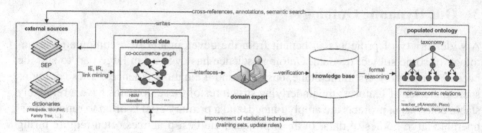

Fig. 1. The InPhO Workflow

little is known about the optimal choices to satisfy specific design goals. Additionally, dynamic ontologists might have questions regarding representational stability. If the domain is itself in flux or controversial, modelers might want to know if they have captured that change. The quantity of feedback may also influence the convergence of a population method to some stable representation. The development of precise metrics about the relationship between an ontology and a domain may be useful in a answering these questions.

The InPhO is a dynamic ontology which models the discipline of philosophy. Our approach leverages expert knowledge by augmenting it with machine reasoning, greatly reducing the need for expensive "double experts". The primary source of text data and domain experts is the Stanford Encyclopedia of Philosophy (SEP)[1]. With over 700,000 weekly article downloads, the SEP is the leading digital humanities resource for philosophy. The corpus consists of over 1,200 articles and 14.25 million words maintained by over 1,600 volunteer authors and subject editors. The tremendous depth of the encyclopedia makes it impossible for any one person to have expertise over the whole domain, necessitating the creation of a useful organization scheme to provide better editorial control and content accessibility. At the same time, the comprehensive richness of the SEP makes it a reasonable proxy for the discipline of philosophy as a whole.

We begin with a small amount of manual ontology construction obtained through collaboration with domain experts. A lexicon is established from SEP article titles, Wikipedia philosophy categories, n-gram analysis and ad hoc additions by the InPhO curators. We then build on this framework using an iterative three-step process of data mining, feedback collection, and machine reasoning to populate and enrich our representation of philosophy (see Figure 1).

First, the SEP is mined to create a co-occurrence graph consisting of several statistical measures. For each term in our lexicon, information entropy is measured, which provides an estimate of relative generality. For each graph edge, we calculate the J-measure, which provides an estimate of semantic similarity. From these measures we are able to generate hypotheses about hypernym/hyponym candidates for sets of terms in the corpus [17]. Second, SEP authors and other volunteers verify these hypotheses by answering questions about relational hypotheses. This reduces the effect of any statistical anomalies which emerge from the corpus. Finally, logic programming techniques are used to assemble these aggregated feedback facts into a final populated ontology [18]. This knowledge base can then be used to generate tools to assist the authors,

---

[1] http://plato.stanford.edu

editors, and browsers of the SEP, through tools such as cross-reference generation engine and context-aware semantic search.

As was mentioned in the introduction, our pragmatic approach recognizes the likelihood that there is no single, correct view of the discipline. However, even if other projects do not agree with our taxonomic projections, our statistical data and expert evaluations may still be useful. By exposing our data from each of the three steps through an easy-to-use API, we encourage other projects to discover alternative ways to construct meaningful and useful representations of the discipline. Additionally, by offering an open platform, we invite other projects to contribute relevant data and expert feedback to improve the quality of the service.

3.1 Raw Materials

In this section we describe the various components of our project which can be exploited for ontology evaluation.

Structure. The core of the InPhO is the *taxonomic representation* marked by the *isa* relations between concepts. *Concepts* in the InPhO may be represented as part of either class or instance relations. *Classes* are specified through the direct *isa* hierarchy of the taxonomy (see below). *Instances* are established between a concept and another concept which is part of the taxonomic structure. *Semantic crosslinks* (hereafter, *links*) can be asserted between two classes to capture the relatedness of ideas deemed mutually relevant by feedback or automatic methods.

Statistics. The InPhO's ontology population and extension techniques rely upon an external corpus (the SEP) to generate hypotheses about similarity and generality relationships. From this corpus we generate a co-occurrence graph $G = (V, E)$ in which each node represents a term in our set of keywords. An edge between two nodes indicates that the terms co-occur at least once.

For each node, the information content (Shannon entropy) is calculated:

$$H(i) = p(i) \log p(i) \tag{1}$$

For each edge, the directed J-measure [19,17] and conditional entropy [20] is calculated bidirectionally. The conditional entropy calculates the information content of a directed edge $i \to j$. This is used as a measure of semantic distance between two terms:

$$H(j \mid i) = p(i, j) \log \frac{p(i)}{p(i, j)} \tag{2}$$

The J-measure calculates the interestingness of inducing the rule "Whenever idea i is mentioned in a fragment of text, then idea j is mentioned as well" [17]. This is used as a measure of semantic similarity between two terms:

$$f(i \to j) = p(j \mid i) \log \frac{p(j \mid i)}{p(j)} \\ + (1 - p(j \mid i)) \log \frac{1 - p(j \mid i)}{1 - p(j)} \tag{3}$$

$$J(i \rightarrow j) = p(i)f(i \rightarrow j) \tag{4}$$

Methods. The taxonomy itself is populated through the use of answer set programming [18]. A population method $M(R, S, F)$ is specified by a set of rules R, a seed taxonomy S, and a set of expert feedback or statistical hypotheses F. Changes in F allow us to measure the impact of groups of expert feedback and to evaluate an ontology extension method. Proposed ruleset changes can be evaluated by maintaining the same set of inputs while testing variations in R. The seed taxonomy is used to reduce the computational complexity of a methodology, and changes to this seed can be used to strengthen the ontology design process. We currently have two years of data collected on nightly repopulation of the published InPhO taxonomy, which is used for evaluation of our ontology extension methods.

3.2 Our Challenges

As hinted above, our dynamic approach to ontology design presents several unique challenges which require that appropriate validation methods be developed to address them. Specifically, there are a variety of different ways that our answer set program could infer a final populated ontology from aggregate expert feedback. For example, there are different ways of settling feedback inconsistencies (e.g. by leveraging user expertise in various ways [18]), by checking for inconsistency between feedback facts (e.g. looking only at directly asserted inconsistencies or by exploring transitivities to look for implied inconsistencies), and by restricting the conditions in which an instance or link relationship can be asserted (e.g. forbidding/permitting multiple classification, forbidding linking to a node when already reachable by ancestry, etc.). It is difficult or impossible to decide which of these design choices is optimal *a priori*, and some precise evaluation metric would be needed to determine which ruleset variations tend to produce better results in certain circumstances.

Furthermore, our current methodology uses a manually-constructed seed taxonomy and populates this taxonomic structure through user feedback. Many options are possible for this initial hand-coded structure, and different experts would produce different conceptualizations; we might want a measure of which basic conceptualization tends to produce representations which best fit the distribution of terms in the SEP. More ambitiously, if we allow the answer set program to use disjunctive branching rules with regards to instantiation (thus creating multiple candidate ontologies from a single set of input), we could produce a large space of possible ontologies consistent with user feedback and a general theory of ontologies; the task would then be to rank these candidates according to their suitability for our metadata goals. Again, a precise evaluation metric which could be used to select the "best" ontology from this space is needed.

Another question concerns the amount of expert feedback needed before we begin to see diminishing returns. For example, we can only collect a limited amount of feedback from volunteer SEP authors and editors before the task becomes onerous; as such, we want to prioritize the collection of feedback for areas of the ontology which are currently underpopulated, or even pay some domain experts to address such sparseness. To optimize efficiency, we would want to estimate the number of feedback facts needed to reach a relatively stable structure in that area.

Finally, given that philosophy is an evolving domain rich with controversies, we might wonder how much our evolving representation of that domain captures these debates as they unfold. One of the alluring applications of dynamic ontology is to archive versions of the ontology over time and study the evolution of a discipline as it unfolds. This is doubly-relevant to our project, as both our domain corpus (the asynchronously-edited SEP) and our subject discipline are constantly evolving. The study of this controversy and the evolution resulting from it could be greatly enhanced by using metrics to precisely characterize change across multiple archived versions of the ontology.

4   Our Scores

By stressing the dynamic nature of philosophy, we do not mean to imply that the sciences lack controversy, or that scientific ontologies do not need ways of managing change. Nevertheless, whereas the sciences typically aim for empirically-grounded consensus, the humanities often encourage interpretation, reinterpretation, and pluralistic viewpoints. In this context, the construction of computational ontologies takes on a social character that makes an agreed-upon gold standard unlikely, and makes individual variation of opinion between experts a permanent feature of the context in which ontology evaluation takes place. Because of the dynamic, social nature of the domain, we do not try to achieve maximal correctness or stability of the InPhO's taxonomy of philosophical concepts in one step. But by iteratively gathering feedback, and improving the techniques used to populate the ontology, we hope to quantify the extent to which a stable representation can be constructed despite controversy among users. Our *volatility* score is designed to provide such a measure.

Many approaches to ontology evaluation, such as our volatility score, focus solely on syntactic (formal) properties of ontologies. These methods provide important techniques for assessing the quality of an ontology and its suitability for computational applications, but stable, well-formed syntax is no guarantee that semantic features of the domain have been accurately captured by the formalism. By using the SEP as a proxy for the domain of philosophy, our *violation* score exploits a large source of semantic information to provide an additional estimate as to how well the formal features of our ontology correspond to the rich source material of the SEP.

4.1   Volatility Score

Volatility measures the structural (in)stability of a set of ontologies or (derivatively) an ontology population method [2] Many in the semantic web community hold that domain ontologies are supposed to be authoritative descriptions of the types of entities in a domain [21]. However, ontology development is often an iterative process [16], especially in dynamic ontology. The volatility score carries with it this assumption that a "final answer" description will not respond to the metadata needs of a dynamic corpus such as the SEP, Wikipedia, or WordNet. Additionally, a domain can undergo wide paradigm shifts, dramatically changing its conceptual landscape [22]. The advent of new theories

[2] We thank Uri Nodelman for early discussion of this idea.

like quantum mechanics or new technologies like computers, for example, radically reshaped the conceptual landscape of philosophy.

A volatility score provides a measure of the amount of change between two or more different versions of a populated ontology. The score fixes the population methodology (e.g., the answer set program), while varying the expert feedback, depending on the assessment task. For *current consensus assessment*, we generate a population of ontologies using random samples of the expert feedback available at a given time. This measure can help determine how much feedback to solicit before seeing diminishing returns with regards to representational stability. For *ongoing controversy assessment*, we generate a population of ontologies using aggregate feedback from different times separated by varying intervals. In the case of the InPhO, controversy assessment is measured using different versions of the populated ontology generated from nightly runs of the answer set program.

Both assessment tasks share the same intuition that the aggregate volatility score should be increased by some amount each time the method "changes its mind" about asserting some particular link in the ontology (e.g., an instance switches from being asserted to not asserted under some class). For example, consider the representation of an ongoing controversy over time: if *behaviorism* is said to be highly related to *philosophy of language* but a handful of expert evaluations indicate otherwise, our model would "change its mind" about asserting a link between *behaviorism* and *philosophy of language*. As other experts choose sides and weigh in on the matter, the volatility continues to increase, further pointing to an area of conflict.

Assumptions and Requirements. Volatility, used to measure both current consensus and ongoing controversy, requires careful examination by domain experts to determine whether representational instability is due to undesirable errors/omissions in feedback or the machine reasoning program, or whether it instead properly highlights controversy within the field. This feedback again avoids the need for expensive "double experts" trained in both the target domain and ontology design. Recommendations from the experts regarding errors in the population methods can guide ontology designers to change the ontology extension methods and be evaluated against the old method. In the case of properly highlighted controversies, this information could be used to inform research in the field. In the case of the InPhO project, this could help facilitate analytic metaphilosophy (see Section 6.1 of Buckner et al. [1]).

Another concern with the violation score is that there are different circumstances which could lead $instance\_of(P, Q)$ to switch from being asserted/non-asserted. One way is for there to be a lack of any feedback facts relevant to that instance which could lead to the assertion of an $instance\_of$ relation (e.g., $more\_specific(P, Q)$ and $highly\_related(P, Q)$); another is due to the resolution of an inconsistency in feedback facts (e.g., in one ontology a connection is asserted between P and Q due to a user's feedback, but not asserted in another because of contrary feedback from another user with a higher level of expertise). In order to isolate these issues we adopt a "conservative" approach to assessing volatility: for any given pair of terms, we will only assess a volatility contribution across the subset of ontologies where at least minimal raw materials are present for asserting an $instance\_of$ relationship. With this consideration, we would want to normalize the volatility contribution such that an $instance\_of$ fact asserted 25 times out of 50 relevant ontologies (i.e., ontologies generated from the relevant raw materials) is more volatile than a *instance_of* fact which shifted 10 times out of 20 relevant ontologies (out of the 50 total generated). Consequently, no violation is assessed for pairs of terms which never have the raw materials for assertion across those random subsets of feedback.

Formalization. The volatility measure treats the entire population of data as a "grab-bag" of raw materials (in the form of expert feedback) and inferred instances. Changes in asserted instances are counted without considering temporal order or duration. For a set of n ontologies generated from k feedback facts, volatility measures the relative proportion of times $instance\_of(P, Q)$ is asserted vs. non-asserted. Thus, for any two terms P and Q, the basic formula for assessing the contribution of that pair to the overall volatility score is

$$v(P, Q) = 1 - \frac{|x - \frac{n}{2}|}{\frac{n}{2}} \tag{5}$$

where x is the number of times that the $instance\_of(P, Q)$ is asserted in the set under consideration. By taking into account our conservative weighting score we arrive at the formula

$$v'(P, Q) = 1 - \frac{|x - \frac{m}{2}|}{\frac{m}{2}} \frac{m}{n} \tag{6}$$

where m is the number of input sets containing the raw materials to assert $instance\_of(P, Q)$. This equation reduces to

$$v'(P, Q) = 1 - \frac{|x - \frac{m}{2}|}{\frac{n}{2}} \tag{7}$$

and yields the following aggregate volatility measure:

$$volatility(z) = \frac{1}{count(P, Q)} \sum_{\forall P, Q} v'(P, Q) \tag{8}$$

Interpretation of Results. The volatility score is intended as a metric to call attention to significant structural changes in the ontology. The raw pairwise volatility score can be used to target feedback solicitation so as to reduce the uncertainty of the ontology population method. For concepts instantiated under multiple classes, the volatility of each class-concept pair can indicate the need to gather additional feedback for the given concepts or classes.

The score may be visualized in many different manners. We propose the use of a *heatmap* (e.g., Figure 2) to highlight areas of disagreement for *ongoing controversy assessment*. "Hotter" areas of the visualization indicate areas of more persistent controversy, while white areas convey that there is widespread agreement about the state of a particular instance assertion. Other alternatives not explored here include a scatter plot of average term volatility for each term X, Y in an $instance\_of(X, Y)$ assertion,

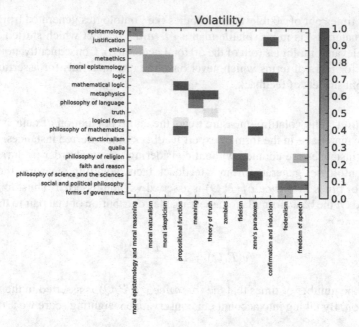

Fig. 2. Heatmap showing terms with average pairwise volatility > 0.45

and a histogram of pairwise volatility, highlighting the distribution of the ontology's controversy and consensus.

Experimental Results. Figure 2 shows a heatmap for the ongoing controversy assessment task at the InPhO Project, using nightly data from October 23, 2008 to April 19, 2011. The heatmap has been filtered to a small selection of the most volatile terms (average pairwise volatility: 0.45). This introspection draws attention to several issues within the InPhO:

- **Specificity.** Many controversies show disagreement in the taxonomic depth at which a term should appear. "Federalism" is a prime example of this, as it is classified under "social and political philosophy" and "forms of government" with moderate degrees of volatility. Both classifications are semantically appropriate, but consensus has not yet been reached as to where the term should be placed.
- **Meta-philosophy.** As discussed in Buckner et al. [1], ontology evaluation can be used to foster meta-studies in the target domain. This type of insight is shown by the concept of "meaning", which is classified under two different parent categories. One shows that meaning is inherent to the structure of the world itself, and therefore belonging in the realm of metaphysics. The other holds meaning to be a function of language, and discussion to fall within the philosophy of language. While many experts in the field could point to this distinction quickly, the evaluation task has highlighted the bifurcation quite simply.
- **Union and Intersection Disambiguation.** Some terms are controversial as a result of polysemous concept labels. "Confirmation and induction" is one such term,

which is instantiated under "justification" and "'logic" with high volatility. Additionally, terms such as "moral epistemology and moral reasoning" find themselves at the intersection of two sections: "ethics" and "epistemology", with the answer set program flipping between the two.

Due to the size of the ontology, instability within these categories would not be immediately obvious to its human curators and users without the aid of software to measure such changes in asserted facts. We are currently working to integrate the average concept volatility into our feedback collection mechanisms, which then permits human expertise and effort to be brought to bear on the specific controversies.

4.2 Violation Score

For a candidate taxonomy, we introduce a "violation score" that is computed by assessing the degree to which its relative placement of terms diverges from statistically generated expectations about those terms relative locations in semantic space (as estimated by their corpus-derived similarity and relative generality measures). Similar to Dellschaft and Staab [6], we consider violation on both a local and the global level. For local violations we only look at parent-child taxonomic relations. For the global violations, we look at the weighted pathwise distance between two terms in a taxonomy.

Assumptions and Requirements. One goal of ontology design is to produce a representation which captures the semantic structure of a domain. In order to have a concrete standard for evaluation, the violation score uses the distribution of terms in corpus (e.g., a reference work in that domain) as a proxy for the domain itself. Evaluation may thus draw upon the statistical measures outlined in Section 3.2.2. However, any metric relating an ontology's taxonomic relations to statistical measures carries with it implicit assumptions regarding the semantic interpretation of the ontology's structural properties, such as the interpretation of edges, pathwise distance, or genealogical depth. In order for the representation to be useful in end user applications (such as visualization, semantic search, and ontology-guided conceptual navigation), we consider several approaches to interpreting ontological structure, which may be adopted with varying degrees of strength:

- **Topic Neutrality.** One might simply wish to regiment all of the vocabulary in a common structure representing only the *isa* relationships that exist among the various terms. The goal of such a taxonomy is simply to enforce a hierarchical structure on all the terms in the language. According to this approach, there is no implied semantic significance to the node depth (aka, genealogical depth) or to path length between pairs of nodes beyond the hierarchical semantics of the *isa* relation itself. For example, if English contains more levels of classificatory terms for familiar animals than it does for relatively unfamiliar organisms, a term such as "dog" may sit at a greater depth in the taxonomy from the root node than terms for other organisms that are similarly specific, but nothing of any semantic significance is implied by this depth (or the distance between term nodes) beyond the existence of the intervening terms in the language.

– **Depth as Generality.** One might desire that all sibling nodes have approximately the same level of generality in the target domain, making node depth (distance from the root node) semantically significant. On this view, the terms *dog* (a species) and *feline* (a family) should not be at the same depth, even if the domain or corpus contains the same number of lexical concepts between *dog* and *thing* as between *feline* and *thing*. Here one expects the entropy of terms at the same depth to be highly correlated.[3]

– **Leaf Specificity.** One might desire that all leaf nodes in the structure represent approximately the same grain of analysis. On this view, regardless of node depth, leaves should have similar entropy. Thus, for example, if *hammerhead shark* and *golden retriever* are both leaf nodes, leaf specificity is violated if these terms are not similarly distributed across the corpus that is standing proxy for the domain.

Choices among these desiderata are central to any argument for edge-based taxonomic evaluation. This is especially true for gold standard approaches which implicitly hold the relations of two candidate ontologies to be semantically equivalent. Additionally, we suspect that most domains have asymmetric taxonomic structures: subtrees of sibling nodes are not typically isomorphic to one another, and this means that even within a given taxonomy, path length between nodes and node depth may not have the same semantic significance.

In our comparison methods we assume that node depth is topic neutral – that is, node depth bears little correlation to specificity or generality on a global level. However, by definition, a child node should be more specific than its parent node. Thus, we measure local violation by comparing the information content of the parent and child nodes. When two terms are reversed in specificity we can count this as a syntactic violation of the taxonomic structure. Additionally, we can expect sibling instances to be closely related to one another and to their parent node by statistical measures of semantic distance. An instance is in violation if it is an outlier compared to the rest of its siblings.

We propose that overall violation is an emergent property from these localized semantic violations. These violations are each weighted by the magnitude of the error (as measured by the difference in semantic distance), ensuring that an ontology with several large mistakes will have greater violation than one with many minute errors.

Formalization. A *generality violation* (g-violation) occurs when two terms are reversed in specificity (e.g., the statistics propose that *connectionism* is more specific than *cognitive science* but the answer set asserts that *cognitive science* is more specific). For two terms S and G, where S is more specific than G, we hypothesize that the conditional entropy will be higher for for G given S than for S given G.

$$H(G \mid S) > H(S \mid G) \qquad (9)$$

---

[3] Edge equality provides a special case of depth as generality. The latter requires only that all edges at a given level represent the same semantic distance, whereas edge equality also requires these distances to be consistent between the different levels (e.g., the movement from a species to a genus represents the same conceptual distance as that between an order and a class).

This makes intuitive sense if one considers the terms dog (S) and mammal (G). The presence of the term *dog* will lend far more certainty to the appearance of *mammal* than the other way around – mentioning *mammal* is not very predictive of *dog.*

If this inequality does not hold, a generality violation (g-violation) is measured:

$$gv(S, G) = H(S \mid G) - H(G \mid S) \tag{10}$$

The mean of the g-violations is then taken to give the overall g-violation.

$$violation_g(O) = \frac{1}{count(S, G)} \sum_{\forall S, G} gv(S, G) \tag{11}$$

A *similarity violation* (s-violation) occurs when an instance's semantic similarity to its parent class is an outlier compared to the rest of its siblings. For example, the entity *(ideas about) federalism* has been observed under both *(ideas about) social and political philosophy* and *(ideas about) forms of government*. However, the siblings of *federalism* under *forms of government* are much closer to their parent node than those under *social and political philosophy*. Therefore, a taxonomy asserting that *federalism* is an instance of *social and political philosophy* will recieve higher violation than one in which *federalism* is an instance of *forms of government*.

Semantic similarity can be measured using a variety of measures reviewed in Jiang and Conrath [23] and Resnik [24]. We use the measure presented in Lin [25]:

$$sim(x_1, x_2) = \frac{2 \times \log P(C)}{\log P(x_1) + \log P(x_2)} \tag{12}$$

Such that x_1 and x_2 are entities in the taxonomy, and C is the most specific class which subsumes x_1 and x_2. As we are simply comparing an instance S to its parent G, we can use:

$$sim(S, G) = \frac{2 \times \log P(G)}{\log P(S) + \log P(G)} \tag{13}$$

The degree of s-violation can be determined by the standard score, which normalizes the values by standard deviation:

$$sv(S, G) = \frac{sim(S, G) - \mu}{\sigma} \tag{14}$$

where x is the raw semantic distance, μ is the mean of the semantic distance to the parent of all sibling nodes and σ is the standard deviation of this population. The final s-violation is calculated as the mean of s-violations.

$$violation_s(O) = \frac{1}{count(S, G)} \sum_{\forall S, G} sv(S, G) \tag{15}$$

Interpretation of Results. The violation score is intended as way to select the best representation of a given set of input parameters. In our methodology, the violation score is used to test variations in ruleset changes or seed taxonomies. This evaluation can be used throughout the ontology design process to perfect methodology. We have used violation to examine changes to the assertion of semantic crosslinks and in the weighting of expert feedback obtained from novice philosophers, undergraduate majors, graduate students, and professors of philosophy.

Additionally, we are able to use the violation score to compare different samples of expert feedback by using the same seed taxonomy and ruleset. The changes in violation scores exposed a steady increase in taxonomic fit from novices to undergraduates to graduate students, before a slight decrease with experts who have published in the area. Further investigation of violations found that our highest-level experts were more likely to go against the statistical prediction in often useful ways. These observations reinforce the importance of our general design strategy of leveraging human expertise against the computational powers of machines. Violation scores may highlight areas which need expert attention, but should not override expert judgment automatically. When comparing different versions of the ruleset, we must carefully reason through whether some particular ruleset change could be subtly biasing the representation towards or against expert feedback (e.g., in the way it settles inconsistency between users and experts).

Experimental Results. Since deploying the initial version of our answer set program (described in Niepert et al. (2008)), we discovered a number of possible improvements, but could not be sure a priori which version of the ruleset would produce better results. The violation score provides us with a way to compare these options in terms of their suitability. We identified three binary parameters along which our program can vary, and have compared the violation scores for each possible combination (resulting in a 2x2x2 matrix). The three parameters are briefly described under their abbreviated names below[4].

- **"plink".** Our original ruleset [18] included non-taxonomic "links" to allow reachability between entities which were semantically related but which, for various reasons, could not be connected taxonomically. To minimize unnecessary taxonomic relations, we added a rule (hereafter, the "nins" rule) which blocked an instance X from being asserted as an instance of a class Y if there was also evidence that X was an instance of class Z and Y was possibly linked ("plink"ed) to Z (since in that case X would already be reachable from Y via the $Y \rightarrow Z$ link). Unexpectedly, we found that this occasionally produced an undesirable "reciprocal plink deadlock" (see Figure 3):

  Whenever links were possible from both $Y \rightarrow Z$ and $Z \rightarrow Y$, the nins rule blocked X from being inferred as an instance of either Y or Z (and thus X often became a taxonomic "orphan"). As such, we created a second version of the program which added a "no plink" restriction to the "nins" rule, preventing this reciprocal plink situation. The "plink" parameter indicates that this restriction was added to the nins rule.

- **"voting".** An important innovation of our project involves the stratification of user feedback into different levels of self-reported expertise and using this information

---

[4] The exact details of the answer set program can be found in Niepert et al. 2009, but are unnecessary for the purposes of our discussion investivating the suitability of different reasoning methods.

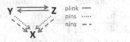

Fig. 3. The reciprocal plink problem

in a two-step process to resolve feedback inconsistencies. The first step in this process involves the application of a "voting filter" which settles *intra*-strata feedback inconsistencies using a voting scheme and can be completed as a preprocessing step before the answer set program is run. The "voting" parameter indicates that this filter was used.

- **"trans".** Much of the information on which our program operates is derived from the transitivity of the "more general than"/"more specific than" feedback predicates. The second step of our method for settling feedback inconsistencies involves settling *inter*-strata inconsistencies, which is completed from within our ruleset. However, transitivities in feedback can be computed either before or after these inter-strata inconsistencies are resolved (the former resulting in many more inconsistencies requiring resolution). The "trans" parameter thus indicates that this version of the ruleset computes transitivities *before* (vs. *after*) our ruleset settles inter-strata inconsistencies.

Each modification was then compared to the current ruleset using both the s-violation and g-violation metrics using corpus statistics and user evaluations from July 24, 2010 (see Figure 4). The number of instances asserted is also included. As we can clearly see, every proposed change decreased both violation scores, with the best results provided by adopting all three changes[5]. The decrease in s-violation can be interpreted as the development of denser semantic clusters subsumed under each class. The decrease in g-violation can be interpreted as movement towards greater stratification in the heirarchy. This is quantitative evidence that the principled design choices outlined above will provide useful additions to the ontology enrichment process.

5   Future Work

With these methods of evaluating ontology structure and function in hand, along with preliminary results on our limited feedback collection, we propose to continue these evaluation experiments as new feedback is rapidly collected from SEP authors. These scores will allow us to pursue a long-desired use of our answer set programming to infer a space of populated ontologies and select an optimal one by ranking them according to violation scores. We can then see how consistent ruleset selection is.

We might also ask how feedback from people with different levels of expertise in philosophy affects the placement of terms in the InPhO. For instance, Eckert et al. [26] have already gathered feedback data from Amazon Mechanical Turk (AMT) users and compared their responses to those of experts. Although we know that as a whole

---

[5] g-violation was lowest when adopting the plink and voting changes, but not trans. However, the result with all three changes was second lowest.

	s-violation		g-violation		instances	
	all-in	voting	all-in	voting	all-in	voting
current	0.8248	0.8214	-0.1125	-0.1170	417	456
plink	0.8111	0.8089	-0.1182	-0.1227	521	568
trans	0.8119	0.8094	-0.1133	-0.1168	452	491
plink, trans	0.8061	0.8031	-0.1153	-0.1188	502	546

Fig. 4. Violation score evaluations on the InPhO using feedback and corpus statistics from July 24, 2010

they differ statistically from experts, we do not yet know how much this matters to the structure that is eventually produced from those feedback facts.

In this paper we have shown an analysis of ongoing controversy using the volatility score. A full analysis of current consensus awaits the collection of sufficient expert feedback to allow rigorous sampling. The integration of pairwise volatility with our feedback collection methods should further this goal.

6 Conclusions

In this paper we have proposed two methods for evaluating the structural and functional aspects of a corpus-based dynamic ontology. Our work focuses on the semantic evaluation of taxonomic relations, rather than the lexical evaluation undertaken by Brewster et al. [8] and Dellschaft & Staab [6]. The violation score gives us a concrete measure of how well an ontology captures the semantic similarity and generality relationships in a domain by examining statistical measures on an underlying corpus. The volatility score exposes areas of high uncertainty within a particular ontology population method, which can be used for many purposes, including current consensus assessment and ongoing controversy assessment, highlighting a dynamic ontology's evolution. We also have examined the considerations necessary to evaluate a taxonomy, and demonstrated how these methods have been used to enhance the enrichment process of the Indiana Philosophy Ontology Project through experiments on ruleset variations, expert feedback stratification and stability.

References

1. Buckner, C., Niepert, M., Allen, C.: From encyclopedia to ontology: Toward dynamic representation of the discipline of philosophy. Synthese (2010)
2. Gangemi, A., Catenacci, C., Ciaramita, M., Lehmann, J.: Modelling Ontology Evaluation and Validation. In: Sure, Y., Domingue, J. (eds.) ESWC 2006. LNCS, vol. 4011, pp. 140–154. Springer, Heidelberg (2006)
3. Guarino, N., Welty, C.A.: An overview of OntoClean. In: Staab, S., Studer, R. (eds.) Handbook on Ontologies, 2nd edn., pp. 151–159. Springer, Heidelberg (2004)
4. Gómez-Pérez, A.: Evaluation of taxonomic knowledge in ontologies and knowledge bases. In: Proceedings of the 12th Banff Knowledge Acquisition for Knowledge-Based Systems Workshop, Banff, Alberta, Canada (1999)
5. Fahad, M., Qadir, M.: A Framework for Ontology Evaluation. In: Proceedings International Conference on Conceptual Structures (ICCS), Toulouse, France, pp. 7–11. Citeseer (July 2008)

6. Dellschaft, K., Staab, S.: Strategies for the Evaluation of Ontology Learning. In: Buitelaar, P., Cimiano, P. (eds.) Ontology Learning and Population: Bridging the Gap Between Text and Knowledge, pp. 253–272. IOS Press (2008)

7. Maedche, A., Staab, S.: Measuring Similarity Between Ontologies. In: Gómez-Pérez, A., Benjamins, V.R. (eds.) EKAW 2002. LNCS (LNAI), vol. 2473, pp. 251–263. Springer, Heidelberg (2002)

8. Brewster, C., Alani, H., Dasmahapatra, S., Wilks, Y.: Data driven ontology evaluation. In: Proceedings of LREC, vol. 2004 (2004)

9. Supekar, K.: A peer-review approach for ontology evaluation. In: 8th Int. Protege Conf., pp. 77–79. Citeseer (2004)

10. Staab, S., Gómez-Pérez, A., Daelemans, W., Reinberger, M.L., Guarino, N., Noy, N.F.: Why evaluate ontology technologies? because it works! IEEE Intelligent Systems 19, 74–81 (2004)

11. Lozano-Tello, A., Gómez-Pérez, A.: Ontometric: A method to choose the appropriate ontology. Journal of Database Management 15, 1–18 (2004)

12. Brank, J., Grobelnik, M., Mladenic, D.: Survey of ontology evaluation techniques. In: Proceedings of the Conference on Data Mining and Data Warehouses (SiKDD) (2005)

13. Velardi, P., Navigli, R., Cucchiarelli, A., Neri, F.: Evaluation of OntoLearn, a methodology for automatic learning of domain ontologies. In: Buitelaar, P., Cimiano, P., Magnini, B. (eds.) Ontology Learning from Text: Methods, Evaluation and Applications. IOS Press, Amsterdam (2005)

14. Porzel, R., Malaka, R.: A task-based framework for ontology learning, population and evaluation. In: Buitelaar, P., Cimiano, P., Magnini, B. (eds.) Ontology Learning from Text: Methods, Evaluation and Applications. IOS Press, Amsterdam (2005)

15. Gruber, T.R.: Toward principles for the design of ontologies used for knowledge sharing. International Journal of Human Computer Studies 43, 907–928 (1995)

16. Noy, N., McGuinness, D.: Ontology development 101: A guide to creating your first ontology (2001)

17. Niepert, M., Buckner, C., Allen, C.: A dynamic ontology for a dynamic reference work. In: Proceedings of the 7th ACM/IEEE-CS Joint Conference on Digital Libraries, p. 297. ACM (2007)

18. Niepert, M., Buckner, C., Allen, C.: Answer set programming on expert feedback to populate and extend dynamic ontologies. In: Proceedings of 21st FLAIRS (2008)

19. Smyth, P., Goodman, R.M.: An information theoretic approach to rule induction from databases. IEEE Transactions on Knowledge and Data Engineering 4, 301–316 (1992)

20. Shannon, C.E.: A mathematical theory of communication. University of Illinois Press, Urbana (1949)

21. Smith, B.: Ontology. In: Luciano, F. (ed.) Blackweel Guide to the Philosophy of Computing and Information, pp. 155–166. Blackwell, Oxford (2003)

22. Kuhn, T.: The Structure of Scientific Revolutions. University of Chicago Press (1962)

23. Jiang, J., Conrath, D.: Semantic similarity based on corpus statistics and lexical taxonomy. In: Proceedings of International Conference Research on Computational Linguistics (ROCLING X), Number Rocling X, Taiwan (1997)

24. Resnik, P.: Semantic similarity in a taxonomy: An information-based measure and its application to problems of ambiguity in natural language. Journal of Artificial Intelligence Research 11, 95–130 (1999)

25. Lin, D.: An information-theoretic definition of similarity. In: Proceedings of the 15th International Conference on Machine Learning, pp. 296–304. Citeseer (1998)

26. Eckert, K., Niepert, M., Niemann, C., Buckner, C., Allen, C., Stuckenschmidt, H.: Crowdsourcing the Assembly of Concept Hierarchies. In: Proceedings of the 10th ACM/IEEE Joint Conference on Digital Libraries (JCDL), Brisbane, Australia. ACM Press (2010)

An Architecture to Support Semantic Enrichment of Knowledge Sources in Collaborative Engineering Projects

Ruben Costa[1] and Celson Lima[2]

[1] UNINOVA, Centre of Technology and Systems, Campus da Caparica
Quinta da Torre2829-516 Monte Caparica, Portugal
[2] Universidade Federal do Oeste do Pará, Instituto de Engenharia e Geociências
Santarém, Brazil
rddc@uninova.pt, celsonlima@ufpa.br

Abstract. This work brings a contribution focused on collaborative engineering projects where knowledge plays a key role in the process, aiming to support collaborative work carried out by project teams, through an ontology-based platform and a set of knowledge-enabled services. We introduce the conceptual approach, the technical architectural (and its respective implementation) supporting a modular set of semantic services based on individual collaboration in a project-based environment (for Building & Construction sector). The approach presented here enables the semantic enrichment of knowledge sources, based on project context. The main elements defined by the architecture are an ontology (to encapsulate human knowledge), a set of web services to support the management of the ontology and adequate handling of knowledge providing search/indexing capabilities (through statistical/semantically calculus), providing a systematic procedure for formally documenting and updating organizational knowledge. Results achieved so far and future goals pursued here are also presented.

Keywords: Collaboration, Knowledge management, Semantic services, Semantic reasoning, Ontology.

1 Introduction

Over the last two decades, the adoption of the Internet as the primary communication channel for business purposes brought new requirements especially considering the collaboration centred on engineering projects. By their very nature, such projects normally demand a good level of innovation since they tackle highly complex challenges and issues. On one hand, innovation often recurs to combination of knowledge (existing, recycled, or brand new) and, on the other hand, it depends on individuals (or groups) holding the appropriate knowledge to provide the required breakthrough.

Engineering companies are project oriented and successful projects are their way to keep market share as well as to conquer new ones. Engineering projects strongly rely

A. Fred et al. (Eds.): IC3K 2010, CCIS 272, pp. 276–289, 2013.

on innovative factors (processes and ideas) in order to be successful. From the organisation point of view, knowledge goes through a spiral cycle, as presented by Nonaka and Takeuchi in the SECI model[1] [1]. It is created and nurtured in a continuous flow of conversion, sharing, combination, and dissemination, where all the aspects and contexts of a given organisation, are considered, such as individuals, communities, and projects.

Knowledge is considered the key asset of modern organisations and, as such, industry and academia have been working to provide the appropriate support to leverage on this asset [2]. Few examples of this work are: the extensive work on knowledge models and knowledge management tools, the rise of the so-called knowledge engineering area, the myriad of projects around 'controlled vocabularies' (i.e., ontology, taxonomies, etc..), and the academic offer of knowledge-centred courses (graduation, master, doctoral).

The quest for innovation to be used a wild card for economic development, growing and competitiveness, affects not only organisations, but also many countries. This demand for innovative processes and ideas, and the consequent pursuit of effectively more knowledge, raise inevitably issues regarding the adoption and use of Knowledge Management (KM) models and tools within organisations.

As relevant literature shows [3]; [4]; [5]; [6], KM does not only comprise creation, sharing, and acquisition of knowledge, but also classification, indexation, and retrieval mechanisms. Knowledge may be classified by its semantic relevance and context within a given environment (i.e., the organisation itself or a collaborative workspace). This is particularly useful to: (i) improve collaboration between different parties at different stages of a given project life cycle; and (ii) assure that relevant knowledge is properly capitalised in similar situations. For example, similar projects can be conducted in a continuously improved way if lessons learned from previous are promptly known when a new (and similar to some previous one) project is about to begin.

Semantic systems utilize an ontology (or a set of ontologies) to encapsulate and manage the collection and representation of relevant knowledge, hence giving information a human-relevant meaning. Semantic description of project resources enhances collaboration through better understanding of document contents (supporting better understanding and extraction of knowledge) [7]. In addition, by introducing ontological reasoning, semantic techniques enable discovery of knowledge and information that was not part of the original use case or purpose of the ontology itself [8].

The work presented here, provides project teams with semantic-enabled services, targeting the improvement of the semantic richness of knowledge sources used/created, during the execution of an engineering project. The work conceptually covers two dimensions, namely collaboration and knowledge engineering, focused on ontology development and knowledge sharing activities [9]. Knowledge, the dimension particularly explored in this paper, relates to the 'currency' being exchanged during a collaborative process, in this case a collaborative engineering process. Technical documents, lessons learned, and expertise, are some examples of such currency.

---

[1] The SECI model covers four transformation processes involving both explicit and tacit knowledge types: Socialisation (from tacit to tacit), Externalisation (from tacit to explicit), Combination (from explicit to explicit), and Internalisation (from explicit to implicit).

This paper is structured as follows: Section 2 defines the problem to be tackled. Section 3 covers the state of practice related to this work. Section 4 introduces the software components handling the knowledge related matters previously introduced. Section 5 gives illustrative examples of the software operation. Finally section 6 concludes the paper and points out the future work to be carried out.

2   Problem Statement

The key question guiding the development of this work is: How to augment the relevance of knowledge sources in collaborative engineering projects in order to support users within problem-solving interactions? The approach adopted here to find a suitable answer is focused on a problem-solution representation, enabling users to keep track of problems occurred and decisions made to solve them, which can be reused whenever necessary to solve new problems. The Building & Construction (BC) sector was found to be a suitable test bed for driving the developments of this work since it is essentially ruled by a project-based delivery paradigm to produce unique products and services.

Though two adjacent buildings may look the same, each has a characteristic of its own when it comes to constructing it [10]. BC projects are characterized by several phases conducted by different teams with different levels of expertise and skills (e.g. architects, engineers, local authorities, etc.). Such teams of professionals vary from phase to phase and have different needs regarding the project goals. For example, a request for information related to a particular issue can produce different results if raised by either a structural engineer or an architect, since different actors likely have different needs regarding the project itself. Summing up, context within B&C projects can be characterized by several features, such as actors, project type, and phase.

The success of collaboration considering an engineering project, where project teams are working together targeting a shared goal, essentially relies on capitalising on the existing knowledge as well as being capable to find innovative solutions to faced problems. Therefore, we can see the instantiation of the SECI model within the collaborative engineering environment towards agile decision making process, where knowledge is: (i) transformed in an evolving way along the time; (ii) managed around problems and solutions in order to be proper capitalised [9]; (iii) better capitalised with the appropriate support of reasoning mechanisms; and (iv) supported by a set of ontology-enabled services to increase semantics.

Collaborative engineering environments are conducted through a series of meetings and every meeting is considered a Decisional Gate (DG), a convergence point where decisions are made, problems are raised, solutions are (likely) found, and tasks are assigned to project participants. Pre-existing knowledge serves as input to the DG, the project is judged against a set of criteria, and the outputs include a decision (go/kill/hold/recycle) and a path forward (schedule, tasks, to-do list, and deliverables for next DG).

Knowledge needs to be shared in order to be proper capitalised during decision making processes. On one hand knowledge sharing is heavily dependent on technical capabilities and, on the other hand, since the social dimension is very strong during collaboration, there is also an increased need to take into account how to support the culture and practice of knowledge sharing. For instance, issues of trust are critical in collaborative engineering projects, since the distribution of knowledge and expertise means that it becomes increasingly difficult to understand the context in which the knowledge was created, to identify who knows something about the issue at hand, and so forth.

Additionally, the following specific questions can be added to that research question in order to better define the frontiers of this work:

1. Which socio-technical framework that represents problem-solving interactions can be adopted?

2. How to define context within collaborative engineering projects for the building & construction sector?

3. Which semantic approach could be adopted to contextualize knowledge sources?

Preliminary answers to these questions are provided here.

3   State of the Art

Ontologies are defined as a shared conceptualization of the knowledge in a certain domain [11]. Based on the scope of the target domain, ontologies can be categorized into domain and application ontologies. Domain ontologies aim at modelling the fundamental concepts in a relatively large domain (for example, Biomedical and Manufacturing). In contrast, application ontologies focus on modelling sub-domains of a major domain. For example, and within the construction domain, one can think of the following application ontologies: cost, productivity and safety.

Ontologies have contributed to the emergence of semantic search engines. Among these, we include the contextual search engines based on domain ontologies, which restrict the search to a well delimited area. We can distinguish two main categories of search engines: ontology-enabled search engines and semantic search engines [12]. We are especially interested in semantic search engines, which can be divided in three groups:

• Context-based Search Engines: the final purpose of these engines is to enhance the performance of traditional search engines (especially precision and recall), by understanding the context of documents and queries. One of their most important parts is the annotator, which is responsible for generating metadata from the crawled pages. These systems are the most practical ones; in fact, they are the next generation of current search engines.

• Evolutionary Search Engines: These search engines are an answer to the famous and well known problem of how to automatically gather information about a specific topic. The main distinguished behaviour of these engines consists in using external metadata. They normally use an ordinary search engine and display augmented information near the original results.

• Semantic Association Discovery Engines: The goal of these systems is to find various semantic relations between input terms and then rank the results based on semantic distance metrics. An upper ontology like WordNet or OpenCyc can be used to evaluate this kind of search engines

The main problems of ontology-based information retrieval are the reformulation of the users' queries and the difficulties to build and manage ontologies. Many ontologies are available on the Web, but it is still difficult to find an ontology for every domain associated to a user's query. However, it is obvious that building ontologies for all domains, so that they can be exploited by semantic search engines, is difficult. So, using previous experiences of users can help to accomplish this mission.

4   Objectives and Relevance of the Work

The main hypothesis supporting this work is that *the relevance of knowledge sources used in collaborative engineering projects can be augmented if they are contextualized and indexed using a semantic approach represented by semantic vectors supported by a reference ontology*. Semantic-enabled systems present a feasible and effective way to support knowledge in collaborative engineering environments due to the fact that such systems allow for better presentation of human knowledge and, through ontologies, provide for efficient means for encapsulating and updating corporate knowledge. i.e., the main advantage of an ontology-based system is that ontologies can be built to encapsulate human knowledge in a machine-readable manner.

The main objective pursued here is related with capturing and reuse of knowledge, by adopting an ontology-based approach using semantic and statistical/proximity based algorithms to better augment the relevance of knowledge sources created/used within collaborative engineering projects. In this sense, the key capabilities to be provided are the following:

• Knowledge documentation and storage: support a consistent approach for documenting lessons learned in ontology-based system that allows semantic retrieval of documents.

• Knowledge classification: knowledge classification is a highly desirable functionality and one having a high priority. Existing tools only allow for the categorisation of knowledge. It is more important to support knowledge item clustering (finding similarities between knowledge items).

• Search for knowledge items: the search, discovery, and ranking of knowledge items are issues of high priority with respect to both the manner in which these are done and in terms of the different types of knowledge items considered (full text search; searches on the basis; and discovery of experts and communities).

It is worth noticing that the Semantic Web is a kind of "convergence point" for the projects and initiatives working in the areas related to semantic matters. As such, the work performed here takes into account the W3C recommendations, such as the language proposed to represent ontologies – the Web Ontology Language (OWL). This seems to be the best currently available format to represent an ontology that aims to become easily used. Therefore, either new ontologies or existing ones are strongly recommended to adopt this representation language, in order to maximise their degree of acceptance and exploitation.

The work aims to provide the best ontological representation for a given Knowledge Source (KS) within a given context, when adding/searching for knowledge. When adding a new knowledge into the knowledge repository, the approach being implemented will extract the best relevant keywords from the KS and calculates their statistical weights. This set of keywords/weights forms the basis of the so-called Semantic Vector (SV), which is analysed against the ontology in order to get the ontological representation of the KS, which is defined by concepts from the ontology. A knowledge representation is then built for the KS and stored into the repository. This is going to be explained more clearly in the following sections.

When searching for knowledge, the system analyses the queries in order to get the appropriate ontological representation. Effectively, the system finds the knowledge representations that best match the concepts in order to get the relevant KS from the knowledge repository for a given query.

Our approach provides three algorithms to perform the process of retrieving the best ontological representation and weight for both the KS and the query. Those algorithms, namely "Lexical Entries based", "Taxonomy based" and "Relation based" work as follows:

• Lexical Entries: Each concept is defined with a list of lexical entries in a different language. The algorithm gets all the lexical entries of all the concepts of the ontology and matches them with all the keywords in the semantic vector. Therefore at the end of the first step, a list of concepts (Lc) matching the semantic vector is built. Further, the weight of the concept C (Wc) is calculated for all the concepts in the list applying the following formula:

$$Wc = \frac{NKm}{NKsm} \tag{1}$$

Where:

Wc: weight calculated to the concept.

NKm: number of keywords that match the concept C.

NKsm: number of keywords in the semantic vector.

• Taxonomy: The algorithm starts from the list "Lc" built in the "Lexical Entries based" algorithm and provides a different way to arrive at the weights. The aim is to try to increase the weight of the concepts which may have received a poor weight in the first stage trying to see if they are close in the taxonomy to a concept that received a good weight in the first stage. The "Lc" list gets the best concepts that match the keywords. A concept is considered a best concept when its weight exceeds the value "best-concept-range" defined in the parameters table. The others are named "worth concepts". For each best concept, the algorithm checks if there are worth concepts nearby concepts of the Lc list in the taxonomy. If this is the case, their respective weights are augmented according the following formula:

$$Wc = Wbc \times Vp \tag{2}$$

Where:

Wc: weight performed for the concept

Wbc : weight of the best concept

Vp: value got in the parameters table depending on the level and the way.

The Vp value depends on the distance between the best concept and the worth concept in the hierarchy of concepts. If the worth concept is a super/sub concept of the best concept then the value of Vp is 1. If they are two levels in the tree between them, Vp is 2.

• Relation: This algorithm will be available in the next version of the system. It aims to integrate the richness of the relations among the concepts in order to provide a more powerful way to represent a KS.

So far, in the process of getting the best concepts that match the keywords, the system provides a list of concepts that match directly the keywords received in the query. All the concepts in the list are indexed by a keyword in the query through one or more lexical entries. At the end of the process (for all algorithms previously described), an additional list of concepts is added to the response. The neighbours, which mean sub/super concepts of the Lc concepts, are added to the list. For those additional concepts, the following formula to calculate their weights is applied:

$$Wa = Wc / Li$$

Wa: weight calculated for an additional concept

Wc: concept weight of the Lc list

Li: Number of level of the Wa concept from the concept of the Lc list.

5   Technical Architecture

The main features of the proposed approach are an ontology (to encapsulate human knowledge) and a set of web services to support the management of ontology (creation, updates), and handling knowledge management requirements (indexing, documentation, retrieval and dissemination). The technical architecture (Fig. 1) that instantiates those features is composed by three main elements:

• Repository: describes and encapsulate a formal representation of domain knowledge to be handled and is composed by three major entities: (i) knowledge representations repository, which is a container that aggregates all the semantic vectors previous calculated and associated with each single knowledge source; (ii) the ontology server represents the abstract conceptual model for construction knowledge, encapsulating main concepts, their relationships and their constraints (axioms); and the knowledge sources repository, responsible for storing all relevant knowledge.

• API: used to manage the ontology, and other semantic services (indexing and search for knowledge sources) which constitute main knowledge enabled services. The API is composed by a set of web-services responsible for managing the ontology and knowledge sources (update, browsing, indexing, access rights).

• User Interface: relates to the front-end with the user via web portal. This interface en enables users to interact with the system.

The adoption of the Web services model also plays a very strategic role regarding openness, interoperability, and integration of the system. We used WSDL to specify the services that form the API layer as well as the services provided by third party tools that have to be integrated into the API layer. Having the WSDL file describing a given web service it is easy to produce the web client able to invoke that service. Thanks to this mechanism, all the public services provided in the API layer are available to any web application in the same way that the system interoperates with any other web application.

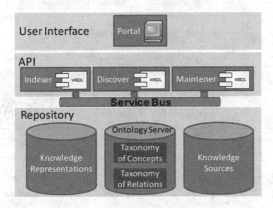

Fig. 1. Technical architecture

The Java language was chosen as it is platform independent and freely available. The repository contains several databases holding the knowledge representations, the ontology and the knowledge sources.

The instantiation of the technical architecture (see Fig. 2), encapsulates the software components and functionalities which are required to support the implementation of what was specified by the technical architecture. While these functionalities are conceptually independent, the access is provided via one single portal, with the distinction being made by means of user access rights. This portal application relies mainly on the JSP technology and uses the apache tomcat as an application server.

Jena technology is used in order to manipulate OWL models. It includes an OWL API that provides support for loading OWL ontologies. The OWL models are manipulated by Jena API which enables to persist OWL ontology into a relational database and programmatically manage the OWL ontology. Within the API layer, the JENA API is the low level component. It loads the ontology model from an OWL file and provides access to all ontological entities.

We use liferay technology as a backend to support the to manage the knowledge sources repository, i.e. liferay acts as a document management system, which provide a set of services to manage the deployment of knowledge sources used/created within the collaborative engineering environment.

MySQL technology is used to implement the repository layer. It implements a relational database which is accessed by a set of services available on the API layer.

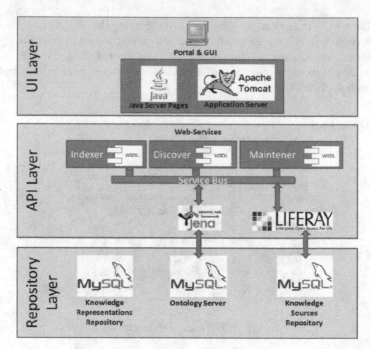

Fig. 2. Instantiation of Technical Architecture

5.1   The Ontology

Knowledge sources strongly rely on ontological concepts, as a way to reinforce their semantic links. The ontology uses a taxonomy of concepts holding two dimensions: on one hand, the knowledge sources themselves are represented in a tree of concepts and, on the other hand, the industrial domain being considered. Instances of concepts (also called individuals) are used to extend the semantic range of a given concept. For instance, the ontological concept of 'Design_Actor' has two instances to represent architect and engineer as roles that can be considered when dealing with knowledge sources (see Fig. 3) related to design (experts, design-related issues/solutions, etc.). Moreover, each ontological concept also includes a list of terms and expressions, called equivalent terms, which may represent synonyms or expressions that can lead to that concept. Ontology support is particularly useful in terms of indexation and classification towards future search, share and reuse.

The ontology is developed to support and manage the use of expressions which contextualize a KS within the knowledge repository. The ontology adds a semantic weight to relations among KS stored into the knowledge repository. Every ontological concept has a list of 'equivalent terms' that can be used to semantically represent such concept. These terms are, then, treated in both statistical and semantic way to create the semantic vector that properly indexes a given KS.

The ontology was not developed from scratch; rather, it has been developed taking into account relevant sources of inspiration, such as the buildingsmart IFD model [13], omniclass[14] , and the e-cognos project[15] .

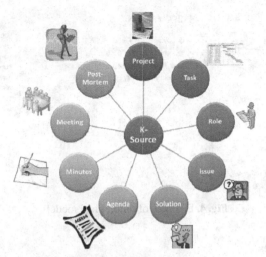

Fig. 3. Knowledge Sources

The basic ontological definition is as follows: *a group of Actors uses a set of Resources to produce a set of Products following certain Processes within a work environment (Related Domains) and according to certain conditions (Technical Topics).*

The ontology is process-centred. Other domains define all relevant process attributes. For example, the Technical Topics domain defines the concepts of productivity, quality standard and duration (see Fig. 4).

All entities (including Process) have three ontological dimensions: state, stage and situation. State concept captures the status of entity development: dormant, executing, stopped, re-executing completed. Stage concept defines various development stages: conceptualization, planning, implementation and utilization. Situation concept refers to planned entities and unplanned entities.

A Project is a collection of processes. It has two types: Brown field projects and Green field projects. It has a project delivery system, a contract, a schedule, a budget, and resource requirements. It also has a set of related aspects that include: start time, a finish time, duration, a quality standard, productivity level, a life cycle and a life cycle cost all of which are defined in the Technical Topics domain.

A Process has an input requirements that include: the completion of all proceeding processes, the availability of required approvals, the availability of required knowledge items (documents, software, etc.), the availability of required Resources (materials, equipment, subcontractors), the availability of required Actors, and the availability of required budget.

A Process has three major sub concepts: Phase, Activity and Task. It also has two major types: engineering process and administrative process. A Process has an output that include: update to a product time-line, an update to the project schedule, and update to the project budget, satisfaction/update to the legal conditions/status of Actors, may result in creating some project incidents (an accident, damage to an equipment, etc.).

A Product (also Actors, Processes and Resources) has attributes, parameters and elements, which are defined in Technical Topics.

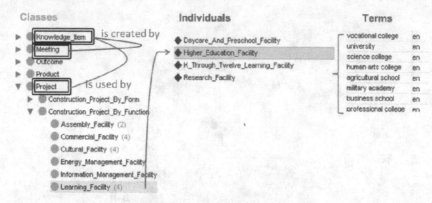

Fig. 4. Example of ontological model

5.2   The Services

The semantic support services that compose the API layer can globally be described as the following:

• Indexing: The service is designed to accept a list of keywords, compare the keywords to ontological concepts, and produce a ranked list of ontological concepts that best matches that list of keywords. For each keyword, it calculates a corresponding weight reflecting its relevance. The set of keyword-weight pairs is the semantic vector of the knowledge source. This vector is then used to assign a hierarchy of relevant metadata to each knowledge source.

• Discover: The service enables the user to perform searches across knowledge elements, is invoked whenever a user requests a search for a set of keywords. The service produces a matching ontological concept for these keywords, and then matches the resulting concept to the metadata of target knowledge source. This ontology-centred search is the essence of semantic systems, where search phrases and semantic vectors are matched through ontological concepts.

• Maintener: This service is responsible for managing the domain ontology enabling the following capabilities:

  o Browse the concepts/relations: this allows navigation through the ontology, showing the description of both concepts and relations;
  o Create new concept: this allows the addition of a new concept into the ontology;
  o Create new relation: this allows the addition of a new relation into the ontology;
  o Create new attribute: this allows the addition of a new attribute to a concept;
  o Import OWL ontology;
  o Remove concept: this allows removal of a concept from the ontology.

6   Indexation Process Using Semantic Vectors

To better understand the indexation process through semantic vectors comparison (Fig. 5), it is necessary to understand how and where these are created and used.

Each semantic vector contains the necessary ontological concepts that best represent a given knowledge source when it is stored into the knowledge repository. These concepts are ordered by their semantic relevance regarding the KS. KS are compared and matched based on their semantic vectors and the degree of resemblance between semantic vectors directly represents the similarity between KS.

Semantic vectors are automatically created using project-related knowledge, using a process which collects words and expressions, to be matched against the equivalent terms which represent the ontological concepts. This produces an inventory of: (i) the number of equivalent terms matched at each ontological concept; and (ii) the total number of equivalent terms necessary to represent the harvested knowledge. This inventory provides the statistical percentage of equivalent terms belonging to each ontological concept represented in the universe of harvested knowledge. This step represents, the calculus of the 'absolute' semantic vector of a given KS, taking into account the equivalent terms-based percentages.

However, the approach presented here also considers a configurable hierarchy of KS relevance, as part of the creation of semantic vectors. This hierarchy is defined using 'relative' semantic factors to all types of KS, which ranges respectively from low relevance (0) to high relevance (1) for the context creation. Both hierarchy and relative semantic factors can be changed if necessary, depending on what KS are considered most relevant for the indexation process. For illustrative purposes only, an example of this hierarchy could be: issues (1), solutions (1), experts (0.7), Post-mortem (0.7), etc.

The final step, which comprehends the semantic evaluation, also includes ontological concepts that are not linked to the knowledge gathered, but have a semantic relationship of proximity with a relevant (heavy) ontological concept. This is done through the definition of a secondary semantic factor to ontological concepts based on their relative distances, inside the ontology tree.

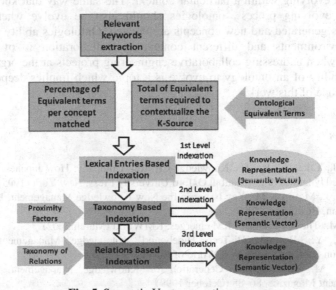

Fig. 5. Semantic Vectors creation process

Summing up, the final calculation of the semantic vector includes: statistical percentages based on the equivalent terms, the hierarchy of relevance for KS, and the weight assigned to the proximity level.

As referred previously, semantic vectors are continuously updated through the project's life cycle, and even in project's post-mortem. This is done in order to maintain the semantic vector's coherence with the level of knowledge available. Semantic vectors are automatically created: (i) whenever a new KS is gathered; and (ii) to help answering queries issued by the users.

7   Conclusions and Future Work

This work brings a contribution focused on collaborative engineering projects where knowledge engineering plays the central role in the decision making process.

Key focus of the paper is the indexation and retrieval of knowledge sources provided by semantic services enabled by a domain ontology. This work specifically addresses collaborative engineering projects from the Construction industry, adopting a conceptual approach supported by knowledge-based services. The knowledge sources indexation process is supported using a semantic vector holding a classification based on ontological concepts. Illustrative examples showing the process are part of this paper.

When addressing collaborative working environments, there is a need to adopt a semantic description of the preferences of the users and the relevant knowledge elements (tasks, documents, roles, etc.). In this context, we foresee that knowledge sources can be semantically enriched when adopting the indexation process described within this work

Ontologies which support semantic compatibility for specific domains should be adaptive and evolving within a particular context. The same way that knowledge by itself is an evolving process, ontologies should also be evolve whenever new knowledge is generated and new concepts are created. Ontologies ability to adapt to different environments and different context of collaboration is of extremely importance, when addressing collaborative engineering projects at the organizational level. The ability of an ontology to evolve is a topic which implies deeper research within the scope of this work.

References

1. Nonaka, I., Takeuchi, H.: The Knowledge-Creating Company: How Japanese Companies Create the Dynamics of Innovation. Oxford University Press, New York (1995)
2. Firestone, J., McElroy, M.: Key Issues in the New Knowledge Management. Butterworth-Heinemann, Burlington (2003)
3. Koenig, M.: The third stage of KM emerges. KMWorld (March 2002)
4. Malhotra, Y.: Beyond 'Hi-Tech Hidebound' Knowledge Management: Strategic Information Systems for the New World of Business (1999)
5. McElroy, M.: The Second Generation of Knowledge Management. Knowledge Management Magazine, 86–88 (October 1999)
6. Dalkir, K.: Knowledge Management in Theory and Practice. Elsevier, Oxford (2005)

7. Zeeshan, A., Chimay, A., Darshan, R., Carrillo, P., Bouchlaghem, D.: Semantic web based services for intelligent mobile construction collaboration. In: ITcon, pp. 367–379 (2004)
8. Lassila, O., Adler, M.: Semantic Gadgets: Device and information interoperability, Cleveland (2003)
9. Costa, R., Lima, C., Antunes, J., Figueiras, P., Parada, V.: Knowledge Management Capabilities Supporting Collaborative Working Environments in a Project Oriented Context. In: European Conference on Intellectual Capital, Lisbon, pp. 208–216 (2010)
10. Kazi, A.S.: Knowledge Management in the Construction Industry: A Socio- Technical Perspective. Idea Group Publishing (2005)
11. Gruber, T.: Toward Principles for the Design of Ontologies Used for Knowledge Sharing. International Journal Human-Computer Studies 43(5-6), 907–928 (1995)
12. Mangold, C.: A survey and classification of semantic search approaches. Journal of Metadata, Semantics and Ontology 2(1), 23–34 (2007)
13. BuildingSMART. IFD Library (December 2010), http://www.ifd-library.org/
14. OCCS. OmniClass (2011), http://www.omniclass.org/
15. Lima, C., Fies, B., El Diraby, T., Lefrancois, G.: The challenge of using a domain Ontology in KM solutions: the e-COGNOS experience. In: International Conference on Concurrent Engineering: Research and Applications, Funchal, pp. 771–778 (2003)

Designing Ontology-Driven System Composition Processes to Satisfy User Expectations: A Case Study for Fish Population Modelling

Mitchell G. Gillespie[1], Deborah A. Stacey[1], and Stephen S. Crawford[2,3]

[1] School of Computer Science, University of Guelph, 50 Stone Rd East, Guelph, ON, Canada
[2] Dept. of Integrative Biology, University of Guelph, 50 Stone Rd East, Guleph, ON, Canada
[3] Chippewas of Nawash Unceded First Nation, RR #5, Wiarton, ON, Canada

Abstract. Ontology-Driven Compositional Systems (ODCSs) are designed to assist a user with semi- or fully automatic composition of a desired system utilizing previously implemented algorithms and/or software. Current research with ODCSs has been conducted around the discovery and composition of web services and a resource management approach. This chapter utilizes the collaboration with a Fish Population Modelling research group to argue that current ODCSs do not fully consider a users' expectations of [a] his/her leverage and acquisition of knowledge from the ODCS, and [b] the trustworthy, high quality, and efficient performance of desired resultant systems. The authors support their argument by acknowledging that the current semantic frameworks have yet to fully represent the knowledge required to make proper discovery, decision-making, and composition. The authors introduce the beginning of their work of utilizing the inheritance of multiple ontologies to fully represent the function, data, execution, quality, trust, and timeline semantics of compositional units within an ODCS. Finally, a case study is utilized to illustrate how a more robust representation model will improve the satisfaction of the user's expectations.

1 Introduction

For many years, stakeholders have utilized previously implemented and modular algorithms/packages to minimize the cost of time and effort for software developers when designing a new software system. This is strongly exemplified in the Open Source community that supports software developers with freely available and downloadable source code [1]. Similarly, the web service and intelligent agents domains allow developers to manually connect (*i.e.* service consumption) to remote services via their own distributed system [2]. Recent research in this area has shifted focus to understand the semantic knowledge requirements and processes to assist with automatic or semi-automatic system composition using a collection of previously developed algorithms, modules, services and packages.

The various semantic knowledge requirements may or may not have an explicit emphasis based on previously implemented software/services being automatically composed. Regardless of this, a mechanism must be utilized to properly represent the semantic knowledge that will drive the discovery, decision-making, and composition processes of an expert system that assists a user. Ontologies are explicit specifications

A. Fred et al. (Eds.): IC3K 2010, CCIS 272, pp. 290–304, 2013.

of a conceptualized body of knowledge, and commonly utilized to understand the sharing of knowledge among people and/or software agents [3]. Thus, ontologies are recognized as a appropriate tool to represent the desired semantic knowledge. Defined within this chapter as *Ontology-Driven Compositional Systems*, current research has focused on a ontology-driven web service composition platform [4], and a resource-management based compositional system using ontologies [5]. Thus far, ODCS research has focused primarily on the discovery and composition of previously developed software, and rarely considers ranking and selection previously developed software to decide which to use. A ranking and selection process requires an understanding of the end-user's expectations to evaluate whether or not a generated resultant system from the ODCS is satisfactory.

Through collaboration with the Integrative Biology Department at the University of Guelph and the Chippewas of Nawash Unceded First Nation fisheries management program [6,7,8], the authors isolated a case study where they can measure user expectations. Why fisheries population modelling? In most cases, fisheries biologists lack the mathematical and computational knowledge to model fish populations properly [6] and ODCSs could provide the appropriate knowledge to allow the successful construction and execution of population models. Furthermore, as a fisheries biologist's knowledge is specialized in population dynamics, their expectations of a composed population modelling system will be more complex than whether or not the computer population models compile and execute. Domain-specific metrics of quality and trust could affect a fisheries biologist's acceptance of certain model components over another.

Utilizing the case study in fisheries population modelling, the authors will illustrate how satisfying user expectations in an ODCS accommodates various research initiatives. Explicitly, compositional systems researchers [9,4,5] will further understand the dynamics involved in the creation of practical applications for ODCS. Also, fisheries management experts will be introduced to an innovative tool to assist with the composition of population models. Implicitly, these population modelling experts will also be embracing a framework for representing instances of population models.

In this paper, current implementations of Ontology-Driven Compositional Systems are assessed, followed by a presentation of the work being conducted by the authors. Section 2 introduces a description of Ontology-Driven Compositional Systems emphasizing the semantic knowledge to be represented in the ontologies and the required processes to provide a functional ODCS. Section 3 presents current implementations while Section 4 argues a general assessment of those implementations. Section 5 presents aspects of User Expectations desired in the design of an Ontology-Driven Compositional System to enhance the quality and trust of the resulting system's outputs. Using the desired user expectations, Section 6 introduces the current work by the authors, including the fisheries population modelling case study. Finally, discussion of future work is presented.

2 Defining Ontology-Driven Compositional Systems

An *Ontology Driven Compositional System (ODCS)* has previously been defined as "a [...] system [...] that demonstrates the power and suitability of using ontologies as the

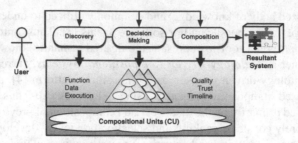

Fig. 1. Discovery, Decision-Making and Composition processes in an Ontology-Driven Compositional System utilizing the semantic knowledge (represented by ontologies) of the Compositional Units available to create a Resultant System for the user to execute. Compositional Units may or may not be locally stored and/or managed (*i.e.* a downloadable package would be locally stored by the compositional system, whereas a web service would not).

main driver for a compositional system" [5]. Simply, it performs *Computer-Aided System Composition*; a process by which an expert system assists a human user with the construction of two or more previously implemented *Compositional Units* (*e.g.* algorithms, packages, services) to create a *Resultant System*. *Compositional Units (CUs)* are defined as functional "black-box" algorithms, packages or services that receive a given set of input, provide a described function or service and sends a set calculated output (see figure 2). The *Resultant System* is the final output of interconnected CUs to be executed by the user, thus providing "yet another functional application" [5].

To satisfy its responsibility, an ODCS must conduct semantic reasoning on a collection of knowledge (*i.e.* ontologies) within a set of processes to construct an outputted resultant system. Within the sub-sections below, we introduce possible semantic knowledge requirements and the set of processes that should be considered for an ODCS (also illustrated in Figure 1).

2.1 ODCS Knowledge: Six Types of Semantics

All ODCS hold some form of semantical knowledge about the compositional units they wish to utilize [10,9]. For web services specifically, work has described four different types of semantical knowledge to be considered by web service compositional systems. This chapter adapts the work of [9] to provide six different knowledge semantics to be considered in all forms of compositional systems, ontology-driven or otherwise:

1. **Function Semantics.** knowledge about the function, features and computational purpose of the compositional units and input/output data.
2. **Data Semantics.** knowledge about the order, format and structure of the input/ output for a given compositional unit.
3. **Execution Semantics.** knowledge about locations, requirements and environment dependancies that need to be utilized or satisfied for a compositional unit to be executed in a resulting system.
4. **Quality Semantics.** knowledge that characterizes performance and other quality-specific metrics of how a given compositional unit performs (adapted from "Quality of Service – QoS Semantics").

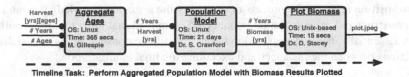

Timeline Task: Perform Aggregated Population Model with Biomass Results Plotted

Fig. 2. A simple example of interconnected compositional units that make up a Resultant System. Five knowledge semantics describe various characteristics of each compositional unit: Function (underlined), Data (purple), Execution (blue), Quality (red), and Trust (green). The overall flow of interconnected compositional units to perform the desired task(Timeline) is emphasized by the dashed process line.

5. **Trust Semantics.** knowledge that distinguishes reputation-based properties of a compositional unit (*e.g.*, name of developer, expert opinion, user ratings, etc.)
6. **Timeline Semantics.** knowledge about the chronological and compositional requirements to create sections or full implementations of a resultant system.

The Functional, Data, Execution, Quality and Trust semantics represent the knowledge about each given instance of a compositional unit, whereas the Timeline semantics are utilized for describing appropriate combinations of compositional units. The collection of the six semantics allow for unique specification of each compositional unit (functional, data, quality & trust, and execution), and combinations of compositional units to perform a given task.

The example illustrated in figure 2 illustrates how a computer-programmed fish population model is a possible instance of a compositional unit (CU). Furthermore, data aggregation routines and data plotting tools are also considered compositional units as they assist the overall statistical methodology of modelling a fish population. Within this illustration, each compositional unit instance is characterized by its unique semantics (Function, Data, Execution, Quality, Trust), and a holistic depiction of the entire modelling process is provided (Timeline).

2.2 ODCS Processes: Discovery, Decision-Making and Composition

Within an ODCS there are three general processes that need to be considered in some manner (adapted from [9]): discovery, decision-making, and composition. Each of the three processes may or may not have a certain emphasis based on the focus of the ODCS.

Discovery refers to the process of accessing the semantic knowledge within the ontologies to *search* and *match* specifications/requirements of a compositional unit instances for a resultant system. For the workflow of compositional units depicted in figure 2, an ODCS would be required to search for "Fish Population Model" compositional units and match the instances that receive a vector of fish harvested by year and outputs a vector of predicted fish biomass over a set number of years.

Decision-making refers to the process of accessing the semantic knowledge within the ontologies to *rank* and *select* discovered compositional units by investigation qualitative/quantitative attributes or metrics that may or may not be based on user preference. Utilizing the figure 2 example again, two different instances of compositional units for data aggregation could execute the same function. Yet, one may perform three times more efficient than the other.

Composition refers to the process of accessing the semantic knowledge to automatically *construct* and *execute* the interconnected compositional units within a resultant system. Essentially, the flow depicted in figure 2 would be constructed by materializing connections between the instances and executing the final configuration.

3   Current Implementations of ODCS

Certain computational characteristics of the compositional units (*e.g.* a web service or downloadable package) affect the specific focus of a compositional system. Two variations of ODCSs are presented: Ontology-Driven Web Service Composition Systems [2,4] and a "Plug-and-Play" Ontology-Driven Composition System [5].

3.1   Ontology Driven Web Service Composition

With the growing popularity of distributed web applications, a strong research focus on the delivery of effective "Semantic Web Services" is under way. From this research, [9] emphasized the importance of ontologies to represent the semantics of web services. This work introduced four knowledge semantics to facilitate semi- or fully-automatic composition, which many research foci have embraced [2,4].

One research initiative considered all processes presented in Section 2.2 with specific application to a winery, web service case study [4]. The system architecture of their ODCS utilized three ontologies to drive the composition: a *Domain ontology*, a *Web Services ontology*, and a *Process ontology* (Figure 3(a)). The *Web Services ontology* utilized work from [12] to represent instances of web services like 'Wine-Searcher' (sub-class of 'Searching Service'). The Domain ontology gave semantic context to the web service input and output, and the *Process Ontology* represented sequences of services that were connected to one another if their interfaces were semantically matched. The quality of web services was briefly considered through proposing the integration of a set of quality criteria [13], however [4] recognized quality as a feature that had to be further developed and researched for their system.

3.2   Resource Management Focused Ontology-Driven Compositional System

A resource-management focus by [5] approached the design of an ODCS in a different manner. The overall goal of their ODCS was to create a "plug & play" hub of syndromic surveillance algorithms (i.e. compositional units) to assist user to explore the discovery, decision-making, and composition of various resultant systems. All algorithms were warehoused on the ODCS system itself to conceptually allow users to download the resultant system of units onto their own executional environment.

The semantic knowledge for this ODCS was represented in two ontologies: an Algorithm ontology and Execution Timeline ontology (See figure 3(b)). The Algorithm ontology provided an extensive representation of syndromic surveillance algorithms (including input/output requirements) and architectural components to help "glue" together input/output of algorithms that could not be connected otherwise. Similar to the Process ontology (from [4]), the Execution Timeline ontology provided semantic descriptions to assist with the composition of events.

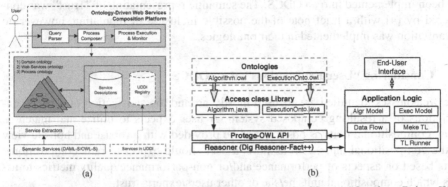

Fig. 3. System Architectures of two current implementations ODCSs: (a) "Ontology Driven Web Service Composition Platform" [4], and (b) Resource Management focused ODCS [11]

4   Assessment of Current Implementations

After a brief investigation of recent implementations of an Ontology-Driven Compositional Systems, the authors identified important strengths and weaknesses:

Strength: Methods of Discovery and Composition. Both implementations [4,5] focused most of their efforts on discovery and composition. As ODCSs are in their youth, a prototype with minimal complexity is optimal. Both ODCSs utilize a unique ontology to assist with defining the combination of compositional units required for a certain resulting system requested by the user. The Input/Output matching algorithms have been implemented successfully, however a larger knowledge base of compositional units, input/output and resulting systems would provide opportunities for more robust matching. [5] accepted this recognition by adding functional semantic knowledge that described "architectural" compositional units to act as *glue* to match more input/output.

Weakness: Representation of Domain-specific Functional Semantic Knowledge.
The functional semantic knowledge of compositional units and their input/output has been limited to the use of a small *domain ontology*, or a minor inheritance relationship in one ontology. Creating a more robust functional semantic knowledge representation of compositional units has been proposed, but not investigated [4]. Furthermore, as the number, variability and specification of different types of compositional units increase, more attributes and descriptors will be necessary to distinguish the difference between which compositional units are desired by the user.

Weakness: Quality & Trust of Compositional Units. Currently, only minor efforts have been attempted to rank and select compositional units in the "Decision-Making" process in an ODCS [5,4]. The ranking of quality was focused mostly on computational performance-based metrics and did not consider domain-specific and qualitative-like performance metrics. Research within the "Semantic Web Services" [9] domain has focused on improving ranking of quality [13,14,15], however that research has not yet been implemented into an ODCS. The semantic representation of "Trust" was considered by [4] with a brief note of the possible inclusion of reputation, however no instantiation was implemented in their ontologies.

4.1   Considering "User Expectations" of an ODCS

The weaknesses identified are focused mostly around the satisfaction of a user's expectations of a resulting composed system. A user expects to utilize and match as many compositional units as possible and be provided with a robust and holistic view of the compositional units. Furthermore, a user should expect to rank compositional units based on aspects of performance and/or non-performance quality metrics to assess which compositional units he/she or other users/experts trust.

5   Design to Satisfy User Expectations of ODCSs

The authors asked themselves, "How may aspects of satisfying a user's expectations of software can be incorporated into the design of an ODCS?". In some fashion, a user must be interacting with any type of ODCS to somehow describe the desired resultant system before it is composed. This interaction provides an ODCS with the opportunity to establish what a user expects of the resultant system. With this understanding the authors consider two facets for "satisfying user expectations": [a] satisfying the expectations of the actual interaction between the user and the interface of an ODCS, and [b] satisfying the expectations a user will have of a resultant system by the incorporation of those expectations in the discovery, decision-making, and composition processes. Within this section, the authors examine a few heuristics to emphasize the importance of the "satisfying user expectations" facets. Obviously, more heuristics should be considered, however for the context of this chapter only a few are exemplified. These examples focus on expectations that are directly related to the characteristics and construction of an ODCS, and not simple user interface usability heuristics.

5.1   Facet 1: User Expectations of Interaction with ODCS User Interface

Leveraging Various Expert Knowledge. In general, users are more satisfied with a system when cognitive processing and knowledge requirements are decreased [12]. Considering the interaction with an ODCS specifically, every user has a varying set/level of predetermined knowledge. For example, one instance of a user could have comprehensive knowledge of programming & software engineering, and would therefore understand execution semantic requirements of compositional units. Alternatively, an instance of a fisheries manager user understands population dynamics of his/her target harvest, however will need to strongly leverage computational, mathematical, and statistical knowledge about the population models, and the other possible interconnected compositional units.

With the examples provided, the authors argue that two tenets of expert knowledge could possibly be leveraged: software/system programming expert knowledge, and domain-specific expert knowledge. In terms of the semantic knowledge presented in Section 2.1, each type of knowledge may or may not need to consider the different tenets of expert knowledge.

The distribution of *software/system programming expert knowledge* should be heavily emphasized in the data, execution, and timeline semantics, moderately in function and quality semantics, and the lightest in trust semantics. In terms of *Discovery* and *Composition* (the most important processes of an ODCS), each discovered compositional unit must have its executional requirements satisfied or it will fail to execute when it is composed. Similarly, if the interconnected (*i.e.* timeline) input/output of compositional units does not follow a certain data semantic structure/format the units will not execute properly. Function and quality semantics do not necessarily effect actual execution, however certain semantic knowledge relates to a software development expert perspective compared to domain-specific (*e.g.*, average performance with specific types of CPUs).

The distribution of *domain-specific expert knowledge* in the different types of semantic knowledge is essentially opposite of the software/system programming expert knowledge. Function, quality, and trust semantics are heavily emphasized, while data and timeline semantics are represented moderately. A very light description of executional semantics would be required for any domain-specific expert knowledge. Considering the more functional focused *Discovery* and *Decision-making* processes, two compositional units may execute similarly with the same input/output requirements, however hold an extreme difference in their functional purpose and sub-features. Also, aspects of quality and trust could be a mutually exclusive representation compared to software/system metrics.

Acquisition of New Knowledge. The main purpose of an ODCS is to allow users to compose previously implemented units together. As a user becomes more comfortable with this type of service, s/he may consciously or unconsciously be acquiring new expert knowledge (software/system programming and domain-specific). By exploring and experimenting with different combinations of compositional units (*e.g.*, a fisheries manager trying different population models) a user's expectations could be satisfied by discovering units they would have otherwise not considered. Overall, this concept is very congruent to "leveraging knowledge".

5.2   Facet 2: User Expectations of Resultant Systems

Acknowledgment of Trust. It has become a general consensus that "if [users] do not trust the new automated tools, they will not use them no matter how useful or efficient they might be" [16]. Within an ODCS, a user would expect to *select* compositional units that have been tested and accepted by software and/or domain-specific experts. Furthermore, if a user favours a specific expert, then compositional units created or recommended by that expert should hold a higher ranking.

Following the same considerations as "leveraging knowledge", different sets of expert semantic knowledge exist. Attributes of trust from software/system programming expert knowledge would consider information like the name of the software developer, number of times it has been executed, reputation of security, minimal number of errors, number of bug fixes, etc.. Whereas, domain-specific expert knowledge would consider trustworthy metrics that are related to its area. For example, a trustworthy fish population modelling metric may be the number of successful simulation studies conducted on a model.

Satisfaction of High Quality and Performance. Ultimately, the resulting system generated from a compositional system is still a piece of computational software. Therefore, most users will still consider performance metrics to be extremely important. Programming specific features like estimated time of execution, amount of memory required or, distance of distributed machines will always be important. Simple metrics and protocols like QoS can be utilized. Domain-specific performance proves slightly more difficult, as the quality of performance could be measured by different numerical representations. For example, many fisheries management population models may only produce a successful modelling simulation some of the time due to factors such as data availability, data quality, or model complexity [17].

6   First Mash-Up of ODCS Motivated by User Expectations

Overall, what has this chapter presented and acknowledged thus far?

- *Describing an ODCS:* an ODCS conducts the discovery, decision-making, and composition of six different types of semantic knowledge (function, data, execution, quality, trust, timeline) to create resultant systems composed of interconnected compositional units.
- *Current ODCSs:* different implementations of ODCS exist with varying levels of automatic processes and semantic knowledge representation. Each also focus on certain types of compositional units (*i.e.* algorithms or web services).
- *Assessment:* the ODCSs presented in this chapter strongly implement their respective forms of discovery and composition, however only a small amount of domain-specific expert knowledge is represented. Methods of ranking and selection (i.e. decision-making) could be improved with the inclusion of more expert knowledge.
- *User Expectaions of ODCS:* the authors propose centring focus on satisfying user expectations of: [a] the user interaction with the ODCS, and [b] the decision-making and composition of the resultant system.

With these perspectives, the authors have begun to investigate the implementation of an ODCS that centres its motivation around satisfying user expectations. With focus on the "expectations" presented in Section 5, the first mash-up design emphasizes the planning and implementation of the semantic knowledge representation required. Through the case study, the authors will briefly illustrate how the proposed semantic knowledge representation will assist to leverage/acquire knowledge and acknowledge trustworthy and high quality resultant systems.

First and foremost, the mash-up presented adapted work from [5] to utilize their "plug and play" design. This allows a user the preference of experimenting with different combinations of compositional units, thus providing opportunity to acquire more

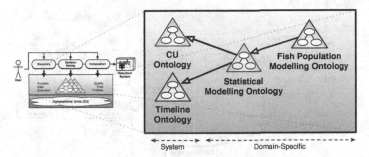

Fig. 4. An illustration of merging and inheritance of a "network" of ontologies for the semantic knowledge representation requirements of software/system programming and domain-specific expert knowledge. The depiction of figure 1 in the background illustrates how the discovery, decision-making, and composition processes could all possibly access the "network" of ontologies. A Statistical Modelling ontology and Fish Population Modelling ontology are exemplified as the domain-specific expert knowledge.

knowledge. Figure 4 presents how the structure of the ontologies (*i.e.* semantic knowledge representation) are engineered in the ODCS mash-up. This illustrates how a "network" of domain-specific ontologies could merge and inherit entities of knowledge from a generalized *Compositional Unit (CU) ontology* and *Timeline ontology*. The "network" of domain ontologies is a proposed attempt to facilitate the ability to leverage the two different forms of expert knowledge; which is further investigated by the case study.

For the remainder of the section, the CU ontology is presented. The Timeline ontology has not yet been adapted by the authors, thus will not be considered for the rest of the chapter. Following this section, the case study investigates how the proposed semantic knowledge structure satisfies user expectations discussed earlier. The chapter is concluded with closing discussions of future work.

6.1   CU Ontology

The CU ontology was adapted from [5] by evolving their Algorithm ontology into what this chapter defines as the Compositional Unit ontology (CU ontology). The change of name to "Compositional Unit" was necessary since the term "Algorithm" does not semantically define all possible units of composition. This ontological representation model is engineered in such a way to modularize the function, data, execution, quality, and trust semantics to describe the holistic context of compositional units. Figure 5 provides an illustration of some of the entities within the CU ontology. As the focus of this mash-up is the satisfaction of user expectations on the discovery and decision-making processes, the execution and timeline semantics are not considered. The remaining four semantics represented in the CU ontology are presented below.

Function Semantic Knowledge. The function semantic knowledge represented in the CU ontology attempts to provide an explanation of the features, elements, and compositional purpose of the compositional units. The term "features" refers to high-level characteristics of a given instance of a compositional unit and the "elements" refers to

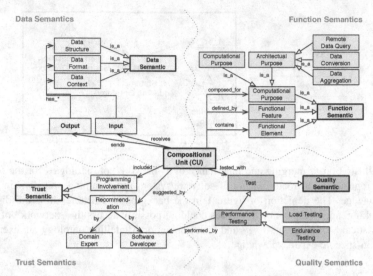

Fig. 5. A few of the essential entities in the prototype Compositional Unit Ontology (CU ontology). The CU Ontology is proposed to be the base system ontology that is utilized along with the Timeline ontology to compose systems. As Figure 4 illustrates, domain-specific ontologies inherit and merge with the CU ontology to allow the classification of specific domain knowledge. Section 7 presents a case study of this paradigm.

functional characteristics *within* an instance of a compositional unit. Finally, "compositional purpose" refers to the reason why compositional units are presented. Similar to [5], most compositional units are "computational" components that provide services described by the "features" and "elements", however some act like *glue* and are semantically represented as "architectural".

Data Semantic Knowledge. Most of the data semantic knowledge is utilized during the matching of input/output (*i.e.* discovery) of the compositional units for a resulting system. Both the input/output could have defined format, structure, and/or type; format refers to string-like protocols that are followed (*i.e.* regular expressions, or string date formats), and structure refers to the method by which the data is represented (*e.g.* matrix, vector, etc.).

Quality and Trust Semantic Knowledge. The quality and trust semantic knowledge in the CU ontology focuses on characteristics that provide the ability to rank and select the compositional unit. Information like positive and negative feedback could dictate whether an end-user would select the given compositional unit. Alternatively, an end-user may not trust a certain expert or developer, therefore any feedback or tests by that expert or developer should be ignored in a ranking process.

7   Case Study: Fisheries Population Modelling

Fisheries managers understand that the anthropogenic effects of harvesting must be studied in a transparent, accountable and scientifically-defensible manner [18].

Fisheries biologists are specifically responsible for collecting and analyzing data to estimate fish population abundance, assess human and non-human sources of mortality, and to ensure sustainable harvest levels [19]. Unfortunately, many fisheries management decisions are based largely on qualitative indicators or expert opinion of fish population status, rather than estimates derived from more complex mathematical modeling tools developed over the past 30 years [20,17]. In some cases, the biologists are constrained by the quantity and/or quality of data required for the mathematical models, while in other cases the biologists are themselves limited in their understanding of how these models function [21].

Through a unique fisheries research collaboration between the Chippewas of Nawash Unceded First Nation and the University of Guelph, the authors have investigated the domain of fisheries population models to understand how fisheries biologists could: [a] leverage and acquire the computational, mathematical, and statistical knowledge to appropriately employ the population models, and [b] enhance their level of trust in the quality and performance of the generated resultant system. The following two sub-sections present how the authors' ODCS could be utilized to achieve the stated objectives. The first sub-section introduces an example of how fisheries managers could leverage a network of expert knowledge through the inheritance and merging of the multiple domain ontologies. The second sub-section provides a specific focus on quality and trust semantic knowledge and how it has been considered through the multiple domain ontologies as well.

7.1 Leveraging Knowledge for Fish Population Models

Fisheries population models are mathematical/statistical models implemented as computer algorithms specifically to focus our understanding on important factors that determine population abundance and condition [22]. Therefore, two levels of domain knowledge need to be considered past the semantic knowledge represented in the Algorithm Ontology. As shown in Figure 6, the first domain of knowledge considered is Statistical Modeling, followed by the domain of Fish Population Modeling itself. Considering functional semantic knowledge in particular, the Statistical Modeling ontology describes high level functional features such as spatial/temporal dimensions or stochastic/deterministic processes and similar functional features of data, such as whether or not given parameters are sampled using normal, log-normal or uniform distributions. The Fish Population Modeling ontology would hold functional features such as mortality and/or recruitment implementations utilized, and functional features of data such as the functional purpose of its input (i.e. mortality, recruitment, etc.). Figure 6 provides an example of how the two domain ontologies in this case study inherit knowledge and merge with the CU ontology.

7.2 Incorporating Quality and Trust

Within each domain presented in this case study, different types of metrics need to be considered to measure and rank aspects of quality and trust specifically. As shown in Figure 6, statisticians would trust a given population model CU more if simulation studies had been conducted and fisheries managers may wish to trust population models that

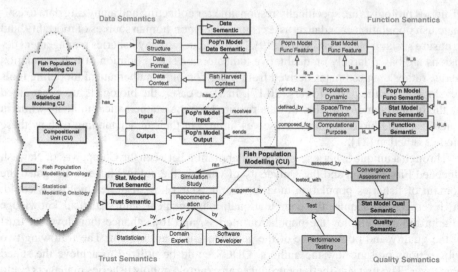

Fig. 6. A case study of fisheries population modelling: an example of how multiple ontologies could inherit the CU ontology and further describe the functional, data, quality and trust of the compositional units within their own domain

have gotten feedback from expert population modelers whom they trust. More complex metrics of trusting quality and performance are also included, however, these are not considered in the examples provided here.

8   Discussion of Future Work

The authors acknowledge that a large amount of work still needs to be dedicated towards their research and in the ODCS academic domain in general. Currently, three different foci of future work are being undertaken by the group: the design of a robust ontological framework to represent more than compositional unit knowledge, a strategy for ontology evaluation, and the construction of a reusable graphical user interface.

8.1   Compositional Unit Ontological Representation Is Not Enough

In recent months, a collaborative research group called the Guelph Ontology Team (GOT) have acknowledged that many more knowledge representation requirements exist for the proper implementation of an ODCS. Recently submitted to academic literature [23], the group illustrated a framework to distinguish five categories of syntactic and semantic knowledge that could be represented: Compositional Units, Workflow, Data Architecture, Human Actors, and Physical Resources. Within each category five subtypes of semantic knowledge exist: function, data, execution, quality, and trust. "Timeline" semantics labelled in this work, is now represented within the Workflow category of knowledge. Relationships between the categories of knowledge are also being investigated. Currently, this framework is being utilized to restructure the ontological structure presented in this chapter. The authors are hopeful to continue collaboration with the Integrative Biology Department and the Chippewas of Nawash Unceded First Nation.

8.2 Strategy for Ontology Evaluation

A structured "golden standard" for ontology evaluation methods is a concept that will likely never be realized. Within ontology engineering methodologies it is very difficult to argue whether or not a given ontological model is properly representing the given knowledge domain [24,25]. With the combination of the knowledge framework with five categories of knowledge (Section 8.1) and ontology evaluation strategies [26], the GOT research group is discussing methods of utilizing a combination of software/system programmers, domain experts, and end-user focus groups to investigate and evaluate the adaptability [27] and flexibility [25] of the ontologies presented in this work.

8.3 Reusable Graphical User Interface

Again, one of the main goals of this chapter is to design an ODCS motivated by satisfying user expectations of his/her interaction with the ODCS and the resultant system. In terms of evaluation, an overall assessment of usefulness and satisfaction for end-users needs to be approached. Currently, a very low level interface is implemented for the current mash-up. The GOT research group has nearly completed the construction of a reusable prototype interface that could be utilized by any knowledge domain given that instances of knowledge are represented within the ontology.

Acknowledgements. The authors would like to acknowledge all of the collaboration with the Integrative Biology Department at the University of Guelph and the Chippewas of Nawash Unceded First Nation, specifically Chief & Council, Scott Lee, Ryan Lauzan, Dan Gillis, and Jasper Tey. Mitchell Gillespie would like to acknowledge the patience, kindness, assistance and opportunities that Dr. Deborah Stacey has provided during his work as a graduate student. Furthermore, Mitchell welcomes the new members to the collaborative research group (Guelph Ontology Team) as he is very excited for future endeavours. Finally, he cherishes all of the support he receives from his partner (Kathryn Marsilio), and his family (Katie, Tom, and Mugz). Deborah Stacey wishes all of her graduate students were so dynamic and productive.

References

1. Feller, J., Fitzgerald, B.: Understanding Open Source Software Development, vol. 84. Addison-Wesley (2002)
2. Meng, X., Junliang, C., Yong, P., Xiang, M., Chuanchang, L.: A Dynamic Semantic Association-Based Web Service Composition. In: Proceedings of the 2006 IEEE/WIC/ACM International Conference on Web Intelligence (2006)
3. Gruber, T.R.: A translational approach to portable ontology specifications. Knowledge Acquisition 5, 199–220 (1993)
4. Arpinar, I.B., Zhang, R., Aleman-Meza, B., Maduko, A.: Ontology-driven Web services composition platform. Information Systems and e-Business Management 3, 175–199 (2005)
5. Hlomani, H., Stacey, D.A.: An ontology driven approach to software systems composition. In: International Conference of Knowledge Engineering and Ontology Development, pp. 254–260. INSTICC (2009)

6. Gillis, D., Tey, J., Gillespie, M., Crawford, S.: Do fisheries biologists have appropriate tools for assessing dynamics of harvested fish populations? (2010)
7. Crawford, S., Gillis, D., Rooney, N.: A review of population level risk assessments for the CANDU Owners Group (2008)
8. Crawford, S., Muir, A., McCann, K.: Ecological basis for recommendation of 2001 Saugeen Ojibway commercial harvest TACs for lake whitefish (Coregonus clupeaformis) in Lake Huron, Report prepared for Chippewas of Nawash First Nation (2001)
9. Cardoso, J., Sheth, A. (eds.): SWSWPC 2004. LNCS, vol. 3387, pp. 1–13. Springer, Heidelberg (2005)
10. Srivastava, B., Koehler, J.: Web Service Composition – current solutions and open problems. In: ICAPS 2003 Workshop on Planning for Web Services (2003)
11. Hlomani, H.: A Bottom-Up Approach to System Compositon using Ontologies. Msc., University of Guelph (2009)
12. Zhang, P., Von-Dran, G.M.: User Expectations and Rankings of Quality Factors in Different Web Site Domains. International Journal of Electronic Commerce 6, 9–33 (2002)
13. Zeng, L., Benatallah, B., Dumas, M., Jayant, K., Sheng, Q.Z.: Quality Driven Web Services Composition. In: Proceedings of the 12th International Conference on the World Wide Web, pp. 411–421. ACM (2003)
14. Tran, V.X.: WSQoSOnto: A QoS Ontology for Web Services. In: 2008 IEEE International Symposium on Service-Oriented System Engineering, pp. 233–238. IEEE (2008)
15. Wang, X., Vitvar, T., Kerrigan, M., Toma, I.: A QoS-Aware Selection Model for Semantic Web Services. In: Dan, A., Lamersdorf, W. (eds.) ICSOC 2006. LNCS, vol. 4294, pp. 390–401. Springer, Heidelberg (2006)
16. Duez, P.P., Zuliani, M.J., Jamieson, G.A.: Trust by design. ACM Press, New York (2006)
17. Methot, R.J.: Stock assessment: operational models in support of fisheries management, pp. 137–165. Springer Science, Netherlands (2009)
18. Walters, C., Martell, S.: Fisheries ecology and management. Princeton University Press, New Jersey (2004)
19. Quinn, T., Deriso, R.: Quantitative fish dynamics. Ofxord University Press, New York (1999)
20. NRC: Improving Fish Stock Assessments. Committee on Fish Stock Assessment Methods, U.S. National Research Council. National Academy Press (1998)
21. Stringer, K., Clemens, M., Rivard, D.: The changing nature of fisheries management and implications for science, pp. 97–111. Springer Science, Netherlands (2009)
22. Hilborn, R., Walters, C.: Quantitative fisheries stock assessment: choice, dynamics and uncertainty. Chapman and Hall, New York (1992)
23. Gillespie, M.G., Hlomani, H., Stacey, D.A.: A Knowledge Identification Framework for Engineering Ontologies in System Composition Processes (Submitted to Conference) (2011)
24. Uschold, M., King, M.: Towards a Methodology for Building Ontologies. In: Workshop on Basic Ontological Issues in Knowledge Sharing, held in conjunction with IJCAI 1995 (1995)
25. Gangemi, A., Catenacci, C., Ciaramita, M., Lehmann, J.: A Theoretical Framework for Ontology Evaluation and Validation. In: Proceedings of SWAP 2005 - Semantic Web Applications and Perspective (2005)
26. Vrande, D.: Ontology Evaluation, pp. 293–313. Springer, Heidelberg (2009)
27. Obrst, L., Ceusters, W., Mani, I., Ray, S., Smith, B.: The Evaluation of Ontologies, pp. 139–158. Springer, Berlin (2007)

Part III

Knowledge Management and Information Sharing

Knowledge Sharing in the First Aid Domain through End-User Development

Daniela Fogli and Loredana Parasiliti Provenza

Dipartimento di Ingegneria dell'Informazione
Università di Brescia, Via Branze 38, Brescia, Italy
fogli@ing.unibs.it, loredana.parasiliti@gmail.com

Abstract. The paper addresses the knowledge sharing needs of an Italian non-profit association for first aid. Their volunteers, and particularly ambulance drivers, need to know the territory to provide first aid quickly and in a safe manner. This knowledge is often tacit and distributed. Paper-based maps are currently the means to spread and share knowledge among volunteers, while training sessions regularly provide information about holdups and fast routes to a place. The paper describes FirstAidMap, a collaborative web mapping system we have designed with volunteers to satisfy their needs. The system beyond supporting the training activity of ambulance drivers provides an interactive space that all volunteers can directly shape to build and share their knowledge about the territory. FirstAidMap integrates proper end-user development functionalities to engage and motivate volunteers to participate in map shaping, thus evolving from passive users to co-designers of map content. Results of a preliminary evaluation are discussed.

Keywords: Collaborative web mapping, Knowledge sharing, End-user development, Participatory design.

1 Introduction

In the era of participatory web or the so-called Web 2.0 [20], digital maps are increasingly the venue where people with different expertise can meet and share knowledge for a specific purpose. In this sense, digital maps can be conceived as social media - as suggested in [14] - where users acting on the map interact with other people directly or indirectly, by sharing and exchanging knowledge related to a territory. Web-based maps, such as Google Maps, are nowadays collaborative web mapping systems that allow users to visually define and shape virtual spaces by choosing what to map according to their own goals, knowledge and practices. The resulting map thus provides a living account of space as a social product of individual embedded knowledge, daily practices, and concerns [10].

In this sense, collaborative web mapping systems are intrinsically end-user development (EUD) environments, whose challenge is "to allow users of software systems who are not primarily interested in software per se to modify, extend, evolve, and create systems that fit their needs" [7]. To effectively address this challenge, collaborative web mapping systems should encompass socio-technical EUD mechanisms for supporting and encouraging user participation in contributing content and mapping space,

A. Fred et al. (Eds.): IC3K 2010, CCIS 272, pp. 307–321, 2013.

especially when the map represents a fundamental knowledge source sustaining users' daily practices. This was the aim we have pursued in the design and development of FirstAidMap, a collaborative web mapping system aimed to satisfy the needs of COSP (Centro Operativo Soccorso Pubblico), an Italian non-profit association for public assistance and first aid. FirstAidMap is a map-based system that allows members of the COSP association to acquire, create and share knowledge about the territory where COSP ensures its assistance. This knowledge is crucial for COSP ambulance drivers to decide, in case of emergencies, how to reach a given place quickly and in a safe manner. However, knowledge of the territory is often tacit and anyway distributed among COSP members, depending on their interests, attitudes, and experiences. FirstAidMap has thus been conceived as a virtual space that users (COSP members) can directly shape and enrich, thus actively building knowledge on the territory and share it within the community they belong to. While interacting with FirstAidMap to make this virtual space evolve, users behave as end-user developers: they modify 'at use time' the system to satisfy their individual and collaborative needs. Moreover, COSP representative users participated 'at design time' in system development. Their contribution was fundamental to build a system that could be not only easy to learn and to use, but also acceptable by the COSP community and trustable by all its members.

The paper is structured as follows. Section 2 introduces the first aid domain by describing the COSP association and its needs. Section 3 illustrates the participatory design activity carried out with representative COSP members: it first presents the requirements emerged for the map and the map-based knowledge that the COSP community needs to share, and then discusses the EUD needs emerged during design sessions and the main ideas for satisfying them. Section 4 illustrates the main characteristics and the EUD activities supported by the resulting application, while Section 5 outlines the results of a preliminary system evaluation. Section 6 compares our work with related literature, and finally, Section 7 concludes the paper.

2   The COSP Association

Centro Operativo Soccorso Pubblico, briefly COSP, is an Italian non-profit association for public assistance and first aid (http://www.cospmazzano.it). The association includes about two hundred volunteers working together in the area of Mazzano near Brescia, in Italy, to provide initial care in case of medical emergencies. To join the association, volunteers are required to attend a certified course for first aid training. Some volunteers are trained to drive the ambulances available at the COSP's offices, and/or to act as specialized rescuers assisting a nurse or an emergency physician from the local hospital in the provision of first aid. Expert drivers usually play the role of driver instructors to train novice drivers, by transferring their knowledge about the territory and their best practice in preparing first aid interventions. In addition to first aid service, COSP volunteers also provide public assistance during sport contexts and demonstrations as well as in transporting patients between places of medical treatment.

In this domain, the role of ambulance drivers is crucial since timely interventions can often save human lives. Even though ambulances are increasingly equipped with navigator satellite systems, such systems are not considered sufficient and satisfactory by COSP volunteers to carefully assist ambulance drivers and the whole emergency crews in bringing medical care to serious patients timely. Current navigator systems do not take into account critical issues when suggesting quickest paths to a place, such as roads with humps or uneven road surfaces (which must not be run in case of patients on board), road yards in progress or weekly open-air markets causing detours that can irreparably delay the provision of first aid. Due to these limitations, ambulance drivers do not generally rely on navigator systems, but they rather prefer trusting in their knowledge and expertise of the territory to decide how to reach a given place quickly and in a safe manner. Therefore, COSP drivers go on using traditional paper maps annotated with their comments and notes in order to encode their knowledge about their territory and share it with the other volunteers. However, due to the rapid topography updates and the perishable nature of paper maps, quick ageing of such traditional maps makes it difficult accessing up-to-date information and sharing such information within the COSP community.

A web-based mapping system tailored to the specific COSP needs may represent a solution to the problem at hand.

3 Participatory Design of FirstAidMap

To meet COSP needs we have designed a web-based mapping system, called FirstAidMap, by involving COSP members according to a participatory approach [21].

During a first meeting, a group of representative COSP volunteers specifically asked for a map-based software system that supports the training activity of new ambulance drivers, who need to know the characteristics of the region where the association provides first aid assistance. A high percentage of interventions are, indeed, performed in the neighbourhood of COSP's offices, which covers a wide territory with about fifteen different villages. As a consequence, drivers often find difficulty orienting themselves, especially to reach villages where interventions are rarely required or when rural areas must be reached. In these cases, a good and up-to-date a priori knowledge of the territory is crucial for guaranteeing fast interventions. To this end, regular training sessions are organized by the COSP association to provide volunteers, and particularly novice drivers, with information about the preferred and the fastest routes to different zones or remote villages of the territory, or about possible danger situations such as temporary holdups on the roads or changes of traffic and road signs. The training activity is usually carried out by using traditional teaching materials, typically by projecting and describing PowerPoint^{TM} slides with annotated maps of the territory. FirstAidMap was therefore intended to support driver instructors during training sessions, as well as ambulance drivers in self-training.

Scenarios and use cases have been used to analyse system requirements with the collaboration of representative ambulance drivers; a task analysis has been carried out to better understand drivers activities, while mock-ups have been prepared and progressively refined to collect feedbacks and suggestions about the map look-and-feel and its interaction possibilities. An iterative approach based on the star-life cycle [12] has been adopted to develop the application.

In the following, we first describe the basic requirements identified at the beginning of the project referred to the training activity of ambulance drivers, and then the needs emerged during system development related to more sophisticated activities for knowledge creation and sharing.

3.1 FirstAidMap Requirements: Map and Information Levels

The activities carried out by drivers and driver instructors require detailed and up-to-date knowledge about the region where first aid interventions could be required.

As previously discussed, the map is the fundamental knowledge source for these volunteers to collect and share knowledge useful to plan first aid interventions. A digital map of different types (road, satellite or hybrid) should be the main component of the FirstAidMap application: its digital nature may increase, on one hand, the ability of COSP volunteers to explore information on the map, on the other, the possibility to encode and represent a richer set of map-based information with respect to the traditional paper-based version. For the specific application domain, there is the need of a digital map whose resolution level is high enough to make roads, but also buildings and houses of interest, recognizable. Additionally, it should be easy to zoom in and zoom out tha map or pan it to better visualize a certain area of the territory. COSP drivers are, indeed, characterized by limited experiences and competencies in information technologies, therefore they need a web mapping system that provides interaction experiences and functionalities suited to their characteristics and needs.

Particular attention should be paid to the knowledge represented on the map. A digital map, although up-to-date and with a high resolution, does not contain all the information the specific community requires about the territory. From the analysis of the application domain, three types of information have been recognized as crucial for the COSP activities: zones, points of interest and notifications.

A *zone* is an area on the map identifying a well-known region of the territory or grouping together several places with common characteristics. An example is a set of roads or neighbourhoods reachable through a same ambulance route from the COSP's offices. A zone is described by a name and eventually characterized by some users' notes and properties.

A *point of interest*, briefly POI, is a fixed and stable element on the territory that acts as a reference point for ambulance drivers to help them to find the way to a place. As in navigator satellite systems, a POI can be a church, a sports ground, a square and so forth. However, it can also be a more specific reference point for an ambulance driver such as a bridge, a dangerous road or other points of interest relevant for first aid activities.

A *notification* is a notice about an alert situation that can interfere with first aid interventions. It aims to notify ambulance drivers about a critical condition occurring in a given place and for a period of time, which can hinder fast interventions, such as work in progress in a specific area of interest or the temporary modification of the road network of a neighbourhood due to a demonstration. Differently from zones and POIs, a notification usually has a limited validity, such as the closing of a motorway tollbooth due to work in progress, or it refers to an event occurring with a certain frequency, e.g. a weekly open-air market taking place in a given square. Therefore, notifications should be displayed on the map in the time frames they are active.

All these types of information contribute to support the activities of COSP drivers. They are all necessary to guide ambulance drivers to the place where medical assistance is needed, thus enriching the geographic map with semantics relevant for the COSP domain. However, they can be a lot of information which altogether can confuse even the most expert users. To avoid information overload, there is the need to properly organize such knowledge. A possible solution is providing FirstAidMap users with all these information organized in four different levels: (a) level 0 with the digital map (street, satellite or hybrid map) retrieved through an available web mapping service; (b) level 1 including the zones identified on the map; (c) level 2 with the POIs; and finally (d) level 3 with the notifications associated with the map.

Ambulance drivers should have the possibility to customize the map level easily, by choosing the type of map displayed (street, satellite or hybrid map) among a set of available web mapping services. Additionally, drivers should be able to access the information of their interest, switching among the four information levels independently, by disabling, if needed, those levels they are not interested in.

3.2   COSP Volunteers: From Passive Users to Co-designers

During the development of a first prototype satisfying the requirements described above, further discussions with representative users led to identify new usage scenarios, beyond driver training. Particularly, a new need emerged: let all COSP volunteers use the system for map consultation to improve their knowledge of the territory as well as a support tool to identify the characteristics of the area around the ambulance destination place, while preparing an emergency intervention.

In this new perspective of FirstAidMap usage, user collaboration to map enrichment is crucial. Therefore, we started to study how 'to transform' COSP volunteers from passive users into co-designers of map content. This should have required to provide users with proper tools to enrich the map with significant and up-to-date information. Moreover, this should have to be achieved without forcing COSP volunteers to become expert neither in information technology nor in cartography, as many commercial geographic information systems require.

To face this problem, the ideas and tools proposed in the end-user development field have been considered. The Network of Excellence on End-User Development, funded by the European Commission during 2002-2003, defined EUD as "the set of methods, techniques, and tools that allow users of software systems, who are acting as non-professional software developers, at some point to create or modify a software artifact" [13]. Particularly, EUD research focuses the attention on those people who use software systems as part of their daily life or daily work, but who are not interested in computers per se [5]. They can be technicians, clerks, analysts and managers who often need to "develop software applications in support of organizational tasks" [2], due to new organizational, business and commercial technologies. The main goal of EUD is therefore studying and developing techniques and applications that empower users to develop and adapt systems themselves [13]. However, the level of complexity of these activities should be appropriate to the users' individual skills and situations, and possibly allow them to easily move up from less to more complex EUD activities. In other words, a "gentle slope of complexity" [17] should be guaranteed, meaning that big steps in complexity should be avoided and a reasonable trade-off between ease-of-use and functional complexity should be always kept in the system. In this way, EUD functionalities should be made available to users progressively, without forcing them to learn soon advanced functionalities. EUD functionalities should not be intrusive nor distract users from their primary task; at the same time, they should encourage users in experimenting system adaptation and modification, by requiring the same cognitive effort necessary for using basic functionalities.

To integrate EUD tools in FirstAidMap while guaranteeing a gentle slope of complexity, different classes of potential end-user developers have been identified within the community of COSP volunteers, by analysing: (i) their current practices and tasks within the association; (ii) their skills and interests in information technologies; (iii) their motivations in collaborating to knowledge sharing. The EUD functionalities the system should offer have been then designed. The next subsection discusses these aspects.

3.3   End-User Developers at COSP

As previously said, the new version of the system must be intended not only for ambulance drivers and driver instructors, but for all COSP volunteers, who could be interested in accessing the knowledge related to their territory, and possibly contributing to its creation. To this aim, we have identified within COSP four different user roles corresponding to different classes of end-user developers: *visitor*, *driver*, *contributor* and *administrator*. For each class, we have designed a suitable interaction experience with FirstAidMap. This classification of end-user developers is based on the following assumptions, which have been discussed and agreed upon during participatory design with representative end users.

All COSP volunteers should be able to access the system easily, without any authentication mechanism[1], as visitor users, just to explore the map-based content, to visualize the map based on their needs and interests, and eventually to point out a danger or a real-time update, which can interfere with first aid interventions, by adding a new notification. As visitor users, they should not be overwhelmed with additional functionalities nor allowed to perform more advanced EUD activities.

Like visitor users, ambulance drivers can access the map and the information associated, visualize active notifications, and eventually insert new ones. They are required to log in FirstAidMap and consult the map before each emergency intervention, in order to check possible alert situations in the route to the emergency site. This organization rule was suggested by the new usage scenario of FirstAidMap, which made the COSP association and its work practice evolve. To support this new practice, a driver role has been added to monitor drivers' accesses to the system.

Some volunteers could be also interested in actively contributing to the updating of map-based information. The role contributor has thus been defined. A volunteer logged in the system as contributor user is provided with advanced EUD tools and functionalities to create and modify zones, POIs and notifications in addition to access and explore the knowledge base as simple visitors.

---

[1] The system is obviously accessible only on the COSP intranet.

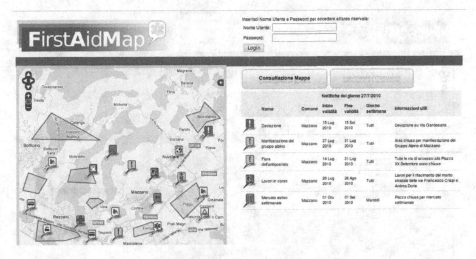

Fig. 1. The FirstAidMap home page

Finally, more active and experienced COSP volunteers should be able to perform advanced EUD activities to let both the content and the whole system evolve according to the COSP community's needs, thus acting as administrator users. An administrator is a power user who manages user profiles, system accesses and all the information associated with the map (POIs, zones and notifications). Furthermore s/he is responsible for configuring the system according to the COSP volunteers' needs.

This classification of the end-user developers is characterized by an increase in the complexity of the EUD activities assigned to them. The usage of the application should motivate and encourage COSP volunteers to become more active in their collaboration to map enrichment. Therefore, after a first period of basic interaction with the system as visitor or driver users, COSP volunteers may wish to become contributor to add and manage zones and POIs, beyond inserting notifications only. In a similar way, a contributor could wish to become an administrator, possibly collaborating with other administrators in the definition of new kinds of POIs and notifications.

4   FirstAidMap: EUD Activities for Knowledge Sharing

Figure 1 illustrates the home page of FirstAidMap. It shows the map with all the zones, POIs and notifications associated with the map, to enable a quick overview and explo ration of the map and its related knowledge. The table on the right contains the list of all the notifications currently active on the map to provide each COSP member with a quick and promptly overview of the critical situations in the territory, as explicitly required during the participatory design activity.

From this page, each volunteer can decide to access as simple visitor, by selecting the button "Consultazione Mappa" (Map consultation) on the top of the table, or to log in the system as a driver, contributor or administrator user by authenticating through the form shown on the top of the screen. The application provides different interaction

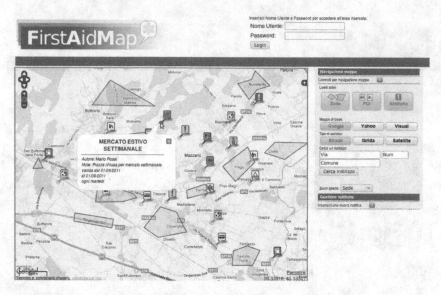

Fig. 2. FirstAidMap environment for map personalization and notification insertion

modalities and supports different kinds of EUD activities according to the user roles logged in the system. These activities can be categorized as follows:

- *Personalization of the map and notification insertion* by filtering the information available, customizing the map, and insert new notifications.
- *Creation and management of new content (zones, POIs, and notifications)*, by adding a multimedia document and a marker (a POI or a notification marker) or defining an area (a zone icon) within the map;
- *System configuration, administration and adaptation* including the modification of the type of content to be added to the map and management of system configuration and user profiles.

In the following subsections, we describe how FirstAidMap enables members of the COSP association to contribute to knowledge creation and sharing at different complexity levels through the EUD activities designed.

4.1 Personalizing the Map and Notifying a Danger

Each visitor user can access FirstAidMap in a consultation mode. In this mode, the user can navigate the map and access content details (see Fig. 2). In particular, the user can interact with the map by clicking on the zoom in/out and pan widgets or using the mouse wheel and left button. S/he can also select an icon on the map, so as a pop-up window appears to display its textual details (in the case at hand the pop-up associated with a notification informs about the closing of a specific square due to an open-air market occurring each Tuesday from June to September 2011). On the right of the map there is a navigation panel to allow the user to customize the map visualization by selecting its type (road, hybrid or satellite map) and the web mapping service (Google, Yahoo,

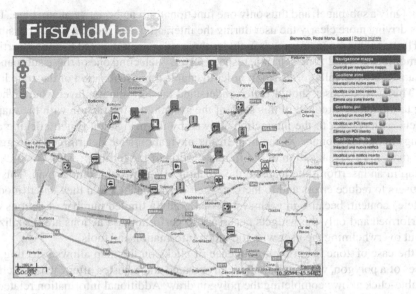

Fig. 3. FirstAidMap environment for content creation

Visual Maps). S/he can also filter the map-related information to be displayed (zones, POIs, notifications) and look for a specific place in the map by specifying its address or immediately points to a relevant place from a list, such as the COSP's offices by choosing the "Sede" item.

Under the navigation panel on the right, there is a notification management panel allowing the user to insert notifications only. By selecting "Inserisci una nuova notifica" · (Insert a new notification), the corresponding panel is enlarged to support the user in inserting a new notification while the only information level displayed on the map is that with active notifications. This allows the user to focus her/his attention on notifications. The visitor user can thus enrich the map-based information by adding a new notification marker on the map and characterizing it through a name, a description, a validity period, a frequency (all days or a given week day) and a type that represents its gravity. The same activities can be carried out with FirstAidMap by driver users who logged in the system. The only difference is the monitoring activity performed by FirstAidMap transparently with respect to the user; this activity, as required by COSP, allows checking a posteriori if drivers consulted the map before their emergency interventions.

4.2 Creating and Managing Zones, POIs and Notifications

More advanced activities can be performed when the user logs in the system as a contributor. In this case, the map view environment is that shown in Fig. 3. As the reader can notice, a richer set of panels is present on the right of the map. This set includes the same navigation panel previously described, and three panels to manage (i.e. insert, modify or delete) zones, POIs and notifications, respectively. The items in each panel can be selected by the user to perform a specific action; the corresponding sub-panel is thus expanded to show all the information necessary to carry out the selected action. Only a sub-panel, and thus only one functionality, can be active at any time. This allows driving more clearly the user during the interaction and reducing error possibilities. The panel for managing zones includes three sub-panels devoted to zone insertion, zone modification and zone deletion, respectively. By selecting one of these sub-panels, the map is automatically refreshed in order to enable the visualization of the zone level only. This way, the user should better understand the information level where s/he is going to operate. Moreover, in this state of the system, the interaction with the map is different with respect to the interaction allowed by the navigation panel: clicking and moving the mouse pointer on the map in the navigation state determines a dragging of the map and a visualization of a different portion of the territory; whilst, a click on the map in an insertion state leads to the creation of a new point on the map. This allows users to reduce errors and to increase their performance when they enter (modify or delete) content, because in each system state only a limited number of actions can be performed and only the widgets necessary to perform those actions are visualized, without overwhelming the user with too much information and tools.

In the case of zone insertion, a sequence of clicks on the map allows selecting the vertices of a polygon, which is automatically created and adjusted after each user click. A double-click allows completing the polygon draw. Additional information related to the zone, such as a name identifying the zone and a detailed description, can then be inserted by filling in the form that is presented in a sub-panel. This form includes also simple instructions that help a non-expert user in perfoming the task. The modification or deletion of a zone can be activated by selecting first the corresponding sub-panel of the zone manager panel, then by clicking on a zone on the map. The shape or the position of the zone can be modified by direct manipulation on the map; while the data associated with the zone, which are automatically loaded and visualized in a form, can be changed by just editing them. FirstAidMap behaves similarly also for managing POIs and notifications. Particularly, for POI insertion, the user can also choose the corresponding icon to be visualized on the map by selecting the POI type (church, soccer field, bridge, etc.).

4.3   Adapting and Managing the System

A separated section can be accessed by administrator users. It does not support a direct interaction with the map, like other roles, but allows creating new kinds of content, managing users, and monitoring user activities.

As a member of the COSP staff, an administrator user will not necessarily be an expert in system administration; s/he will be a power user, with some deeper knowledge in information technologies with respect to the other volunteers. The administrator user should therefore be supported in performing administration activities by easy-to-use tools and user-oriented terminology. For this reason, we classify also administration activities within the EUD activities supported by FirstAidMap. An interesting EUD activity at the hands of an administration user is concerned with the application configuration. Fig. 4 shows the page devoted to this activity. At the top of the page the user can select the base map to be loaded at the application start. Then, s/he can manage the types of POIs and notifications by defining new ones or changing the existing ones. The administrator can define a new type by inserting a name and selecting an icon from

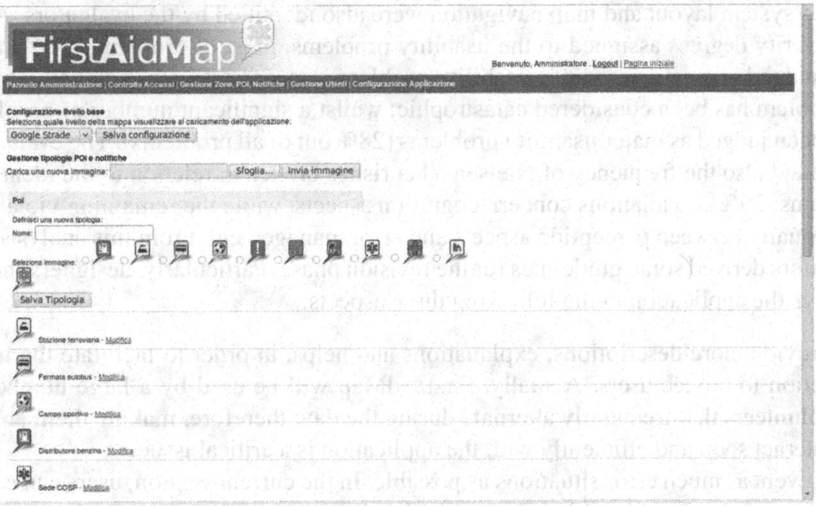

Fig. 4. FirstAidMap environment for system adaptation

those avalaible in a group of radio buttons. If the user does not find a suitable icon, s/he can upload a new image, which will then be added to the available icons. The types of POIs and notifications already existing in the application are shown as a list in the bottom part of the page; each one is represented by its icon and name, whilst the link "Modifica" (modify) allows the user to access the tool for changing the icon or the name.

5   Preliminary Evaluation of FirstAidMap

Before releasing the system and going on with a systematic experiment with COSP volunteers, a low-cost heuristic evaluation has been carried out to identify most usability problems of the resulting prototype.

The evaluation has been carried out following the well-known Nielsen's heuristics [19] by two outside evaluators, who have a background in information engineering, usability theory and practice, and a prior experience with usability evaluations. Each evaluator has carried out two individual evaluation sessions, each one lasting two hours. During the evaluation, usability problems were documented by annotating: (i) the problem title, (ii) an extended problem description, (iii) the Nielsen's heuristics broken, and (iv) a screenshot illustrating the problem. During the debriefing session (lasting three hours and a half) the usability problems identified by the two evaluators were reviewed and consolidated, and a severity degree was assigned to each problem to facilitate the subsequent prioritization of potential solutions to the usability problems identified. Following Nielsen [18], the severity scale comprises five values, from 0 - denoting no usability problem - to 4 meaning a usability catastrophe.

The usability evaluators detected 60 usability problems. The majority of usability problems refers to the EUD activities devoted to content creation and system administration (50% and 28% respectively). A limited number of problems (13) related to the general system layout and map navigation were also identified by the evaluators. As to the severity degrees assigned to the usability problems, most of problems (72%) have been judged cosmetic or minor usability problems (severity degree equal to 1 or 2); no problem has been considered catastrophic; whilst a significant number of problems have been judged as major usability problems (28% out of all problems). The evaluators calculated also the frequency of Nielsen's heuristics broken in relation to the identified problems: 39% of violations concern cognition aspects, while the remaining violations split equally between perception aspects and error management. From this analysis, the evaluators derived some guidelines for the revision phase. Particularly, designers should improve the application in the following three aspects:

1. Provide more descriptions, explanations and helps, in order to facilitate the interaction to novice users. Actually, FirstAidMap will be used by a large number of volunteers that frequently alternate during the day; therefore, making them able to interact soon and efficiently with the application is a critical issue.
2. Prevent as much error situations as possible. In the current version, users can easily make mistakes without being notified by the system. Therefore, error situations must be thoroughly analysed and proper mechanisms for their prevention or for suggesting users how to recover from them must be implemented.
3. Improve consistency. Even though none of consistency problems identified are severe, improving consistency may decrease the learning curve and users' cognitive overload.

Designers are currently fixing these problems, to go on with an in-depth code testing with the aim to ensure correct and secure operations with the system in critical situations.

6   Related Work

EUD techniques have been used for many years in commercial software, such as macro recording in word processors, formula composition in spreadsheets or filter definition in e-mail clients. However, on the one hand, they are far to be used extensively by a large community of end users, and, on the other hand, there exists the potential for employing EUD techniques in many other application domains and with different levels of complexity [7]. Particularly, EUD-based solutions are advocated in cooperative domains, similar to that considered in this paper. For example, in [3] an EUD approach is proposed to facilitate document-mediated cooperative work in the healthcare domain. As far as the technical solutions are concerned, component-based approaches for EUD are proposed in the computer-supported cooperative work field [15]. Myers et al. focus instead on natural programming languages and environments that permit people to program by expressing their ideas in the same way they think about them [16]. Annotation mechanisms and visual programming through direct manipulation are the main EUD techniques implemented in software shaping workshops [4]. A lightweight visual design paradigm is also proposed in [22], where the approach allows business users to create enterprise widgets. Moving from domain-oriented systems to more general web-based applications, the approach presented in [6] is based on the definition of a meta-model of web applications and a set of form-based tools that can be used by end users to customize and shape their applications. A form-based approach is also proposed in [8], as a way to support the development of e-government services on behalf of administrative employees, who do not have any competences in information technology and neither are interested in acquiring them. Anslow and colleagues [1] propose the adoption of Web 2.0 technologies, such as wikis, to support end users not only in contributing content, but also in performing computational tasks, such as the development of business queries. In this line, it has been observed that also mashup makers include much support for EUD [11], which usually adopts a model-based on composition.

In this work, we have capitalized on these proposals by adopting direct manipulation and form-based interaction as basic means to implement EUD functionalities in a collaborative web mapping system. Actually, other systems are available on the Web having many similarities with ours. For example, Google Maps enables users to create personalized maps and share them with relatives and friends. Particularly, users can create their own maps by using place markers, shapes, and lines to define a location, an entire area, or a path. However, the interaction with tools for map personalization is much more programmer-oriented than in our system, with terminology and interaction style resulting to be sometimes intimidating for our classes of users (especially drivers and contributors). Other systems, such as WikiMapia (www.wikimapia.org) or OpenStreetMap (www.openstreetmap.org) are more oriented toward the creation of a social network rather than of a virtual place where to accumulate and share knowledge for a specific and common purpose. WikiMapia allows registered users to select interesting places by drawing polygons and add text notes about the places, as well as images. Other users can see places, read annotations and add comments. Registered users can also vote for an annotation. If the annotation has more than one vote against it, it is deleted. This parameter, which WikiMapia uses to control the correctness of annotations, is a typical feature of social networks. YourHistoryHere (www.yourhistoryhere.com) is another map-based wiki based on the Google Maps API. It is similar to WikiMapia, since it enables users to mark a place with a flag and to add textual annotations telling the history of the specific place. Other logged users can comment on the history. In all these examples, users who comment on or edit annotations constitute informal groups, characterized by common interests or common knowledge about a same place. With respect to such systems, FirstAidMap has been designed in a participatory way in collaboration with its intended users, and therefore functionalities for knowledge creation and sharing are tailored to users' skills and experiences, and aimed at satisfying the specific needs of the community. For example, the notion of levels and the different kinds of information to be made available at users' pace emerged from the discussions with representative users, as well as the distinctions between POIs and notifications. We argue that these characteristics may better sustain user participation in knowledge creation and sharing, and thus increase the usefulness and meaningfulness of the application. This paper extends the work presented in [9] by providing more details on the case study and the results of a preliminary system evaluation.

7 Conclusions

In this paper, we have described FirstAidMap, a collaborative web mapping system we have developed to satify the needs of COSP, an Italian non-profit association for public assistance and first aid. The system allows COSP volunteers to create, manage and share territorial knowledge that can be useful in case of emergencies. Representative volunteers participated in the design of FirstAidMap and in the study of EUD functionalities that could make all volunteers able to contribute their own knowledge and share it with the other volunteers. The resulting system aims to engage and motivate members of the COSP association to participate in map shaping with the goal of reinforcing the sense of community and individual awareness. A preliminary evaluation has been carried out. This evaluation provided us with interesting indications to improve the application.

As to future work, we plan to go on with the system evaluation and organize a systematic experiment with a sample of COSP volunteers. We will study the integration in FirstAidMap of more advanced EUD functionalities, such as the possibility for users to create new information levels. Furthermore, we plan to re-engineer the application to make it flexible enough to be easily adaptable to other application domains. Finally, a portable version of FirstAidMap endowed with a real-time data updating based on a Global Positioning System for ambulances will be studied.

Acknowledgements. The authors wish to thank the volunteers of COSP Mazzano for their collaboration. We are also grateful to Francesca Facchetti and Paolo Melchiori for their contribution to the prototype design and implementation. Finally, we would like to thank Maddalena Germinario and Annamaria Percivalli for the usability evaluation of the application presented in this paper.

References

1. Anslow, C., Rielhe, D.: Towards End-User Programming with Wikis. In: Proc. WEUSE 2008, Leipzig, Germany (2008)
2. Brancheau, J.C., Brown, C.V.: The Management of End-User Computing: Status and Directions. ACM Computing Surveys 25(4), 437–482 (1993)
3. Cabitza, F., Simone, C.: LWOAD: A Specification Language to Enable the End-User Development of Coordinative Functionalities. In: Pipek, V., Rosson, M.B., de Ruyter, B., Wulf, V. (eds.) IS-EUD 2009. LNCS, vol. 5435, pp. 146–165. Springer, Heidelberg (2009)
4. Costabile, M.F., Fogli, D., Mussio, P., Piccinno, A.: Visual Interactive Systems for End-User Development: a Model-based Design Methodology. IEEE Transactions on Systems Man and Cybernetics, part A - Systems and Humans 37(6), 1029–1046 (2007)
5. Cypher, A.: Watch What I Do: Programming by Demonstration. The MIT Press, Cambridge (1993)
6. Da Silva, B., Ginige, A.: Modeling Web Information Systems for Co-Evolution. In: Proc. ICSOFT 2007, Barcelona, Spain (2007)
7. Fischer, G.: End-User Development and Meta-Design: Foundations for Cultures of Participation. Journal of Organizational and End User Computing 22(1), 52–82 (2010)
8. Fogli, D.: End-User Development for E-Government Website Content Creation. In: Pipek, V., Rosson, M.B., de Ruyter, B., Wulf, V. (eds.) IS-EUD 2009. LNCS, vol. 5435, pp. 126–145. Springer, Heidelberg (2009)

9. Fogli, D., Parasiliti Provenza L.: End-User Development for Knowledge Sharing: A Collaborative Web Mapping Application in the First Aid Domain. In: Proc. of KMIS 2010 - International Conference on Knowledge Management and Information Sharing, pp. 5–14 (2010)
10. Giaccardi, E., Fogli, D.: Affective Geographies: Towards Richer Cartographic Semantics for the Geospatial Web. In: Proc. ACM Int. Conf. AVI 2008, Naples, Italy, pp. 173–180 (May 2008)
11. Grammel, L., Storey, M.: An End User Perspective on Mashup Makers. Technical Report DCS-324-IR, Department of Computer Science, University of Victoria (September 2008)
12. Hix, D., Hartson, H.R.: Developing User Interfaces: Ensuring Usability through Product & Process. John Wiley (1993)
13. Lieberman, H., Paternò, F., Klann, M., Wulf, V.: End-User Development: An Emerging Paradigm. In: Lieberman, H., Paternò, F., Wulf, V. (eds.) End-User Development, pp. 1–8. Kluwer Academic Publishers, Dordrecht (2006)
14. Marcante, A., Parasiliti Provenza, L.: Social Interaction through Map-based Wikis. PsychNology Journal 6(3), 247–267 (2008)
15. Morch, A., Stevens, G., Won, M., Klann, M., Dittrich, Y., Wulf, V.: Component-Based Technologies for End-User Development. Communications of the ACM 47(9), 59–62 (2004)
16. Myers, B.A., Pane, J.F., Ko, A.: Natural Programming Languages and Environments. Communications of the ACM 47(9), 47–52 (2004)
17. Myers, B.A., Smith, D.C., Horn, B.: Report of the 'End-User Programming' Working Group, Languages for Developing User Interfaces, pp. 343–366. Jones and Bartlett, Boston (1992)
18. Nielsen, J.: Usability Engineering. Academic Press, San Diego (1993)
19. Nielsen, J.: Heuristic evaluation. In: Nielsen, J., Mack, R.L. (eds.) Usability Inspection Methods. Wiley, New York (1994)
20. What is Web 2.0 - Design Patterns and Business Models for the Next Generation of Software. O'Reilly (2006),
   http://oreilly.com/web2/archive/what-is-web-20.html
21. Schuler, D., Namioka, A.: Participatory Design, Principles and Practice. Lawrence Erlbaum Ass. Inc., Hillsday (1993)
22. Spahn, M., Wulf, V.: End-User Development of Enterprise Widgets. In: Pipek, V., Rosson, M.B., de Ruyter, B., Wulf, V. (eds.) IS-EUD 2009. LNCS, vol. 5435, pp. 106–125. Springer, Heidelberg (2009)

Knowledge Management Tools for Terrorist Network Analysis

Uffe Kock Wiil, Jolanta Gniadek, Nasrullah Memon,
and Rasmus Rosenqvist Petersen

Counterterrorism Research Lab, The Maersk Mc-Kinney Moller Institute
University of Southern Denmark, Campusvej 55, 5230 Odense M, Denmark
ukwiil@mmmi.sdu.dk

Abstract. A terrorist network is a special kind of social network with emphasis on both secrecy and efficiency. Such networks (consisting of nodes and links) needs to be analyzed and visualized in order to gain a deeper knowledge and understanding that enable network destabilization. This paper presents two novel knowledge management tools for terrorist network analysis. CrimeFighter Investigator provides advanced support for human-centered, target-centric investigations aimed at constructing terrorist networks based on disparate pieces of terrorist information. CrimeFighter Assistant provides advanced support for network, node, and link analysis once a terrorist network has been constructed. The paper focuses primarily on the latter tool.

Keywords: Social network analysis, Terrorist network analysis, CrimeFighter investigator, CrimeFighter assistant.

1 Introduction

A terrorist network is a special kind of social network with emphasis on both secrecy and efficiency. Such networks are intentionally structured to ensure efficient communication between members without being detected [1], [2], [3], [4], [5].

Knowledge about the structure and organization of terrorist networks is important for both terrorism investigation and the development of effective strategies to prevent terrorist attacks. Theory from the knowledge management field plays an important role in dealing with terrorist information. Knowledge management processes, tools, and techniques can help intelligence analysts in various ways when trying to make sense of the vast amount of data being collected in relation to terrorism [6]. The collected data needs to be analyzed and visualized in order to gain a deeper knowledge and understanding of the terrorist network.

A terrorist network can be modeled as a generalized network (graph) consisting of nodes and links. Nodes are entities (people, places, events, etc.) and links are relationships between the entities. Techniques from social network analysis (SNA) and graph theory [7] can be used to identify key nodes in the network, which is helpful for network destabilization purposes. Taking out key nodes will decrease the ability of the terrorist network to function normally [8].

A. Fred et al. (Eds.): IC3K 2010, CCIS 272, pp. 322–337, 2013.

Previous research on terrorist network analysis (TNA) has to a large degree focused on analysis of nodes. Links are seldom first class objects in the terrorism domain models with the same properties as nodes. This is in contrast to the fact that the links between the nodes provide at least as much relevant information about the network as the nodes themselves [9].

A terrorism domain model with both nodes and links as first class objects will allow for a better balance between analysis of nodes and analysis of links, which will result in more precise knowledge about the terrorist network. This paper presents two novel knowledge management tools for TNA that supports construction, analysis, and visualization of terrorist networks with a balanced focus on network, node, and link measures to address the above issue.

Section 2 briefly describes our overall knowledge management approach to counterterrorism called CrimeFighter. In Section 3, we briefly present the CrimeFighter Investigator tool. Section 4 presents the CrimeFighter Assistant tool. In Section 5, we demonstrate the use of the CrimeFighter Assistant tool through a case study of the 2002 Bali bombing. Section 6 compares our approach with related work. Section 7 concludes the paper and discusses future work.

2 Knowledge Management for Counterterrorism

The CrimeFighter toolbox for counterterrorism (see Figure 1) is a novel approach to TNA. The goal is to provide a number of desktop tools that are grouped into three overall software packages each containing knowledge management tools and services relevant to counterterrorism [6]. These tools and services are designed and implemented to enable them to interoperate and exchange information.

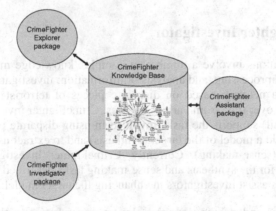

Fig. 1. The CrimeFighter toolbox for counterterrorism

The Explorer and Investigator packages each support different knowledge management processes that result in generation of terrorist networks consisting of nodes and links. These terrorist networks are stored in the knowledge base. The

Assistant package provides various features to analyze and visualize networks – as generated by the Explorer and Investigator packages.

The research on CrimeFighter can be divided into four overall areas:

1. **CrimeFighter Explorer** is a software package with various services aimed at acquiring data from open sources and extracting valuable information from the data by processing it in various ways (filtering, mining, etc.).

2. **CrimeFighter Investigator** is a software package that provides various services that enables an intelligence analyst to work with emergent and evolving structure of terrorist networks to uncover new relationships between people, places, events, etc.

3. **CrimeFighter Assistant** is a software package with various services that supports analysis and visualization of terrorist networks. TNA is aimed at finding new patterns and gaining a deeper knowledge and understanding about terrorist networks. Terrorist network visualization deals with the complex task of visualizing the structure of terrorist networks.

4. **CrimeFighter toolbox architecture.** In order for the developed tools and services to be able to interoperate and exchange information, the overall software architecture of the toolbox must enable a service in one package to use a service in another package. For instance, the structure generated by the services of the Investigator package must be able to use the analysis and visualization services available in the Assistant package.

This paper focuses on describing the various TNA and visualization techniques available in the CrimeFighter Assistant. As mentioned, the starting point for TNA is the existence of a network structure. Hence, much knowledge management work needs to take place prior to network analysis. CrimeFighter Investigator is briefly described in Section 3 to demonstrate how this type of knowledge management work can be supported.

3   CrimeFighter Investigator

Terrorist investigations involve a number of complex knowledge management tasks such as collection, processing, and analysis of information; investigations are aimed at piecing together a network based on disparate pieces of terrorist information that tends to dribble in over time from various sources. CrimeFighter investigator (see [10] for additional detail) supports the tasks involved in using disparate pieces of terrorist information to build a model of the target (synthesis) and to extract useful information from the model (sense-making). Currently, CrimeFighter Investigator focuses on providing support for the synthesis and sense-making tasks described below.

Synthesis tasks assist investigators in enhancing the target model:

• Creating, editing, and deleting **entities.** Investigators basically think in terms of people, places, things, and their relationships.

• Creating, editing, and deleting **relations.** The impact of link analysis on investigative tasks is crucial to the creation of the target model. Descriptive relations between entities helps discover similarities and ultimately solve investigation cases.

• **Re-structuring.** During an investigation, information structures are typically emerging and evolving, requiring continuous re-structuring of entities and their relations.

• **Grouping.** Investigators often group entities using symbols like color and co-location (weak), or they use labeled boxes (strong). Groupings can be used to highlight and emphasize particular entities and their relations.

• **Collapsing and expanding** information is essential since the space available for manipulating information is limited physically, perceptually, and cognitively. Zooming is a way to visually collapse or expand information in the space; however, depending on the zooming degree, it facilitates information overview at the expense of information clarity.

• **Brainstorming** is often used in the early phases of an investigation to get an initial overview of the target and the investigation at hand. Brainstorming is an example of a task that involves both synthesis and sense-making activities. Brainstorming is often supported by different types of mind mapping tools that allows the generated information elements to be organized in a hierarchical manner.

Sense-making tasks assist investigators in extracting useful information from the synthesized target model:

• **Retracing the steps.** Investigative teams often retrace the steps of their investigation to see what might have been missed and where to direct resources in the continued investigation. Walking through an existing recorded investigation is used by new team members to understand the current status of the investigation and for training purposes.

• **Creating hypotheses.** Generating hypotheses and possibly competing hypotheses is a core task of investigation that involves making claims and finding supporting and opposing evidence.

• **Creating alternative interpretations.** Investigators use both fact- and inference-based reasoning to rationalize about their beliefs either in a top-down or bottom up manner. This results in different interpretations of the information at hand (sequences of information, thought experiments, alternative stories, etc.).

• **Prediction.** The ability to determine the presence or absence of relationships between and groupings of people, places, and other entity types is invaluable when investigating a case.

• **Exploring perspectives.** To reduce the cognitive biases associated with a particular mind set, exploring different perspectives (views) of the information is a key investigative task.

• **Decision-making.** During an investigation, decisions have to be made such as selecting among competing hypotheses and selecting among alternative interpretations of the information.

Figure 2 shows a screenshot of the CrimeFighter Investigator from a case study of the Daniel Pearl investigation (Daniel Pearl was kidnapped by terrorists in Karachi, Pakistan in 2002). At this point in the investigation, a small terrorist network has been pieced together from various sources. Once a network like this has been constructed it can be further analyzed by the CrimeFighter Assistant tool.

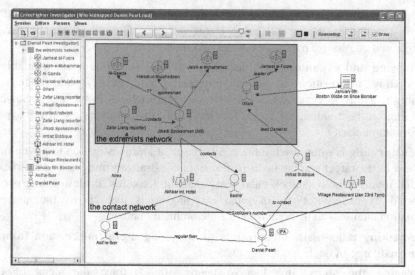

Fig. 2. A screenshot of CrimeFighter Investigator

4   CrimeFighter Assistant

The goal from an intelligence analysis perspective is to support the analysts in making informed decisions regarding possible actions to destabilize the network by determining the most important nodes and links in the network. Looking at the diversity of terrorist groups (e.g., al-Qaeda, ETA, or Liberation Tigers of Tamil Eelam), the way they work, their goals, and their means are different. Therefore, using just one strategy to counter them is impossible. To gain the best possible knowledge and understanding about a terrorist network, one should analyze the network as a whole together with the properties of its nodes and the properties of its links. Various questions might be asked in this process, such as:

Network measures:

- How covert is the network?
- How efficient is the network?
- What is the density of the network?
- What is the trade-off between secrecy and efficiency in the network?

Node measures:

- Who are the central (important) persons in the network?
- What makes the person important?
- What role does a particular person have?
- How is the network affected after removal of a particular node?

Link measures:
- What links are important for communication in the network?
- How important is a particular link in relation to network efficiency and secrecy?
- What is the information backbone of the network?

Answering the above questions without any tools to support the task would be very time consuming. CrimeFighter Assistant provides various TNA features that can support intelligence analysts in answering the above questions. In the following sections, the system architecture and analysis and visualization features are presented.

4.1   System Architecture

The overall system architecture of CrimeFighter Assistant is shown in Figure 3.

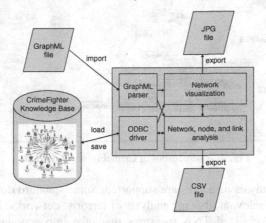

Fig. 3. CrimeFighter Assistant system architecture

CrimeFighter Assistant provides two primary features: network, node, and link analysis and network visualization. Networks can be loaded from the knowledge base or from GraphML files. GraphML is extensively used in SNA applications. Therefore, CrimeFighter Assistant supports this file format as an interchange format to provide interoperability with other SNA and TNA tools. Network data stored in GraphML files can be loaded into the workspace and the same analysis and visualizations can be performed as for network data stored in the CrimeFighter knowledge base.

Analysis results can be exported to CSV (comma-separated values) format to be used in other applications such as Microsoft Excel. Visualized networks can be exported to a printable format (JPEG format). Visualization is based on the JUNG (Java Universal Network/Graph) library [11]. The entire package is coded in Java.

4.2   Analysis and Visualization

A screenshot of CrimeFighter Assistant is shown in Figure 4. The panel to the left is used for visualizing the network, while the panel to the right is used for displaying network, node, and link analysis results. If the user clicks on a node or a link in the analysis results part, the corresponding node or link in the network visualization part will be highlighted in red.

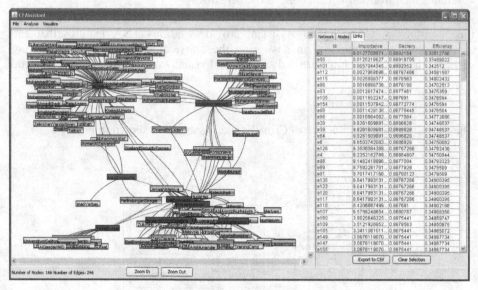

Fig. 4. A screenshot of CrimeFighter Assistant

A number of analysis measures are supported. Some standard domain independent SNA measures are relevant also for analysis of terrorist networks. However, there is also a need for specialized TNA measures that take into account the specifics of terrorist networks. The measures listed below in black font color are standard SNA measures [7], while the measures listed in red font color are specific TNA measures (secrecy, efficiency, performance, position role index, and link importance).

A few definitions are needed regarding graphs to explain the analysis measures.

A graph G consists of two sets of information: a set of nodes, $N = \{n_1, n_2, ..., n_n\}$, and a set of links $L = \{l_1, l_2, ..., l_l\}$ between pairs of nodes. There are n nodes and l links. In a graph, each link is an unordered pair of distinct nodes, $l_k = \{n_i, n_j\}$.

Additional relevant definitions are:

• **Size** is defined as the number of nodes (n) in the network.

• **Nodal degree** is defined as the number of links that are incident with the node.

• A **cluster** is a part of the graph with high density of nodes and links between them.

• The **average shortest path** is the average length of the geodesic between two nodes.

Network Analysis. The following network measures are supported:

• **Density** is the number of links (l) in proportion to the number of links that are possible in G (if all nodes where connected to each other).

• **Diameter** is the maximum distance between any pair of nodes in the network (calculated using the shortest path).

• According to [2], **secrecy** is a measure which is defined by two parameters: the exposure probability and the link detection probability. The exposure probability applies to individual nodes and depends on the location in the structure. It is defined as the probability of a member of the network to be detected as a terrorist. Link detection probability represents the chance of exposure of a part of the network if a member is detected. The secrecy depends on the number of links, the number of nodes, and their degree. The higher the degree of nodes, the lower the secrecy is in the network.

• According to [5], **efficiency** is a measure to quantify how efficiently the nodes of a network can exchange information. To calculate the efficiency of a network, all the shortest path lengths between any pair of nodes in the graph must be calculated. The assumption is made that every link can be used to transfer information in the network. The efficiency is calculated in two parts: (1) the inverse of the sum of the shortest paths between any pair of nodes are calculated; (2) the result from (1) is divided by the possible number of pairs of nodes to find the average efficiency of the network.

• According to [2], **performance** is a measure of the overall performance of a network calculated as the product between secrecy and efficiency. This measure is used to assess the performance of the network in the light of the goals of terrorist network to reach a balance between secrecy and efficiency.

Lindelauf, Borm, and Hamers [2] use the term information performance instead of efficiency. Information performance is calculated in a manner similar to efficiency as proposed by Latora and Marchiori [5].

Node Analysis. The following node measures are supported:

• **Degree centrality**. A node is central when it has many ties (links) to other nodes in the network. This kind of centrality is measured by the degree of the node. The higher the degree, the more central the node is.

• **Closeness centrality** indicates that a node is central when it has easy access to other nodes in the network. This means that the average distance (calculated as the shortest path) to other nodes in the network is small.

• **Betweenness centrality**. Usually, not all nodes are connected to each other in a network. Therefore, a path from one node to another may go through one or more intermediate nodes. Betweenness centrality is measured as the frequency of occurrence of a node on the geodesic connecting other pairs of nodes. A high frequency indicates a central node.

• **Eigenvector centrality** is like a recursive version of degree centrality. A node is central to the extent that the node is connected to other nodes that are central. A node that is high on eigenvector centrality is connected to many nodes that are themselves connected to many nodes.

• According to [12], **position role index** (PRI) is a measure aimed at making a distinction between the gatekeeper and follower roles. PRI evolved from testing efficiency of a network based on the assumption that a network without followers has a higher efficiency as followers are less connected within the structure. PRI is measured as the change of network efficiency after removal of a node. A high PRI value indicates a large loss of efficiency, if a particular node is removed.

Link Analysis. The following link measures are supported:

• **Link betweenness** measures the frequency of link occurrence on the geodesic connecting pairs of nodes [13]. Link betweenness indicates how much information flows via a particular link. The assumption is that communication flows along the shortest path. A high frequency indicates a central link.

• According to [14], **link importance** measures how important a particular link is in a terrorist network by measuring how the removal of the link will affect the secrecy and efficiency (performance) of the network. A high loss of efficiency (when removing the link) indicates an important link.

Visualization. CrimeFighter Assistant can visualize networks using various visualization layouts [15]:

• Fruchterman-Reingold layout
• Kamada-Kawai layout
• Spring layout
• Radial layout
• Self-organizing map layout
• Tree layout

The user decides which layout is the most appropriate for a given network by selecting a menu item in the "Visualize" menu. It is possible to switch between different layouts at any time by simply selecting a different menu item.

In network visualizations with many nodes, vertices might overlap. This might make the graph somewhat unclear. To cope with this issue, a zooming feature has been added.

5   Case Study: 2002 Bali Bombing

At 23:05 on October 12, 2002 an electronically-triggered bomb blew apart Paddy's Bar, a popular night spot in Kuta on the Indonesian island of Bali. Seconds later, as the terrified and injured customers fled, another more powerful bomb hidden in a white Mitsubishi minivan detonated in front of the Sari Club across the street. 202 victims died in the explosions and more than 200 were injured [16].

Members of the South East Asian militant network Jemaah Islamiah were responsible for the attack. It is believed that Riduan Isamuddin (a.k.a. Hambali) ordered a new strategy of hitting soft targets such as nightclubs and bars. Hambali, who is currently in US custody in Guantanamo Bay, is believed to have been the South East Asian contact for Osama Bin Laden's al-Qaeda network. But he is not thought to have played an active part in the Bali plotting. Instead, 43-year-old Islamic teacher Mukhlas (a.k.a. Huda bin Abdul Haq) was convicted as the overall coordinator of the attacks. He also recruited two of his younger brothers, Amrozi and Ali Imron, to play key roles in the attack.

Important roles were also played by Imam Samudra (a.k.a. Abdul Aziz), Azahari Husin (a Malaysian who was alleged to be Jemaah Islamiah's top bomb-making expert and to have helped assemble the Bali bombs; he was killed by police in eastern Indonesian in November 2005) and alleged bomb-maker Noordin Mohammad Top (killed during a police raid in Solo, Central Java in September 2009) [17].

Additionally, Khalid Sheikh Mohammed (leading member of 9/11 attacks) confessed during his hearing at Guantanamo Bay on March 10, 2007 to have been the leader of the Bali bombing plot.

The perpetrators mentioned above were not the only ones involved in planning and carrying out the attack. Therefore based on known facts, a terrorist network for the Bali bombing can be built. The dataset used in this case study is based on the work in [12].

CrimeFighter Assistant will be used to analyze the network, nodes, and links in order to be able to answer the various questions raised in the previous section.

After loading the Bali bombing data set, the status bar shows basic information about the network (number of nodes and links/edges) and the various options for network analysis and visualization become active. Figure 5 shows the status bar after loading the Bali bombing network.

Number of Nodes: 166 Number of Edges: 246

Fig. 5. Status bar after loading the Bali bombing network

5.1   Network Analysis

The result of the network analysis is shown in the right side panel of CrimeFighter Assistant (see Figure 6).

Network	Nodes	Links
Name	Value
Density	0.01796276
Diameter	4.0
Efficiency	0.34907728
Secrecy	0.8874098
Performance	0.30977460741996765

Fig. 6. Network analysis results

The Bali bombing network is sparsely connected: for 166 nodes only 246 links (edges) exists. Hence, the density of the network is low (0.0179): only 1.8 % of all possible links between nodes exist.

The structure of the network consists of stars (clusters of people) that are loosely connected with each other. Three overall clusters can be identified: centered on al-Qaeda, Khalid Sheikh Mohammed (the leader of the plot), and the people directly responsible for the attack, respectively. The only dense segment is formed by the people directly responsible for the attack (see Figure 7).

The diameter of the network is 4 meaning that the largest distance between any pair of nodes is 4. Taking the diameter and the star structure (clusters) into account, information does not need to travel very far in the network.

The structure has a direct impact on the secrecy, efficiency, and performance of the network. The terrorists had to work with a high level of secrecy, which is reflected in the structure of the network. A star structure (excluding the center node) is resistant for uncovering since the other nodes only know one other member. The secrecy value for the network is 0.89, which means that the structure of the network provided a high level of covertness. The high level of secrecy had an impact on the communication possibilities. The efficiency of the network is 0.35. This is however still a high value taking under consideration the conditions in which the terrorist group operated. The overall performance of the network is 0.31 – measured as the trade-off between secrecy and efficiency.

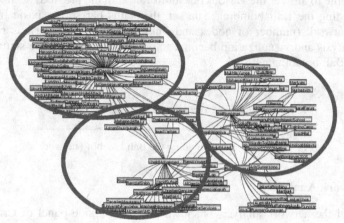

Fig. 7. Bali bombing network structure

The similar values for the 9/11 network is a density of 0.08, a diameter of 5, a secrecy of 0.86, an efficiency of 0.34, and an overall performance of the network of 0.29. Hence, the Bali bombing network managed to have a good trade-off between secrecy and efficiency due to the star-like structure combined with a more densely connected cluster of people taking directly part in the attack.

5.2   Node Analysis

The results of analyzing the nodes in the Bali bombing network are shown in Table 1. The five different node centralities (degree, eigenvector, closeness, betweenness, and PRI) explained in the previous section have been calculated. The table shows the results of the most important nodes ordered according to the degree centrality (highest at the top). Khalid Shaikh Mohammed has the highest score in all the centrality measures. Thus, the analysis strongly indicates that he is the most important person in the network. According to his confession mentioned earlier, he was in fact the leader

Table 1. Node analysis results from the Bali bombing network

Name	Degree	Eigenvector	Closeness	Betweenness	PRI
Khalid Sheikh Mohammed	30	0,060020	2,000000	3118,940177	0,064867
Riduan Isamuddin	23	0,046400	2,457831	1610,549559	0,027781
Huda bin Abdul Haq	12	0,024307	2,566265	169,652490	0,000803
Yazid Sufaat	12	0,024291	2,578313	1441,260256	0,030670
Wan Min Wan Mat	11	0,022286	2,596386	366,862572	0,004090
Imam Samudra	9	0,018222	2,686747	118,365560	0,000254
Azahari Husin	9	0,018231	2,710843	176,577028	0,000249
Amrozi Nurhasyim	8	0,016209	2,698795	201,889424	0,000229
Noordin Mohammad Top	8	0,016194	2,704819	163,530717	0,000218
Ali Imron	7	0,014199	2,698795	160,470996	0,000193
Agus Dwikarna	6	0,011954	2,740964	235,378571	0,003376
Aris Sumarsono	5	0,010080	2,746988	372,804052	0,006535
Osama Bin Laden	3	0,006208	2,349398	55,888300	0,000852
Aafia Siddiqui	2	0,004173	2,530120	0,000000	0,000575
Adnan Shukrijumah	2	0,004173	2,530120	0,000000	0,000575
Mohamed Atta	2	0,004173	2,530120	0,000000	0,000575
Abual Zarqawi	2	0,004173	2,530120	0,000000	0,000575
Hasan Ghul	2	0,004173	2,530120	0,000000	0,000575
Khalid AlHajj	2	0,004173	2,530120	0,000000	0,000575

of the plot. Also, Riduan Isamuddin (believed to be responsible for strategy) and Huda bin Abdul Haq (coordinator of the attack) are both ranked very high according to the centrality measures. The PRI values suggest that Khalid Shaikh Mohammed, Riduan Isamuddin, and Yazid Sufaat were sources of information and gatekeepers. Yazid Sufaat is believed to be the supplier of explosives. Removal of those nodes would lead to the highest decrease in network efficiency.

5.3   Link Analysis

The results of analyzing the links in the Bali bombing network are shown in Table 2. The link analysis measures described in the previous section have been calculated. The table shows link betweenness and link importance for the most important links. The influence of each link in relation to secrecy and efficiency has also been calculated. The secrecy and efficiency columns show how these values will be affected in case the link is removed. The links are ordered according to their link importance values (highest at the top).

The three most important links (e2, e56, and e101) connect the three overall clusters in the network centered on al-Qaeda, Khalid Sheikh Mohammed, and the Bali bombing actors. Other important links connect the individual members that were directly responsible for the attack. Figure 8 shows the Bali bombing network with the 10 most important links highlighted in red. The most important links points out the information backbone of the network. Important communication takes place between the three clusters and inside the cluster directly responsible for the attack.

Table 2. Link analysis results from the Bali bombing network

Link id	Betweenness	Importance	Secrecy	Efficiency
e2	3063,266036	0,012770897	0,8892154	0,33512786
e56	1662	0,012521663	0,88919705	0,33489022
e101	1037,131532	0,005704435	0,8882053	0,342512
e112	889,328022	0,00279687	0,88787466	0,34581587
e115	366	0,002569838	0,8878563	0,34603432
e96	249,8333333	0,001689674	0,8878196	0,34702513
e93	335,6526446	0,001261747	0,8877461	0,3475359
e105	160	0,001165225	0,887691	0,3476594
e154	206,6336996	0,001153794	0,88772774	0,3476594
e88	270,5785714	0,001142814	0,88776445	0,3476594
e86	175,0131868	0,001088408	0,8877094	0,34773886
e29	177,4738095	9,93E-04	0,8886828	0,34746537
e39	177,4738095	9,93E-04	0,8886828	0,34746537
e64	177,4738095	9,93E-04	0,8886828	0,34746537
e6	335,9455458	9,65E-04	0,8886828	0,34750062

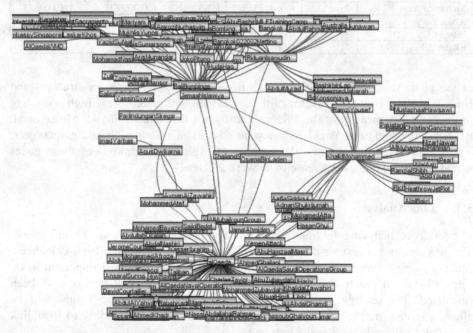

Fig. 8. The 10 most important links in the Bali bombing network

The results of the link analysis point in the same direction as the results of the node and network analysis. Important links are to a high degree connected to what was found to be important nodes (further indicating the importance of those nodes). Also, important links connect the three overall clusters of the network (further emphasizing the use of clusters to structure the network to achieve a good trade-off between secrecy and efficiency).

5.4   Summary

A case study of the 2002 Bali bombing was used to show that the network, node, and link analysis features of CrimeFighter Assistant can provide significant help in answering the important questions related to destabilization of terrorist networks.

6   Comparison to Related Work

We have studied various existing software packages for SNA and TNA to see what features they include: 1. Network Workbench [18]; 2. Social Networks Visualizer [19]; 3. UCINET [20]; 4. Visione [21]; 5. VisuaLyzer [22]; 6. Pajek [23]; 7. NetMiner [24]; 8. Analyst's Notebook 8.5 [25]; 9. *iMiner* [12]; and 10. CrimeFighter Assistant. We have compared the software packages against some of the network, node, and link analysis features available in CrimeFighter Assistant. Table 3 summarizes our results.

Table 3. Comparison of analysis features in SNA and TNA software packages

	1	2	3	4	5	6	7	8	9	10
Secrecy	–	–	–	–	–	–	–	–	–	+
Efficiency	–	–	–	–	–	–	–	–	+	+
Performance	–	–	–	–	–	–	–	–	–	+
Degree centrality	–	+	+	+	+	+	+	+	+	+
Closeness centrality	+	+	+	+	+	+	+	+	+	+
Betweenness centrality	+	+	+	+	+	+	+	+	+	+
PRI	–	–	–	–	–	–	–	–	+	+
Link betweenness	–	–	+	+	–	+	+	+	–	+
Link importance	–	–	–	–	–	–	–	–	–	+

A minus (–) indicates that the feature is not supported. A plus (+) indicates that the feature is supported. All the examined software packages support visualization of network structures – some more advanced than others. The software packages for SNA (1 to 7) as well as Analyst's Notebook (8), a commercial tool for analysis and visualization that is widely used by law enforcement and intelligence agencies, support to a varying degree the ordinary SNA features, but do not support the domain specific TNA features (secrecy, efficiency, performance, PRI, and link importance). *iMiner* [12] which is also a TNA tool supports both SNA and TNA features, but lack some of the latest TNA features that were reported in the research literature after the tool was developed (secrecy, performance, and link importance). On the other hand, some of the software packages (1 to 8) provide many features not currently supported in CrimeFighter Assistant such as detecting communities, k-plex, k-core, clustering coefficients, etc.

Additional TNA features have been proposed in the literature by Memon [12]:

• **Detecting hidden hierarchy**. This method aims to identify hidden hierarchical structures in horizontal networks. The method uses SNA measures and graph theory to indicate parent-child relationships of nodes in the network.

• **Subgroup detection**. A terrorist network can often be partitioned into cells (subgroups) consisting of individuals who interact closely with each other. This method uses SNA measures and graph theory to indicate clusters (subgroups) in relation to a particular node and the diameter from that node.

Rhodes [26] proposed the use of Bayesian inference techniques to **predict missing links** in a covert network, demonstrated through a case study of the Greek terrorist group November 17. The assumption is that during the analysis of terrorist networks it is unlikely that the intelligence analysts have an overview of the full terrorist network. Prediction of missing links can be a useful method to gain deeper understanding and conduct detailed analysis of the terrorist network.

CrimeFigther Assistant provides many of the typical SNA features as well as features dedicated for TNA. Some of the latest TNA features are so far only implemented in CrimeFigther Assistant, thus making the tool unique in certain aspects. However, there are still a number of SNA and TNA features (detecting hidden hierarchy, subgroup detection, link prediction, k-plex, etc.) that can be implemented in future versions to make CrimeFigther Assistant a more complete tool for TNA.

7   Conclusions and Future Work

This paper described the CrimeFigther Assistant knowledge management tool for TNA. The network, node, and link analysis features of the tool were demonstrated based on a case study of the 2002 Bali bombing. CrimeFigther Assistant provides the following contributions:

• The tool supports a balanced approach to TNA focusing on network, node, and link analysis as an attempt to support intelligence analysts in making informed decisions regarding possible actions to take to destabilize a terrorist network.

• The tool provides the first implementation of the link importance measure proposed by Wiil, Gniadek, and Memon [14].

• The tool also provides the first implementation of the secrecy, efficiency, and performance measures proposed by Lindelauf, Borm, and Hamers [2].

Future work will focus on further developing the tool in various ways:

• We plan to include new algorithms for TNA including detecting hidden hierarchy, subgroup detection, and link prediction.

• We are currently looking into how link weights can be incorporated, since not all links are equally important. We believe that incorporation of link weights will result in more precise link analysis measures.

• We wish to optimize the existing TNA algorithms to perform more efficient to be able to analyze large networks of thousands of nodes and links.

• We are currently including additional data sets to test and evaluate the usefulness of the tool more thoroughly.

Acknowledgements. This paper is an extended version of a paper previously published at the International Conference on Knowledge Management and Information Sharing (KMIS 2010) [27].

References

1. Baccara, M., Bar-Isaac, H.: Interrogation methods and terror networks. Mathematical Methods in Counterterrorism, 271–290 (2009)
2. Lindelauf, R., Borm, P., Hamers, H.: The influence of secrecy on the communication structure of covert networks. Social Networks 31, 126–137 (2009)

3. Enders, W., Su, X.: Rational terrorists and optimal network structure. Journal of Conflict Resolution 51(1), 33 (2007)
4. Baker, W.E., Faulkner, R.R.: The social organization of conspiracy: illegal networks in the heavy electrical equipment industry. American Sociological Review, 837–860 (1993)
5. Latora, V., Marchiori, M.: How the science of complex networks can help developing strategies against terrorism. Chaos, Solitons and Fractals 20(1), 69–75 (2004)
6. Wiil, U.K., Memon, N., Gniadek, J.: Knowledge management processes, tools and techniques for counterterrorism. In: Proceedings of the International Conference on Knowledge Management and Information Sharing, pp. 29–36. INSTICC Press (2009)
7. Wasserman, S., Faust, K.: Social network analysis: methods and applications. Cambridge University Press (1994)
8. Carley, K.M., Lee, J.S., Krackhardt, D.: Destabilizing networks. Connections 24(3), 31–34 (2001)
9. Gloor, P.A., Zhao, Y.: Analyzing actors and their discussion topics by semantic social network analysis. In: Information Visualization (IV 2006), pp. 130–135 (2006)
10. Petersen, R.R., Wiil, U.K.: Hypertext structures for investigative teams. Accepted for the Proceedings of the Conference on Hypertext. ACM Press (2011)
11. O'Madadhain, J., Fisher, D., Smyth, P., White, S., Boey, Y.B.: Analysis and visualization of network data using JUNG. Journal of Statistical Software VV, 1–35 (2005)
12. Memon, N.: Investigative data mining: mathematical models for analyzing, visualizing and destabilizing terrorist networks. PhD thesis, Aalborg University, Denmark (2007)
13. Girvan, M., Newman, M.E.J.: Community structure in social and biological networks. Proceedings of the National Academy of Sciences 99(12), 7821–7826 (2002)
14. Wiil, U.K., Gniadek, J., Memon, N.: Measuring link importance in terrorist networks. In: Proceedings of the International Conference on Advances in Social Networks Analysis and Mining, Odense, Denmark, pp. 225–232. IEEE CS Press (August 2010)
15. Di Battista, G., Eades, P., Tamassia, R., Tollis, I.G.: Algorithms for drawing graphs: an annotated bibliography. Computational Geometry-Theory and Application 4(5), 235–282 (1994)
16. Wise, W.: Indonesia's war on terror. United States-Indonesia Society, 1–107 (2005)
17. BBC News, The Bali bombing plot (2008), http://news.bbc.co.uk/2/hi/asia-pacific/3157478.stm
18. NWB Team, Network Workbench Tool. Indiana University, Northeastern University, and University of Michigan (2006), http://nwb.slis.indiana.edu
19. SocNetV (2011), http://socnetv.sourceforge.net
20. UCINET (2011), http://www.analytictech.com/ucinet
21. Brandes, U., Wagner, D.: Analysis and visualization of social networks. Graph Drawing Software, 321–340 (2003)
22. VisuaLyzer (2011), http://www.mdlogix.com/solutions
23. Batagelj, V., Mrvar, A.: Pajek – Program for large network analysis (2011), http://pajek.imfm.si
24. NetMiner (2011), http://www.netminer.com
25. i2 (2011), http://www.i2group.com
26. Rhodes, C.J.: Inference approaches to constructing covert social network topologies. Mathematical Methods in Counterterrorism, 127–140 (2009)
27. Wiil, U.K., Gniadek, J., Memon, N.: CrimeFighter Assistant: a knowledge management tool for terrorist network analysis. In: Proceedings of the International Conference on Knowledge Management and Information Sharing, pp. 15–24. INSTICC Press (2010)

Optimization Techniques for Range Queries in the Multivalued-partial Order Preserving Encryption Scheme

Hasan Kadhem[1], Toshiyuki Amagasa[1,2,3], and Hiroyuki Kitagawa[1,2]

[1] Graduate School of Systems and Information Engineering
[2] Center for Computational Sciences, University of Tsukuba
1-1-1 Tennodai, Tsukuba, Ibaraki 305-8573, Japan
[3] Institute of Space and Astronautical Science, Japan Aerospace Exploration Agency
3-1-1 Yoshinodai, Chuo-ku, Sagamihara, Kanagawa 252-5210, Japan
hsalleh@kde.cs.tsukuba.ac.jp
{amagasa,kitagawa}@cs.tsukuba.ac.jp

Abstract. Encryption is a well-studied technique for protecting the privacy of sensitive data. However, encrypting relational databases affects the performance during query processing. Multivalued-Partial Order Preserving Encryption Scheme (MV-POPES) allows privacy preserving queries over encrypted databases with reasonable overhead and an improved security level. It divides the plaintext domain into many partitions and randomizes them in the encrypted domain. Then, one integer value is encrypted to different multiple values to prevent statistical attacks. At the same time, MV-POPES preserves the order of the integer values within the partitions to allow comparison operations to be directly applied on encrypted data. However, MV-POPES supports range queries at a high overhead. In this paper, we present some optimization techniques to reduce the overhead for range queries in MV-POPES by simplifying the translated condition and controlling the randomness of the encrypted partitions. The basic idea of our approaches is to classify the partitions into many supersets of partitions, then restrict the randomization within each superset. The supersets of partitions are created either based on predefined queries or using binary recursive partition. Experiments show high improvement percentage in performance using the proposed optimization approaches. Also, we study the affect of those optimization techniques on the privacy level of the encrypted data.

Keywords: Encryption, Order-preserving, Range queries, Binary recursive partition.

1 Introduction

Encryption can provide strong security for sensitive data against inside and outside attacks. The primary interest in database encryption results from the recently proposed "database as service" (DAS) architecture [1]. In DAS or database outsourcing, a database owner outsources its management to a "database service provider", which provides

A. Fred et al. (Eds.): IC3K 2010, CCIS 272, pp. 338–353, 2013.

online access mechanisms for querying and managing the hosted database. At the same time, the service provider incurs most of the server management and query execution load. Clients would like to take advantage of the provider's robust storage, but in many cases they cannot trust the provider. Specifically, the provider should be prevented from observing any of the outsourced database contents. Encryption is a common technique used to protect the confidentiality and privacy of stored data in the DAS model. However, encrypting relational databases affects the performance during query processing.

Preserving the order of the encrypted values is a useful technique to perform quires over the encrypted database with a reasonable overhead. For instance, given three integers $\{a,b,c\}$, such that $(a<b<c)$, then the encrypted values are $(E^K(a)<E^K(b)<E^K(c))$. Here, $E^K(v)$ denotes ciphertext value of v with encryption key K. The order preserving encryption scheme OPES proposed firstly by [2]. The strength and novelty of OPES is that comparison operations, equality and range queries as well as aggregation queries involving MIN, MAX and COUNT can be evaluated directly on encrypted data, without decryption. Another encryption scheme proposed by [3,4], where a sequence of polynomial functions is used to encrypt integer values while preserving the order. The decryption is made by solving the inverses of each polynomial function in the sequence in reverse order.

Unfortunately, the previous order preserving encryption (OPE) schemes are not secure against known plaintext attacks and statistical attacks. In those attacks, it is assumed that the attacker has a prior knowledge about plaintext values or statistical information on plaintext domain. Here, the attacker who has access to the encrypted values and has knowledge about the plaintext can map both the plaintext and the encrypted values and make use of them to obtain the key. This is because the OPE schemes preserve the order of all integers in the domain, so the order of encrypted values is exactly the order of plaintext values (we called those schemes as full OPE).

In [5], we proposed a database encryption scheme called MV-POPES (Multivalued-Partial Order Preserving Encryption Scheme), which divides the plaintext domain into many partitions and randomizes them in the encrypted domain. It allows one integer to be encrypted to many values using the same encryption key while preserving the order of the integer values within the partitions. Here, we can still get benefit from the partial order preserving in the encrypted data to perform queries directly at the server without decrypting data. At the same time, we can prevent attackers from inferring individual information from the encrypted database even if they have statistical and special knowledge about the plaintext database. The reason is that the encrypted values are totally in different order compared with the plaintext values. The results from an implementation of MV-POPES show that security for sensitive data can be achieved with reasonable overhead in performing different types of queries.

However, MV-POPES supports range queries over encrypted database at a high overhead compared with the full OPE schemes. The overhead of range queries in MV-POPES is mainly dependent on the number of partitions. Large number of partitions will lead to high overhead on performing range queries over encrypted database. That is because the condition in the WHERE clause becomes complex with many sub conditions connected with boolean connectors (AND and OR).

In this paper, we present some optimization techniques to reduce the overhead for range queries in MV-POPES by simplifying the translated condition and controlling the randomness of the encrypted partitions. The basic idea of our approaches is to classify the partitions into many supersets of partitions, then we restrict the randomization within each superset. The supersets of partitions are created either based on predefined queries or using Binary Recursive Partition (BRP). Experiments show high improvement percentage in performance using the proposed optimization approaches. Also, we study the affect of those optimization techniques on privacy level of the encrypted data.

1.1 Organization of the Paper

The rest of paper is organized as follows. We first discuss related work in Section 2. Section 3 gives a brief overview of the MV-POPES. Section 4 describes merging conditions technique. Section 4 discusses the multilevel partitioning based on predefined queries and binary recursive partition. Section 6 reports the experimental results. Section 7 analyzes the security effects of the optimization techniques. Section 8 introduces the privacy-performance trade-off. We conclude with a summary and directions for future work in Section 9.

2 Related Work

Many full OPE schemes [2,3,4,6,7] have been proposed, but no work discussed the security of the encryption schemes against known plaintext attack and statistical attack. The authors in [8] proposed a new encryption scheme (Chaotic Order Preserving Encryption (COPE)). COPE [8] hides the order of the encrypted values by changing the order of buckets in the plaintext domain. It is secure against known plaintext attack. However, COPE can be used just on trusted server where the encryption keys are used to perform many queries such as join and range queries. The overhead of range queries over encrypted database is much higher than the overhead of range queries over plaintext database. In addition, it uses many keys to change the order of buckets and in some cases that may lead to have duplicated values. Another drawback in COPE is the encryption and decryption cost. That is because of the computation complexity to randomize the buckets and assign the correct order within each bucket.

The bucketing approach [9,10,11,12,13] is closely related to our scheme in sense of dividing the plaintext domain into many partitions (buckets). The encrypted database in the bucketing approach is augmented with additional information (the index of attributes), thereby allowing query processing to some extent at the server without endangering data privacy. The encrypted database in the bucketing approach contains etuples (the encrypted tuples) and corresponding bucket-ids (where many plaintext values are indexed to same bucket-id). In this scheme, executing a query over the encrypted database is based on the index of attributes. The result of this query is a superset of records containing false positive tuples. These false hits must be removed in a post filtering process after etuples returned by the query are decrypted. Because only the bucket id is used in a join operation, filtering can be complex, especially when random mapping is used to assign bucket ids rather than order preserving mapping. The number of false positive records depends on the number of buckets involved. Using a small number of buckets will hide the real values within the bucket index, but the filtering overhead can become excessive. On the other hand, a large number of buckets will reduce the filtering overhead, but the scheme will be vulnerable to estimation exposure. In bucketing, the projection operation is not implemented over the encrypted database, because a row level encryption is used. In addition, updating attributes in the bucketing approach requires that two attributes be updated, the bucket-id and the etuple. This means that all attributes in the row must be re-encrypted, thereby increasing overhead for the update query.

Many works [14,15,16,17] studied the challenges in balancing query efficiency and data secrecy using bucketing based encryption. The basic idea of these researches is to design a bucketization algorithm based on predefined queries that minimizes the false positive tuples while ensuring the privacy of data. It assumed that the data and queries are not frequently updated. While in real world most databases systems are frequently uploaded and updated.

3   Preliminaries

3.1   An Overview of MV-POPES

We begin with the brief overview of the MV-POPES. When encrypting plaintext values in a column having values in the range $[D_{min}, D_{max}]$, first, we divide the domain into n partitions and assign for each partition a random number from 1 to n. This number will be the order of partitions in the encrypted domain. We change the order of partitions to hide the original order of plaintext values. Then, boundaries are generated for all integers in all partitions using an order preserving function. The order is preserved within the partition to be able to evaluate queries efficiently on encrypted database. The generated boundaries identify the intervals. For instance, interval I_i is identified by $[B_i, B_{i+1})$. We then generate the encrypted values for integer i as random values from the interval I_i, so one plaintext value is encrypted to many different values. This will change the frequencies of the plaintext values to prevent the encrypted database against statistical attack.

3.2   Partitioning and Metadata

Here, we explain the partitioning function for each attribute's domain and what is stored in the metadata for each domain. We first divide the plaintext domain of values $[D_{min}, D_{max}]$ into partitions $\{p_1, \ldots, p_n\}$, such that these partitions cover the whole domain and there is no overlap between them. Then, we assign for each partition a unique random number in the range of $[1, n]$. This number is the new order of partitions in the encrypted domain.

As an example, Figure 1 shows the partitions metadata for the domain $[1,100]$. The domain is divided into 5 partitions: $[1,20], [21,40], [41,60], [61,80], [81,100]$. (F), (L) are the first and the last number in the partition. The partition identifier (PID) represents the original order of partitions in plaintext domain. The encrypted partition identifier $(EPID)$ represents the order of partition in the encrypted domain. $(PREV)$ and $(NEXT)$ are the

PID	EPID	PREV	F	L	NEXT
1	3	80	1	20	81
2	1	-	21	40	61
3	5	100	41	60	101
4	2	40	61	80	1
5	4	20	81	100	41

Fig. 1. Partitioning metadata

previous and the next number in the encrypted domain for a partition. For instance, the partition [61,80] where $PID=4$, and $EPID=2$, the $(PREV)$ will be the last number for the partition with $EPID=1$ (which is 40), and the $(NEXT)$ will be the first number for the partition with $EPID=3$ (which is 1).

3.3 Partition Identification Functions

The partition identification functions will be used to translate conditions that contain comparison operations. Let A be an attribute, v be a value in the domain and i be a partition identifier. Table 1 shows the partition identification functions. Using the running example, $PID_A^<(50)=\{1,2\}$ and $PID_A^>(50)=\{4,5\}$.

Table 1. Partition Identification Functions

PID_A	set of PID for attribute A
$PID_A(v)$	PID to which value v belongs in the domain of A
$PID_A^<(v)$	set of PID for attribute A that are less than the partition that contains v
$PID_A^>(v)$	set of PID for attribute A that are greater than the partition that contains v
$PID_A^{<i}$	set of PID for attribute A that are less than i
$PID_A^{>i}$	set of PID for attribute A that are greater than i

3.4 Translation of Range Query

In this section, we explain the translation of range query using the condition **Attribute < Value** as an example. Since the MV-POPES preserves the order of the encrypted values within the partition ($v_i < v_j \rightarrow E^K(v_i) < E^K(v_j)$, v_i and v_j belong to the same PID), the translation of $(A<v)$ is as follows:

$$(A^E < B_v \wedge A^E \geq B_{F(PID_A(v))}) \bigvee \bigvee_{i \in PID_A^<(v)} (A^E \geq B_{F(i)} \wedge A^E > B_{NEXT(i)})$$

where $F(i)$ is the first integer in the partition i, and $NEXT(i)$ is the next integer of the partition i. Simply, the result contains all encrypted values that are less than the left boundary (B_v) of the interval (I_v) within the partition that contains v. In addition, all partitions whose PID are less than the partition of v are included in the result. For example, the translation for the condition $(A<55)$ is:

$$(A^E < B_{55} \wedge A^E \geq B_{50}) \vee (A^E \geq B_1 \wedge A^E < B_{81}) \vee (A^E \geq B_{21} \wedge A^E < B_{61})$$

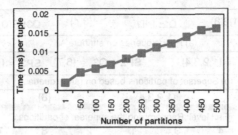

Fig. 2. Range query overhead in MV-POPES

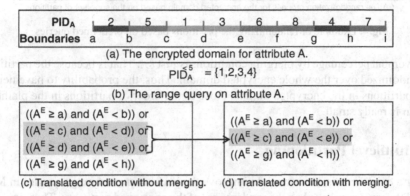

(a) The encrypted domain for attribute A.

$PID_A^{\leq 5} = \{1,2,3,4\}$

(b) The range query on attribute A.

((A^E ≥ a) and (A^E < b)) or
((A^E ≥ c) and (A^E < d)) or
((A^E ≥ d) and (A^E < e)) or
((A^E ≥ g) and (A^E < h))

((A^E ≥ a) and (A^E < b)) or
((A^E ≥ c) and (A^E < e)) or
((A^E ≥ g) and (A^E < h))

(c) Translated condition without merging.   (d) Translated condition with merging.

Fig. 3. Example of merging conditions in MV-POPES

Figure 2 shows the range query (selectivity=50%) execution times on MV-POPES over 10^5 encrypted tuples that picked randomly from a uniform distribution between 1 and 10^5. We can clearly see that the overhead is sharply increasing with increasing the number of partitions, because the condition becomes more complicated. Thus, some optimization techniques are needed to reduce the overhead for range queries in MV-POPES by simplifying the translated condition.

4   Merging Conditions

The first solution to reduce the overhead of range queries in MV-POPES is to merge two or more conditions into one condition. Thus, the query condition in the WHERE clause will be less complex. This can be done in the query translation by searching for neighbor encrypted partitions that are included in the range query. For instance, the original condition shown in Figure 3 (c) over encrypted domain (Figure 3 (a)) consists of 8 sub conditions that specify the first and the last values for each encrypted partition. Knowing that partitions with ids 1 and 3 are adjusted to each other in the encrypted domain, the four sub conditions related to those partitions are simplified to two sub conditions (Figure 3 (d)). Using partitioning metadata, the neighbors' partitions in the encrypted domain can be discovered easily and then merged in the translated condition.

The improvement in performance by merging conditions is based on the number of neighbors' partitions in the range query. The experiments show that the maximum

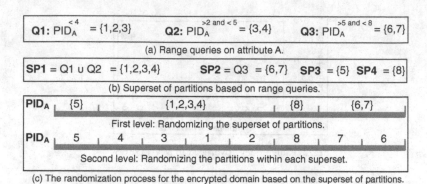

Fig. 4. Example of randomizing the partitions based on predefined queries

improvement percentage by merging conditions is just 5%. That is because the partitions are randomized over the whole encrypted domain. Thus, the probability to have neighbors' partitions in the encrypted domain close to the order of partitions in the plaintext domain is really small.

5   Multilevel Partitioning

Merging conditions is not enough to improve the performance of range queries in MV-POPES because it is based on the randomness of the encrypted partitions. Thus, in order to reduce the overhead, we need to control the randomness of the encrypted partitions.

5.1   Predefined Queries

The first approach to control the randomness of the encrypted partitions is based on a set of predefined range queries. Given a set of range queries $Q=\{q_1,...,q_k\}$ on an attribute A, such that $q_i=[l,h]$ where $l<h$ and $l,h \in PID_A$. The supersets of the partitions are created by merging all queries that have a partition in common. So, for queries q_i,q_j, such that $q_i \cap q_j \neq \phi$, a superset of partitions $SP=q_i \cup q_j$ is created. A separate superset is created for each partition that is not included in any queries. Then, the supersets of partitions are randomized at a first level of randomization. The second level of randomization is done within each superset by randomizing the set of partition for each query. At the final level, the partitions in each set are randomized restrictly within each set of partitions. Figure 4 shows an example of controlling the randomness of 8 encrypted partitions based on 3 range queries. Figure 4 (a) shows the predefined range queries on attribute A, Figure 4 (b) shows the supersets of partitions based on the predefined range queries and Figure 4 (c) shows the process of randomizing the partitions in the encrypted domain.

Using such a technique, we can clearly see that the translated condition will be approximately as simple as the condition on the plaintext domain. That is because most of the partitions that are included in a range query q_i will be neighbors in the encrypted domain even if they are in a different order. So, merging conditions technique now is effective. The overhead of performing range queries over encrypted domain will be much closer to the overhead on the plaintext domain. However, this technique is complicated

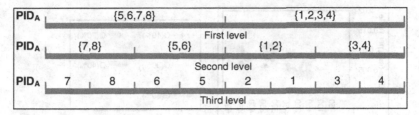

Fig. 5. Example of randomizing the partitions based on BRP

to implement. In addition, when the set of range queries changes, all the data in the attribute need to be re-encrypted based on the new supersets of partitions. Thus, this technique is not good when the data and queries are changing frequently.

5.2 Binary Recursive Partitioning (BRP)

The second approach to control the randomness of the encrypted partitions is based on the binary recursive partitioning BRP [18]. BRP is an iterative process of splitting the data into partitions, and then splitting it up further on each of the branches. There are three characteristics for the BRP. First, the data are partitioned into homogeneous subgroups based on values or rules (Partitioning). Second, data are always partitioned into two groups (Binary). Finally, after data have been partitioned once, more partitions are created within the original partitions (Recursive).

Here, BRP is used to control the randomness of the encrypted partitions by dividing the partitions into homogeneous subgroups or subsets of partitions. Using BRP, the partitions are divided recursively into two subsets till they reach the predefined level or predefined rules. There are many types of rules that decide the splitting position. For simplicity of analysis, we will restrict our attention to splitting the subsets of partitions based on equi-width method. Given a set of partition $PID_A = \{1,...,n\}$ for attribute A and a predefined level pl to stop the partitioning process. At each level in the partitioning process, and for each subset of partitions that consists of $[l,h]$ where $l < h$ and $(l,h \in PID_A)$, the splitting point is $\frac{l+h}{2}$. For each level, the new subsets are randomized within the superset position. At the last level, the partitions are randomized within each subset's position they belong. For instance, Figure 5 shows an example for 3 levels of binary recursive partition applied on 8 partitions.

6 Experiments

We have conducted experiments to examine the validity and the effectiveness of the BRP for performing range queries in MV-POPES. We generated 100,000 records picked randomly from a uniform distribution using [1,100000] as an input domain. The evaluations were performed on a various number of partitions ({50,100,...,500}) using different BRP levels ({1,2,3,4}). The queries used in those experiments have different selectivity {10%,20%,30%,40%,50%}. Figure 6 (a) shows the execution time for the set of range queries performed on non-optimized partitions. The overhead is sharply increasing with

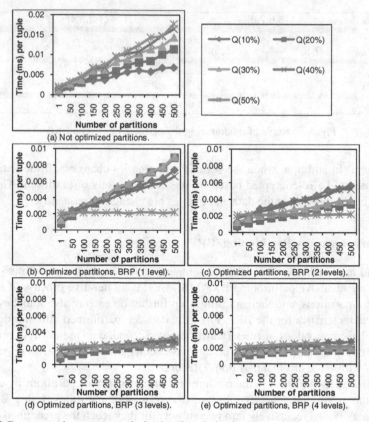

Fig. 6. Range queries on non-optimized and optimized encrypted partitions (using BRP)

increasing the number of partitions. As we discussed before, that is because the condition in the WHERE clause becomes more complicated using large number of partitions.

Figure 6 (b, c, d, e) shows the execution time of the same set of range queries performed on an optimized encrypted domain (using different BRP levels). The figures show that the overhead in the optimized encrypted attribute is less than the non-optimized domain. Also, we notice from those figures that the overhead is decreased by increasing the level in BRP. That is because the translated condition can be simplified more in advance BRP level by merging the sub conditions for neighbor partitions. Figures 6 (b, c, d) show that the high selectivity queries (50%) take less time than the low selectivity queries (10%), especially with the large number of partitions ([400–500]). The reason behind that is that high selectivity queries consist of a large number of partitions, thus the probability for merging partitions into one condition is high. On the other hand, in the case of low selectivity queries such as 10%, the queries consist of small number of partitions; so the probability for merging neighbor partitions is small. Thus, the overhead becomes higher when large number of partitions is used to encrypt the plaintext domain. Overall, in advanced BRP level, the overhead using large number of partitions (MV-POPES) is slightly more than the overhead using one partition (full OPE scheme).

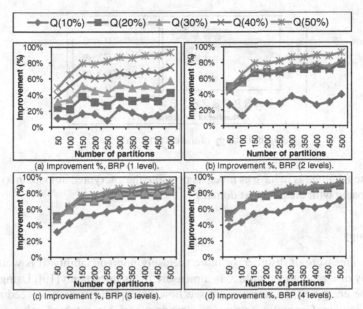

Fig. 7. Improvement percentage in execution time for range queries using different BRP levels

Figure 7 shows the improvement percentage in the execution time for range queries using optimized encrypted partitions (different BRP levels) compared with non-optimized domain. The figure shows that the improvement is increasing with increasing the number of partitions. Also, we can notice that the improvement percentage is increasing with increasing in level for BRP. The improvement reaches approximately 92% in case of 50% selectivity with 500 partitions in the case of 4 BRP levels. The overhead in optimized partitions is slightly increasing using large number of partitions, while in non-optimized partitions the overhead is sharply increasing by using large number of partitions. Thus, the improvement using large number of partitions such as 500 is higher than the improvement using smaller number of partitions such as 50.

7   Security Analysis

The techniques proposed above to control the randomness of the encrypted partitions affect the privacy level of the encrypted domain. That is because those techniques restrict the position probability for a partition in the encrypted domain. Without optimization techniques the probability for a partition to be in a position in the encrypted domain is $\frac{1}{n}$, where n is the number of partitions. On the other hand, using BRP, the probability for a partition to be in a position in the encrypted domain just after the first level is $\frac{1}{(n/2)}$, this number becomes smaller in advanced level of BRP. In this section, we analyze the privacy of the encrypted partitions in both cases; with optimization technique (BRP) and without optimization. We study the **Entropy (H)** and **Variance (V)** of the distribution of encrypted partitions as measures of privacy. We base our choice of entropy and variance on the security definitions presented in [5]. The basic idea to ensure privacy of

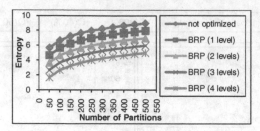

Fig. 8. Entropy of the encrypted domain

the encrypted domain is to have as much as possible the encrypted domain in a different order compared with the plaintext domain.

7.1 Entropy

Entropy is a measure of disorder, or more precisely unpredictability. It is well-known that entropy of a random variable X is a measure of its uncertainty[19]. Using the entropy of the encrypted domain we can see how encrypted partitions spread over the encrypted domain. Generally, entropy of a random variable X taking values $x_{i=1,...,n}$ with corresponding probabilities p_i ,$i=1,...,n$ is given by:

$$Entropy(X){=}H(X){=}-\sum_{i=1}^{n} p_i \times log_2(p_i)$$

Given the partitions in the plaintext domain $PID{=}\{1,...,n\}$, the corresponding partitions in the encrypted domain $EPID{=}\{\forall i{\in}PID, \exists! j{\in}EPID, (1{\leq}j{\leq}n)\}$ and the corresponding probabilities p_i , the entropy of the encrypted partitions $H(EPID)$ without controlling the randomness can be written as:

$$H(EPID) = -\sum_{i=1}^{n} p_i \times log_2(p_i)$$

$$Since\ p_i\ for\ all\ partitions\ (p_i{=}\tfrac{1}{n})$$

$$= -\sum_{i=1}^{n} \tfrac{1}{n} \times log_2(\tfrac{1}{n})$$

$$= -n \times \tfrac{1}{n} \times log_2(\tfrac{1}{n})$$

$$= log_2(n)$$

The entropy in the encrypted domain without optimization depends on the number of partitions. A large number of partitions has more entropy than a small number of partitions which means it is more secure. Using multilevel partitioning to control the randomness of the encrypted partitions, the entropy depends on the number of super partitions and the number of partitions on each super partition. That is because the entropy for a set of partitions (SP_i, $i{=}\{1,...,m\}$) is the summation of entropy in each super partition $H(SP_i)$ multiplied by the probability p_{SP_i} for each super partition (p_{SP_i} is the number of elements in each partition divided by the total number of elements). Given the partitions in the plaintext domain $PID{=}\{1,...,n\}$ and super partitions SP_i, $i{=}1,...,m$, the entropy of the encrypted partitions $H(EPID)$ can be written as:

$$H(EPID) = \sum_{i=1}^{m} \tfrac{|SP_i|}{n} \times H(SP_i)$$

in which $|SP_i|$ is the number of partitions in SP_i. The entropy for the encrypted domain $H(EPID)$ in this case will be less than the entropy without optimization. As an example, we analyze the entropy of the optimized encrypted domain using binary recursive partition. Given the partitions in the plaintext domain $PID=\{1,...,n\}$, the level pl of the BRP and super partitions SP_i, $i=1,...,2^{pl}$, the entropy of the encrypted partitions $H(EPID)$ using BRP can be written as:

$$H(EPID) = \sum_{i=1}^{2^{pl}} p_{SP_i} \times H(SP_i)$$

$$Since\ SP_i\ has\ same\ size\ (p_{SP_i} = \tfrac{1}{2^{pl}})$$

$$= \sum_{i=1}^{2^{pl}} \tfrac{1}{2^{pl}} \times H(SP_i)$$

$$= 2^{pl} \times \tfrac{1}{2^{pl}} \times H(SP_i)$$

$$= H(SP_i)$$

$$= log_2(\tfrac{n}{2^{pl}})$$

Here, the entropy is the summation of entropy for each super partition $H(SP_i)$ multiplied by the probability for each super partition p_{SP_i}. Since all super partitions have same size and same probability, the entropy of the encrypted domain with BRP after pl levels is $log_2(\tfrac{n}{2^{pl}})$. Thus, the entropy will be smaller by increasing the level in BRP. The privacy also decreases in an advanced level of BRP. That is because the randomness in an advanced level of BRP will be less effective to hide the order of the encrypted partitions. Figure 8 shows the entropy of the encrypted domain in both cases; (a) without optimization and (b) optimized domain (BRP) using different levels.

7.2 Variance

The variance and the closely-related standard deviation are measures of how spread out a distribution is. In other words, they are measures of variability. In this section, we study two types of variance; **Horizontal Variance (HV)** and **Vertical Variance (VV)**. The horizontal variance used to measure the difference of orders between plaintext and encrypted domain. While vertical variance used to measure the partial order in the encrypted domain. Given the partitions in the plaintext domain $PID=\{1,...,n\}$ and the corresponding partitions in the encrypted domain $EPID=\{\forall i \in PID, \exists! j \in EPID, (1 \leq j \leq n)\}$, the horizontal variance HV can be written as:

$$HV = \tfrac{1}{n} \sum_{i=1}^{n} \left(\left(PID(i) - EPID(i)\right) - HM \right)^2$$

where HM is the horizontal mean and can be calculated as follows:

$$HM = \tfrac{1}{n} \sum_{i=1}^{n} \left(PID(i) - EPID(i)\right)$$

Note that the horizontal mean HM will be 0 for all permutation. Thus, the horizontal variance can be rewritten as:

$$HV = \tfrac{1}{n} \sum_{i=1}^{n} \left(PID(i) - EPID(i)\right)^2$$

The horizontal variance describes how far the order of the encrypted partitions lies from the order of partitions in the plaintext domain. The minimum value for HV is 0

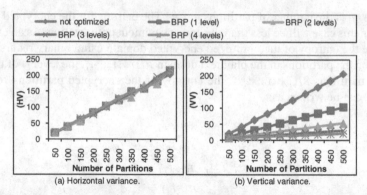

(a) Horizontal variance.   (b) Vertical variance.

Fig. 9. Horizontal and vertical variance in non-optimized and optimized (BRP) encrypted domain

when both PID and $EPID$ have same order. High value for HV means high difference of order between partitions in plaintext domain and partitions in encrypted domain, which leads to high privacy level. However, horizontal variance dose not measure the partial order in the encrypted domain. We can have high HV when encrypted partitions are in descending order which is not secure at all. Thus, we need another measure to describe the order in the encrypted domain. Here, we calculate the variance within the encrypted domain (vertical variance) by taking the difference between each two sequenced partitions. The vertical variance VV for the encrypted partitions $EPID$ is given as follows:

$$VV = \tfrac{1}{n-1} \sum_{i=1}^{n-1} \left(\left(EPID(i) - EPID(i+1) \right) - VM \right)^2$$

where VM is the vertical mean and can be calculated as follows:

$$VM = \tfrac{1}{n-1} \sum_{i=1}^{n-1} \left(EPID(i) - EPID(i+1) \right)$$

The vertical variance describes the degree of partial order in the encrypted domain. The minimum value for VV is 0 when the encrypted partitions are totally ordered either in ascending or descending order. High value for VV means the degree of partial ordering in the encrypted domain is low, which leads to high privacy level.

Figure 9 shows the effect of BRP (levels={1,2,3,4}) on horizontal and vertical variance of encrypted partitions in MV-POPES. The horizontal variance (Figure 9 (a)) using different BRP levels is approximately same as the HV in non-optimized encrypted domain. That is because the encrypted partitions can be in different order compared with the original order in the plaintext domain, even when we restrict the randomness options by using different BRP levels. Figure 9 (b) shows that the vertical variance is reduced because of BRP. That is because the partial order in the encrypted domain using BRP is much more than the non-optimized domain. Also, we can clearly see from the figure that the vertical variance is decreasing by increasing the level of BRP. The results are expectable since the process of controlling the randomness for encrypted partitions in BRP keeps the encrypted partitions somehow ordered within each superset of partitions. Thus, increasing the level of BRP will increase the partial order level in the encrypted domain, and that helps to perform range queries over the encrypted domain efficiently. However, increasing the level of BRP will decrease the privacy level.

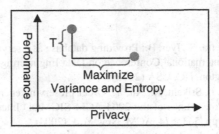

Fig. 10. Privacy-performance trade-off

8   The Privacy-Performance Trade-off

The optimization techniques proposed in this paper lead to better performance for range queries in MV-POPES, for a given number of partitions. However, those techniques also lead to reducing the level of privacy (that is the encrypted partitions might not have a large enough entropy and variance). The research issue here is how to re-randomize the encrypted partitions, starting with the optimized performance for range queries and allowing a bounded amount of performance degradation, in order to maximize the two measures related to the privacy (entropy and variance) simultaneously (Figure 10). We formalize the problem being addressed below:

Trade-off Problem. Given a domain $D=[D_{min}, D_{max}]$ for attribute A, and an initial set of encrypted partitions $\{EPID_1, ..., EPID_n\}$ in an optimized order, re-randomize and/or divide the encrypted partitions into new order $\{C\ EPID_1, ..., C\ EPID_w\}$ such that no more than a factor T of performance degradation is introduced and the entropy and vertical variance for the encrypted partitions are simultaneously maximized.

This problem required advanced analysis and algorithms to maximize the entropy and variance within T factor of performance degradation by swapping and subdividing the partitions. We will study this problem in future work.

9   Conclusions and Future Work

MV-POPES supports range queries at a high overhead compared with the full OPE schemes. In this paper, we presented some optimization techniques to reduce the overhead for range queries in the MV-POPES by simplifying the translated condition and controlling the randomness of the encrypted partitions. The basic idea of our approaches is to classify the partitions into many supersets of partitions, then we restrict the randomization of the partitions within each superset. The supersets of partitions are created either based on predefined queries or using binary recursive partition BRP. Experiments show high improvement percentage in performance using the proposed optimization approaches. However, those techniques also lead to reducing the level of privacy. In the future, we will study the privacy-performance trade-off in the MV-POPES.

Acknowledgements. This work was supported by MEXT Grant-in-Aid for Scientific Research on Innovative Areas (#21013004) and JSPS Grant-in-Aid for Young Scientists (B) (#21700093).

References

1. Hacigumus, H., Mehrotra, S., Iyer, B.: Providing database as a service. In: ICDE 2002: Proceedings of the 18th International Conference on Data Engineering, pp. 29–38. IEEE Computer Society, Washington, DC, USA (2002)
2. Agrawal, R., Kiernan, J., Srikant, R., Xu, Y.: Order preserving encryption for numeric data. In: SIGMOD 2004: Proceedings of the 2004 ACM SIGMOD International Conference on Management of Data, pp. 563–574. ACM, New York (2004)
3. Chung, S.S., Ozsoyoglu, G.: Anti-Tamper databases: Processing aggregate queries over encrypted databases. In: ICDEW 2006: Proceedings of International Conference on Data Engineering Workshops, p. 98. IEEE Computer Society, Washington, DC, USA (2006)
4. Ozsoyoglu, G., Singer, D.A., Chung, S.S.: Anti-tamper databases: Querying encrypted databases. In: 17th Annual IFIP Working Conference on Database and Applications Security, pp. 4–6. Estes Park (2003)
5. Kadhem, H., Amagasa, T., Kitagawa, H.: A secure and efficient order preserving encryption scheme for relational databases. In: KMIS 2010: Proc. of International Conference on Knowledge Management and Information Sharing, pp. 25–35 (2010)
6. Wang, H., Lakshmanan, L.V.S.: Efficient secure query evaluation over encrypted XML databases. In: VLDB 2006: Proceedings of the 32nd International Conference on Very Large Databases, pp. 127–138. VLDB Endowment (2006)
7. Kadhem, H., Amagasa, T., Kitagawa, H.: Mv-opes: Multivalued-order preserving encryption scheme: A novel scheme for encrypting integer value to many different values. IEICE Transactions 93-D, 2520–2533 (2010)
8. Lee, S., Park, T., Lee, D., Nam, T., Kim, S.: Chaotic order preserving encryption for efficient and secure queries on databases. IEICE Transactions on Information and Systems 92, 2207–2217 (2009)
9. Hacigümüş, H., Iyer, B., Li, C., Mehrotra, S.: Executing SQL over encrypted data in the database-service-provider model. In: SIGMOD 2002: Proceedings of the 2002 ACM SIGMOD International Conference on Management of Data, pp. 216–227. ACM, New York (2002)
10. Damiani, E., di Vimercati, S.D.C., Finetti, M., Paraboschi, S., Samarati, P., Jajodia, S.: Implementation of a storage mechanism for untrusted dbmss. In: International IEEE Security in Storage Workshop, p. 38 (2003)
11. Hacigümüs, H., Iyer, B.R., Mehrotra, S.: Ensuring the integrity of encrypted databases in the database-as-a-service model. In: DBSec 2003: Seventeenth Annual Working Conference on Data and Application Security, pp. 61–74. Kluwer (2003)
12. Damiani, E., De Capitani di Vimercati, S., Foresti, S., Jajodia, S., Paraboschi, S., Samarati, P.: Metadata Management in Outsourced Encrypted Databases. In: Jonker, W., Petković, M. (eds.) SDM 2005. LNCS, vol. 3674, pp. 16–32. Springer, Heidelberg (2005)
13. Ceselli, A., Damiani, E., Vimercati, S.D.C.D., Jajodia, S., Paraboschi, S., Samarati, P.: Modeling and assessing inference exposure in encrypted databases. ACM Trans. Inf. Syst. Secur. 8, 119–152 (2005)
14. Tang, Y., Yun, J., Zhou, Q.: A multi-agent based method for reconstructing buckets in encrypted databases. In: Proceedings of the IEEE/WIC/ACM International Conference on Intelligent Agent Technology, IAT 2006, pp. 564–570. IEEE Computer Society, Washington, DC, USA (2006)
15. Li, J., Omiecinski, E.: Efficiency and Security Trade-Off in Supporting Range Queries on Encrypted Databases. In: Jajodia, S., Wijesekera, D. (eds.) Data and Applications Security 2005. LNCS, vol. 3654, pp. 69–83. Springer, Heidelberg (2005)

16. Hore, B., Mehrotra, S., Tsudik, G.: A privacy-preserving index for range queries. In: Proceedings of the Thirtieth International Conference on Very Large Data Bases, VLDB 2004, vol. 30, pp. 720–731. VLDB Endowment (2004)
17. Damiani, E., Vimercati, S.D.C., Jajodia, S., Paraboschi, S., Samarati, P.: Balancing confidentiality and efficiency in untrusted relational dbmss. In: Proceedings of the 10th ACM Conference on Computer and Communications Security, CCS 2003, pp. 93–102. ACM, New York (2003)
18. Breiman, L., Friedman, J.H., Olshen, R.A., Stone, C.J.: Classification and Regression Trees. Wadsworth (1984)
19. Cover, T.M., Thomas, J.A.: Elements of information theory. Wiley Interscience, New York (1991)

miKrow: Enabling Knowledge Management One Update at a Time

Guillermo Álvaro, Víctor Penela, Francesco Carbone, Carmen Córdoba
Michelangelo Castagnone, José Manuel Gómez-Pérez, and Jesús Contreras

iSOCO, Avda. del Partenón 16-18, 1-7, 28042, Madrid, Spain
{galvaro,vpenela,fcarbone,ccordoba,
mcastagnone,jmgomez,jcontreras}@isoco.com
http://lab.isoco.net/

Abstract. Knowledge Management technologies have been around for a while, and while their growth has been increasing lately, it still lacks the current traction that other related technologies are experiencing. One of the hardest bottlenecks Knowledge Management systems currently face are the hurdles that users face, reducing their interest and finally discouraging them for being involved. This proposal showcases the benefits that extending the current Enterprise 2.0 approach can provide. The key evolutions proposed for this lightweight Knowledge Management system are adding (1) a semantic back-end, making the system more intelligent both internally with the use of domain ontologies, and externally by leveraging the Linked Data paradigm, and (2) a simple and smooth microblogging front-end, that improves user experience and makes users more comfortable by taking advantage of a familiar environment they can relate to, in this case. A current implementation and evaluation are also discussed, as well as different boosting techniques that are being studied and deployed.

Keywords: Microblogging, Enterprise 2.0, Semantic web.

1 Introduction

Knowledge Management (KM) within enterprises is a discipline that comprises a set of techniques and processes which pursue the following objectives: i) identify, gather and organize the existing knowledge within the enterprise, ii) facilitate the creation of new knowledge, and iii) foster innovation in the company through the reuse and support of workers' abilities.

Arguably, there is already a wide range of tools in the market that address and support KM processes within enterprises. However, in most of the cases, the potential of those tools gets compromised by an excessive complexity that prevents end-users from getting deeply involved with the system. This leads to end-users not following the protocols, and eventually to a loss of the knowledge that the tools are supposed to capture. Additionally, the integration of complex KM systems within the infrastructures of large organizations is both effort and time-consuming.

On the opposite end of the spectrum, one can find the Web 2.0 paradigm, where end-user involvement is fostered through lightweight and easy-to-use tools. These techniques are increasingly penetrating into the context of enterprise solutions, in a paradigm

A. Fred et al. (Eds.): IC3K 2010, CCIS 272, pp. 354–367, 2013.

usually referred to as Enterprise 2.0. In particular, the trend of microblogging -of which Twitter[1] is the most prominent example- based on short messages and the asymmetry of its social connections, has been embraced by a large number of companies as the perfect way of easily allowing its workers communicate and actively participate in the community, as demonstrated by successful examples like Yammer[2], which has implemented its microblogging enterprise solution into more than 100.000 businesses worldwide.

Our proposal is to apply the Web 2.0 principles and in particular the microblogging approach to the Knowledge Management processes, hence creating an easy-to-use tool that wouldn't prevent users from using it. One of the main characteristics of the proposal is its external simplicity -the only input parameter from the end-user would be used both for capturing his experience and for retrieving suggestions from the system-, though it is supported by complex processes underneath. In fact, our contribution is enriched by semantics, though hidden to the users, in order to support the Knowledge Management processes. Firstly, internally the system is supported by a domain ontology related to the particular enterprise, which can capture the different concepts relating to the company knowledge, and secondly, externally by making use of Linked Data resources available on the Web.

This paper is structured in three main sections: we describe the State of the Art regarding Knowledge Management and Microblogging in 2, we introduce our theoretical contribution in 3 and finally we cover implementation details and evaluation results in 4.

2   State of the Art

2.1   Knowledge Management

The value of Knowledge Management relates directly to the effectiveness[1] with which the managed knowledge enables the members of the organization to deal with today's situations and effectively envision and create their future. Because of the new features of the market like the increasing availability and mobility of skilled workers, the growth of the venture capital market, external options for ideas sitting on the shelf, and the increasing capability of external suppliers, knowledge is not anymore proprietary to the company. It resides in employees, suppliers, customers, competitors, and universities. If companies do not use the knowledge they have inside, someone else will.

In recent years computer science has faced more and more complex problems related to information creation and fruition. Applications in which small groups of users publish static information or perform complex tasks in a closed system are not scalable and nowadays are out of date. In 2004, James Surowiecki introduced the concept of "The Wisdom of Crowds"[2] demonstrating how complex problems can be solved more effectively by groups operating according to specific conditions, than by any individual of the group. The collaborative paradigm leads to the generation of large amounts of content and when a critical mass of documents is reached, information becomes unavailable. Knowledge and information management are not scalable unless formalisms are adopted. Semantic Webs aim is to transform human readable content into machine

---

[1] Twitter http://www.twitter.com

[2] Yammer: http://www.yammer.com

readable[3]. With this goal data interchange formats (e.g. RDF/XML, N3, Turtle, N-Triples), and languages such as RDF Schema (RDFS) and the Web Ontology Language (OWL) have been defined.

The term "Computer Supported Cooperative Work" (CSCW) was defined by Grief and Cashman[4] in 1984 to designate the discipline whose aim is the study of the influence of technology on work. Over the years, CSCW researchers have identified a number of basic dimensions of collaborative work:

– Awareness: people working together should be able to produce a certain level of shared knowledge about the activities of others[5].
– Articulation work: people who cooperated in some way must be able to divide the work into units and dividing them among themselves and finally rebuilding them[6][7].
– Appropriation or Tailorability: technology can be adapted as needed in a particular situation[8][9][10].

A common problem with existing platforms is their limited ability to "capture knowledge" [11]: the channels are not accessible to all and platforms do not allow interaction and only store the final result of a process that has required collaboration and exchange knowledge.

Computer supported collaborative work research analyzed the introduction of Web 2.0 in corporations: McAfee[12] called "Enterprise 2.0", a paradigm shift in corporations towards the 2.0 philosophy: collaborative work should not be based in the hierarchical structure of the organization but should follow the Web 2.0 principles of open collaboration. This is especially true for innovation processes which can be particularly benefited by the new open innovation paradigm[13]. In a world of widely distributed knowledge, companies do not have to rely entirely on their own research, but should open the innovation to all the employees of the organization, to providers and customers.

In a scenario in which collaborative work is not supported and members of the community can barely interact with others, solutions to everyday problems and organizational issues rely on individual initiative. Innovation and R&D management are complex processes for which collaboration and communication are fundamental. They imply creation, recognition and articulation of opportunities, which need to be evolved into a business proposition in a second stage. The duration of these tasks can be drastically shortened if ideas come not just from the R&D department. This is the basis of the open innovation paradigm which opens up the classical funnel to encompass flows of technology and ideas within and outside the organization. Ideas are pushed in and out the funnel until just a few reach the stage of commercialization. Technologies are needed to support the opening of the innovation funnel, to foster interaction for the creation of ideas (or patents) and to push them through and inside/outside the funnel. In a Web 2.0 environment, it is easier to edit and create content, collaboration provides automatic filtering and every member has a simple way to track proposals evaluation. Microblogging model covers the dimensions identified in CSCW, is accessible to all employees, and records all interactions fostering collaboration.

Nowadays solutions like folksonomies, collaborative tagging and social tagging are adopted for collaborative categorization of contents. In this scenario we have to face

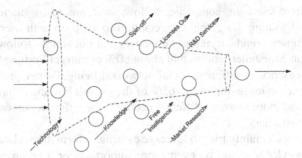

Fig. 1. Open Innovation Funnel

the problem of scalability and interoperability[14]: making users free to use any keyword is very powerful but this approach does not consider the natural semantic relations between the tags. Semantic Web can contribute introducing computer-readable representations for simple fragments of meaning. As we will see, an ontology-based analysis of a plain text provides a semantic contextualization of the content, supports tasks such as finding semantic distance between contents and helps in creating relations between people with shared knowledge and interests.

Moreover, a reward system is necessary to involve people in the innovation process. Money is not a sole motivating factor and there may be other factors such as prestige and ego. A company could collaborate in another firms innovation process as a marketing strategy, in order to have a public recognition as an "innovative partner". Technology has to support the innovation process in this aspect as well, helping decision makers in the enterprise to evaluate the ideas and to reward the members of the community.

2.2 Semantic Technologies on Social Networks

Microblogging is one of the recent social phenomena of Web 2.0, being one of the key concepts that has brought Social Web to more than merely early adopters and tech savvy users. The simplest definition of microblogging, a lite version of blogging where messages are restricted to less than a small number of characters, does not make true judgment of the real implications of this apparently constraint. Its simplicity and ubiquitous usage possibilities have made microblogging one of the new standards in social communication.

Although several microblogging networks have been built, Twitter is currently and by far the most extended, counting more than 100 million users in April of 2010. With its ease of use and the countless number of mobile and desktop applications built over its API, Twitter has been able to grow from a mere tool to a key way of communication.

One of Twitter's key strategies has been its public by default attitude in terms of tweets and basic user information. This approach, although quite interesting from a social point of view, rises several issues in terms of privacy [15], particularly in a work related environment where most of the information could be highly confidential: sharing company information in a public social network could lead to unintended leaks, misappropiation of internal know-how and problems with property rights.

Obviously, where users go, companies follow, so it was just a matter of time for companies to start joining the global conversation to keep up with user's comments, opinions and with new trends, trying to be leaders and not simply followers. A recent study from Burson-Marsteller[3] shows that about 80% of current Fortune 50 companies have an online presence in different social networks, being Twitter probably the one where their presence is more important -65% of the overall Fortune companies according to the study- and more relevant -different accounts for different purposes with direct interaction with customers.

While this approach mainly tries to leverage external information related to the company, internal knowledge could be even more important for a company: what their employees know, which are their opinions on company issues,... Yammer enters the microblogging scene as the first social network with a clear enterprise orientation. Its products, as simple as Twitter high level design could be (status updates as plain text), has reached a huge success counting more than 70.000 companies from all kind of sizes and fields as their clients. However, Yammer does not really offer more than a simple evolution from current chat tools, evolving into a Web 2.0 approach, not providing with any of the benefits of the knowledge management sciences, thus relying only in syntactic analysis.

3 An Intra-enterprise Semantic Microblogging Tool as a Micro-knowledge Management Solution

In this section, we describe our theoretical contribution towards Knowledge Management, addressing the processes involved in order to benefit from the microblogging approach, and how they are enriched by the use of semantics.

3.1 General Description

Unlike powerful yet complex Knowledge Management solutions which expose a broad range of options for the end-user, we propose a Web interface with a single input option for end-users, where they are able to express what are they doing, or more precisely in a work environment, what are they working at. We explain how this single input, which follows the simplicity idea behind the microblogging paradigm, can still be useful in a Knowledge Management solution while reducing the general entry barriers of this kind of solutions.

The purpose of the single input parameter where end-users can write a message is twofold: Firstly, the message is semantically indexed so it can be retrieved later on (see section 3.2), as well as the particular user associated to it; secondly, because the content of the message itself is used to query the same index for relevant messages semantically related to it (section 3.3), as well as end-users associated to those messages ("experts", section 3.4).

Supporting the process of indexing and retrieving relevant information, domain ontologies are used so messages can be associated even if they do not contain the same

---

[3] Burson-Marsteller: http://www.burson-marsteller.com/

expressions. The domain ontology is also used in order to identify the areas in which the system will identify experts.

In addition to the domain ontology, the system takes advantage of the Linked Data paradigm[16] as an efficient manner of accessing structured data already available via Web, thus enriching the system with external information (see section 3.5).

Finally, the system uses contextual information in order to enrich the interactions of end-users with the system. This way, the location information is also stored in the semantic index, so it can be used in the querying step to improve the suggestions (see section 3.6).

3.2 Message Indexing

When a user interacts with the system and a new status message is created, this is indexed into a status repository, permitting its efficient retrieval in the future. Similarly, a repository of experts is populated by relating the relevant terms of the message with the particular author.

Technically, messages that users post to the system are groups of terms T (both key-terms T^K, relevant terms from the ontology domain, and normal terms) $\bigcup T$. The process of indexing each message results in a message repository that contains each document indexed by the different terms it contains, as shown in figure 2.

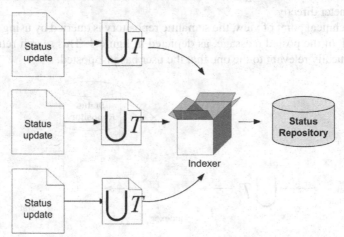

Fig. 2. Message repository creation

Additionally, the process of indexing a message is followed by the update of a semantic repository of experts. In this case, each user can be represented by a group of key-terms (only those present in the domain ontology) $\bigcup T^K$. This way, the repository of experts will contain the different users of the systems, that can be retrieved by the key-terms. Figure 3 illustrates this experts repository.

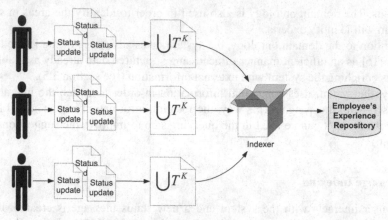

Fig. 3. Employees expertise repository creation

3.3 Message Search

As stated in 3.2, the posting of a new message by a user subsequently triggers a search over the semantic repository. This is performed seamlessly behind the scenes, i.e., the user is not actively performing a search, but the current status message is used as the search parameter directly.

From a technical point of view, the semantic repository is queried by using the group of terms $\bigcup T$ of the posted message, as depicted in figure 4. This search returns messages semantically relevant to the one that the user has just posted.

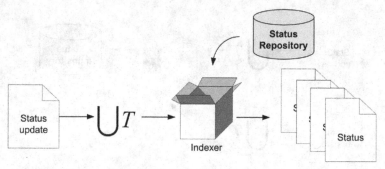

Fig. 4. Detection of related statuses

It is worth noting that the search process in the repository is semantic, therefore the relevant messages might contain some of the exact terms present in the current status message, but also terms semantically related through the domain ontology.

3.4 Expert Search

Along with the search for relevant messages, the system is also able to extract experts associated with the current status message being posted. As stated before, the

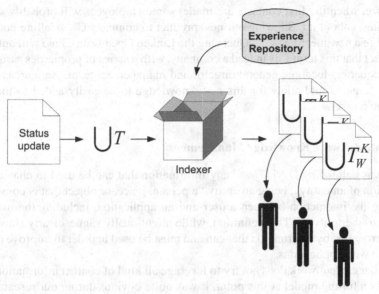

Fig. 5. Expert identification

experts have been identified by the terms present in the messages they have been writing previously.

In this case, the search over the semantic repository of experts is performed by using the key-terms contained in the posted message $\bigcup T^K$, as depicted in figure 5.

3.5 Linked Data Boost

One of the issues of the previous approach is the need of a global ontology that models as close as possible the whole knowledge base of an enterprise, which, depending on the size and the diversity of the company, may differ from difficult to almost impossible (new knowledge concepts being generated almost as fast as they can be modeled).

As an open approach to solve this issue we propose to take advantage of information already available in a structured way via the Linked Data paradigm, providing with an easy and mostly effortless mechanism for adding new knowledge to the system knowledge base. Each new message posted will be processed with NLP methods against the distributed knowledge base that the Linked Data Cloud could be seen as. New concepts or instances extracted from that processing will be added to a temporary knowledge base of terms that could be used to add new information to the system's ontology. These terms would be semiautomatically added to the knowledge via algorithms that weighs the instance usage and the final input of a ontology engineer that decides whether the proposed terms are really valid or is a residue from common used terms with no further meaning to the company.

The main advantage of this approach is that it allows the whole system to adapt to its real usage and to evolve with an organic growth alongisde the evolution of the company knowhow. That way, when a new client starts to make business with the company (or even before, when the first contacts are made) some employees will probably start to post messages about it ("Showing our new product to company C", "Calling company C to arrange a new meeting",...). Querying the Linked Open Data Cloud will automatically detect that this term C is indeed a company, with a series of propierties associated to it (headquarters location, general director and management team, main areas of expertise,...), and would allow for this new knowledge to be easily added to the base knowledge dataset.

3.6   Context-Aware Knowledge Management

Context was defined by Dey[17] as "any information that can be used to characterize the situation of an entity", being an entity "a person, place, or object that is considered relevant to the interaction between a user and an application, including the user and applications themselves". This definition, while intentionally vague, clearly shows that user is surrounded by information that can and must be used in order to improve his/her interaction with applications.

While our current work does not try to leverage all kind of context information or to even apply a formal model at this point, it was quite obvious during our research and particularly during the testing phase that, although users have a clear perception of how a tool like this can be improved by exploiting information about themselves, answers are usually vague in terms of which information do they really find relevant for this kind of application.

For testing purposes we experimented with different kinds of context awareness trying to narrow down which ones where more useful in a work environment like this, and particularly for knowledge management purposes. Location was obviously the first variable that provides useful information, with most users preferring a experts rank where user closeness was positively weighed. As well, a first step into leveraging the dimension of social context was taken into account, by weighing up experts that where somehow close socially (working in the same area or having contacts in common) and thus more easily reachable.

4   Current Implementation: miKrow

The theoretical contribution covered in the previous section has been implemented as a prototype, in order to be able to evaluate and validate our ideas. The nickname chosen for this prototype, miKrow, is based on our micro-Knowledge Management approach. In the following subsections, we address the implementation details and the evaluation performed.

4.1   miKrow Implementation

Figure6 depicts the Web page of the last implementation of miKrow used within iSOCO[4]. The interface features a new stream of messages relevant to the one just

---

[4] Further information on this prototype at http://mikrow.isoco.net/about

Fig. 6. miKrow implementation snapshot

posted, and the Linked Data terms found in it. Experts about relevant terms in the domain ontology, in this case "GLOCAL", are highlighted as well. This last evolution new features include the recommendation of different types of elements from ACTIVE platform[18], being one of its goals to showcase how easy adding external knowledge sources to miKrow can be, and the deployment of the whole system on a Cloud Computing environment[5], particularly Google App Engine[6], reducing substantially the cost and effort of deploying a solution like this in a new enterprise.

Microblogging Engine. miKrow implements Jaiku [7] as the microblogging network management layer, relying on it for most of the heavy lifting related to low level transactions, persistence management and, in general, for providing with all the basic needs of a simple social network. Jaiku was one of the first microblogging social networks available, even earlier than the now omnipresent Twitter, and, after being bought by Google, its source code was released under an open source license.

---

[5] A demo is available and access can be provided on request.

[6] Google App Engine: https://appengine.google.com/

[7] Jaiku: http://www.jaiku.com

Using Jaiku as an starting point reduced the burden of a huge part of the implementation that should have been devoted to all the middleware and infrastructure needed for even the simplest process such as create a new user or post a new status update to be functional, thus allowing the new development to be completely focused in evolving the current microblogging state of art from a simple tool for post and reading to an intelligent knowledge management semantically-enabled environment.

The choice of Jaiku over other possibilities available is based essentially in its condition of having been extensively tested and the feasibility of being deployed in a cloud computing infrastructure[19] such as Google App Engine, thus reducing both the IT costs and the burden of managing a system that could have an exponential growth. Current versions of miKrow have moved from a classic server deployment to a cloud-powered environment, allowing for a faster and cheaper deployment, and thus being able to reach customers easily.

Semantic Engine. The semantic functionalities are implemented in a three layered architecture as shown in figure 7: ontology and ontology access is the first layer, keyword to ontology entity mapping is the second one and the last layer is semantic indexing and search.

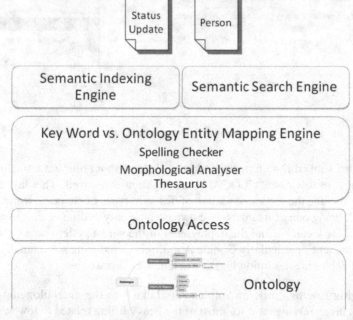

Fig. 7. Semantic Architecture

For each company, an ontology modeling the specific business field needs to be implemented in RDF/RDFS. Knowledge engineers and domain experts worked together to define concepts and relations in the ontologies. Ontologies are accessed through the OpenRDF framework[8].

---

[8] OpenRDF: http://www.openrdf.org

An engine to map keywords to ontology entities has been implemented in order to detect which terms (if any) in the text of an idea are present in the ontology. For this task we consider: morphological variations, orthographical errors and synonyms (for the terms defined in the ontology). Synonyms are manually defined by knowledge engineer and domain experts as well.

The indexes and the search engine are based on Lucene[9]. Two indexes have been created: user activities index and experts index. Each index contains terms tokenized using blank space for word delimitation and ontology terms as single tokens. When we look for related activities to a given one the following tasks are executed:

- extraction of the text of the idea for using it as a query string;
- morphological analysis;
- ontology terms identification (considering synonyms);
- query expansion exploiting ontological relations.

If a synonym of an ontology term is detected, the ontology term is added to the query. If a term corresponding to an ontology class is found, subclasses and instances labels are used to expand the query. If an instance label is identified, the corresponding class name and sibling instance labels are added to the query. Different boosts are given to the terms used for each different query expansion.

For expert detection, semantic search results are filtered with statistical results about related activities.

In order to minimize the maintenance of the ontology, we have added a system based on Linked Data in order to identify relevant terms in the contents created by the users. When a concept doesn't belong to the ontology, Named Entity Recognition services such as Open Calais[10] will be used in order to extend the current knowledge base with external information.

4.2 miKrow Evaluation

An evaluation of the miKrow implementation was carried in-house inside iSOCO where the application was developed. iSOCO has currently around 100 employees distributed in 4 different cities all around Spain, being this important geographical distribution as well as their different knowledge backgrounds and experience a common issue for sharing knowledge between different employees and branches of the company. The miKrow online application was made available for the workers to participate. Additionally, they were asked to rank the suggestions the system made in each occasion, and some of them also provided feedback.

Qualitatively, some conclusions extracted from the evaluation process:

- The system was more useful and provided better suggestions after an initial period of adaptation where the messages were training the system. Arguably, the integration of such a system could be enhanced by the incorporation of previous existing knowledge into the system, e.g., predefined experts that would be substituted gradually through the interactions with the system.

---

[9] Lucene: http://lucene.apache.org/
[10] Open Calais: http://www.opencalais.com/

- Users were significantly more pleased with the suggestions that involved semantics, when they were presented with suggestions and experts with different words than the ones they used, because they perceived some sort of "intelligence" in the system.
- Misleading suggestions were often caused by stop-words that should not be considered, for instance some initial activity gerunds (e.g., "working", "preparing"). A system such as this one should consider them to avoid providing wrong suggestions.

From a quantitative point of view, during the evaluation period there was a considerable increase of interactions of the workers with new tool, in comparison with the previous existing systems such as the intranet. One has to take into account, though, that this increase is related to the context in which the new system was introduced (as it was a project developed in-house). A more consistent evaluation will be carried out if the prototype evolves and is introduced in an external-client.

5   Conclusions

We have presented the concept of a semantic microblogging tool to be used within an enterprise as a lightweight method for Knowledge Management, applying Web 2.0 concepts in order to lower down the entrance barriers for these kinds of systems, thus fostering participation and increasing the utility of the system. We have also described an implementation of a tool that follows these ideas, miKrow, and the evaluation tests that have been possible thanks to it.

References

1. Bellinger, G.: Systems thinking-an operational perspective of the universe. Systems University on the Net 25 (1996)
2. Surowiecki, J., Silverman, M., et al.: The wisdom of crowds. American Journal of Physics 75, 190 (2007)
3. Fensel, D., Hendler, J., Lieberman, H., Wahlster, W.: Spinning the semantic Web: Bringing the World Wide Web to its full potential. MIT Press (2003)
4. Grudin, J.: Computer-supported cooperative work: History and focus. Computer 27, 19–26 (1994)
5. Patterson, J., Hill, R., Rohall, S., Meeks, S.: Rendezvous: an architecture for synchronous multi-user applications. In: Proceedings of the 1990 ACM Conference on Computer-Supported Cooperative Work, p. 328. ACM (1990)
6. Greenberg, S., Marwood, D.: Real time groupware as a distributed system: concurrency control and its effect on the interface. In: Proceedings of the 1994 ACM Conference on Computer Supported Cooperative Work, pp. 207–217. ACM (1994)
7. Nardi, B., Whittaker, S., Bradner, E.: Interaction and outeraction: instant messaging in action. In: Proceedings of the 2000 ACM Conference on Computer Supported Cooperative Work, pp. 79–88 (2000)
8. Hughes, J., Randall, D., Shapiro, D.: Faltering from ethnography to design. In: Proceedings of the 1992 ACM Conference on Computer-Supported Cooperative Work, pp. 115–122 (1992)

9. Tang, J., Isaacs, E., Rua, M.: Supporting distributed groups with a montage of lightweight interactions. In: Proceedings of the 1994 ACM Conference on Computer Supported Cooperative Work, p. 34 (1994)
10. Neuwirth, C., Kaufer, D., Chandhok, R., Morris, J.: Issues in the design of computer support for co-authoring and commenting. In: Proceedings of the 1990 ACM Conference on Computer-Supported Cooperative Work, p. 195 (1990)
11. Davenport, T.: Thinking for a Living. Harvard Business School Press, Boston (2005)
12. McAfee, A.: Enterprise 2.0: The dawn of emergent collaboration. MIT Sloan Management Review 47, 21 (2006)
13. Chesbrough, H., Vanhaverbeke, W., West, J.: Open innovation: Researching a new paradigm. Oxford University Press, USA (2006)
14. Graves, M.: The relationship between web 2.0 and the semantic web. In: European Semantic Technology Conference, ESTC 2007 (2007)
15. Humphreys, L., Gill, P., Krishnamurthy, B.: How much is too much? privacy issues on twitter. In: Conference of International Communication Association (2010)
16. Bizer, C., Heath, T., Idehen, K., Berners-Lee, T.: Linked data on the web (ldow 2008). In: WWW 2008, pp. 1265–1266 (2008)
17. Dey, A.K.: Understanding and using context. Personal and Ubiquitous Computing 5, 4–7 (2001)
18. Warren, P., Kings, N., Thurlow, I., Davies, J., Bürger, T., Simperl, E., Ruiz, C., Gómez-Pérez, J., Ermolayev, V., Ghani, R., Tilly, M., Bösser, T., Imtiaz, A.: Improving knowledge worker productivity – the active approach. BT Technology Journal 26, 165–176 (2009)
19. Armbrust, M., Fox, A., Griffith, R., Joseph, A., Katz, R., Konwinski, A., Lee, G., Patterson, D., Rabkin, A., Stoica, I., et al.: Above the clouds: A berkeley view of cloud computing. EECS Department, University of California, Berkeley, Tech. Rep. UCB/EECS-2009-28 (2009)

A Pervasive Approach to a Real-Time Intelligent Decision Support System in Intensive Medicine

Filipe Portela, Manuel Filipe Santos, and Marta Vilas-Boas

Departamento de Sistemas de Informação, Universidade do Minho
Campus de Azurém 4800-058 Guimarães, Portugal
{cfp,mfs,mvb}@dsi.uminho.pt

Abstract. The decision on the most appropriate procedure to provide to the patients the best healthcare possible is a critical and complex task in Intensive Care Units (ICU). Clinical Decision Support Systems (CDSS) should deal with huge amounts of data and online monitoring, analyzing numerous parameters and providing outputs in a short real-time. Although the advances attained in this area of knowledge new challenges should be taken into account in future CDSS developments, principally in ICUs environments. The next generation of CDSS will be pervasive and ubiquitous providing the doctors with the appropriate services and information in order to support decisions regardless the time or the local where they are. Consequently new requirements arise namely the privacy of data and the security in data access. This paper will present a pervasive perspective of the decision making process in the context of INTCare system, an intelligent decision support system for intensive medicine. Three scenarios are explored using data mining models continuously assessed and optimized. Some preliminary results are depicted and discussed.

Keywords: Real-time, Pervasive, Remotely access, Knowledge discovery in databases, Intensive care, INTCare, Intelligent decision support systems.

1 Introduction

Intensive care units (ICU) are a particular environment where a great amount of data related to the patients' condition is daily produced and collected. Physiological variables such as heart rate, blood pressure, temperature, ventilation and brain activity are constantly monitored on-line [1]. Due to the complex condition of critical patients and the huge amount of data, it can be hard for physicians to decide about the best procedure to provide them the best health care possible. The human factor can lead to errors in the decision making process; frequently, there is not enough time to analyse the situation because of stressful circumstances; furthermore, it is not possible to continuously analyse and memorize all the data [2] .

Care of the critically ill patients requires fast acquisition, registering and availability of data [3].

Accordingly, rapid interpretation of physiological time-series data and accurate assessment of patient state is crucial to patient monitoring in critical care. The data

A. Fred et al. (Eds.): IC3K 2010, CCIS 272, pp. 368–381, 2013.

analysis allows supporting decision making through prediction and decision models. Algorithms that use Artificial Intelligence (AI) techniques have the potential to help achieve these tasks, but their development requires well- annotated patient data [4, 5].

We are deploying a real-time and situated intelligent decision support system, called INTCare[1], whose main goal is to improve the health care, allowing the physicians to take a pro-active attitude in the patients' best interest [6, 7].

INTCare is capable of predicting organ failure probability, the outcome of the patient for the next hour, as well as the best suited treatment to apply. To achieve this, it includes models induced by means of Data Mining (DM) techniques [6], [8-11].

Further improvements include the adjustment of the system to new requirements in order to make it pervasive and ubiquitous [12]. This allows the system to be universal, i.e. can be used anywhere and anytime, eliminating any sort of barrier be it time or place.

This paper is organized as follows. Section 2 presents some background related to Intelligent Decision Support Systems (IDSS), Knowledge Discovery in Databases (KDD), Intensive Medicine and the Pervasive Computing. In next sections the INTCare system is presented, focusing on its features (section 3), the information architecture (section 4), pervasive approach (section 5) and the latest DM models developed (section 6). Section 7 and 8 conclude this paper, presenting a discussion, a conclusion and pointing to future work.

2   Background

2.1   Intelligent Decision Support Systems

According to Turban [13], a Decision Support System (DSS) is an interactive, flexible and adaptable information system, developed to support a problem solution and to improve the decision making. These systems usually use AI techniques and are based on prediction and decision models that analyse a vast amount of variables to answer a question.

The decision making process can be divided in five phases: Intelligence, design, choice, implementation and monitoring [13]. Usually it is used in the development of rule based DSS [14]. However, these DSS are not adaptable to the environment in which they operate. To address this fault, Michalewicz [15] introduced the concept of Adaptive Business Intelligence (ABI). The main difference between this and a regular DSS is that it includes optimization that enables adaptability. An ABI system can be defined as "the discipline of using prediction and optimization techniques to build self-learning decisioning systems. ABI systems include elements of data mining, predictive modelling, forecasting, optimization, and adaptability, and are used to make better decisions." [15].

As it is known, predictive models' performance tends to degrade over time, so it is advantageous to include model re-evaluating on a regular basis so as to identify loss of accuracy [8] and enable their optimization.

There is a particular type of DSS, the real-time DSS. Ideally, the later includes adaptive behaviour, supporting the decision making in real-time.

To achieve real-time DSS, there is a need for a continuous data monitoring and acquisition systems. It should also be able to update the models in real time without human intervention [6]. In medicine, most systems only use data monitoring to support its activities, without predictive behaviour and with poor integration with other clinical information.

2.2   Knowledge Discovery from Databases

KDD is one of the approaches used in Business Intelligence (BI). According to Negash [16], BI systems combine data gathering, data storage, and knowledge management with analytical tools to present complex and competitive information to planners and decision makers. KDD is an interactive and nontrivial process of extracting implicit and previously unknown and potentially useful and understandable information from data [17].

The KDD process is divided in 5 steps: Selection, pre-processing, transformation, data mining and interpretation/evaluation [18]. This process starts with raw data and ends with knowledge.

The automation of the knowledge acquisition process is desirable and it is achieved by using methods of several areas of expertise, like machine learning [9]. The knowledge acquisition takes advantage of KDD techniques, simplifying the process of decision support [8].

Knowledge discovery is a priority, constantly demanding for new, better suited efforts. Systems or tools capable of dealing with the steadily growing amount of data presented by information system, are in order [19].

2.3   Intensive Medicine

Intensive medicine can be defined as a multidisciplinary field of the medical sciences that deals with prevention, diagnosis and treatment of acute situations potentially reversible, in patients with failure of one or more vital functions [20].

These can be grouped into six organic systems: Liver, respiratory, cardiovascular, coagulation, central nervous and renal [21].

ICU are hospital services whose main goal is to provide health care to patients in critical situations and whose survival depends on the intensive care [22], [23].

In the ICU, the patients' vital signs are continuously monitored and their vital functions can be supported by medication or mechanical devices, until the patient is able to do it autonomously [22].

Clinical intervention is based on the degree of severity scores like the SOFA (Sequential Organ Failure Assessment) score, that allow the evaluation of the patient's condition according to a predefined set of values [24].

The assessment of these severity scores are based on several medical data acquired from bedside monitors, lab results and clinical records.

2.4   Real-Time

A system that aims to support decision making must analyse many parameters and output in short real-time and consider online monitoring [25]. It is known that in the ICU setting, there is a huge amount of noisy, high dimensional numerical time series data describing patients. Consequently, such systems must go beyond classical medical knowledge acquisition, since they have to handle with high dimensional data in real-time.

Data acquisition in real-time implies the need for a system responsible for collecting the relevant data to the DSS. This process can be divided in two phases: monitoring and acquisition and storage. Initially, the required data (variables) for the project is identified for further being monitored by sensors or other technology. Subsequently, data is acquired and stored in DB.

This is a critical phase, for technical, human and environment factors are involved and may condition the quality of the data acquired by a gateway, for example, and its storage on a DB. Usually, the monitoring is continuous and there is a small percentage of failures. Although they may occur, they are relatively easy to correct. The biggest problem occurs in the communication between the monitoring system and the storage system.

In conclusion, monitoring in real-time is relatively easy; usually, problems arise in the data storage process.

2.5   Pervasive Computing

Satyanarayanan [26] characterizes the pervasive computing (ubiquitous) as an evolutionary step resulting from the harmonization of the fields of distributed computing and mobile computing (Fig. 1).

This involves not only issues related with saturated environments of communication and user interaction, but also in the support of user's mobility.

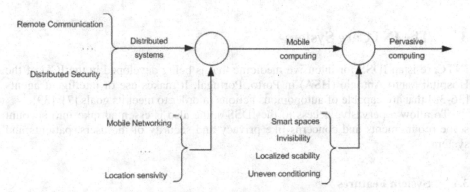

Fig. 1. Pervasive Computing [26]

This type of computing has evolved with the emergence of the Internet. Nowadays, the Internet is easily accessible in all developed countries, schools, public organizations, at home, inside or outside of buildings.

The appearance and ease of access to wireless networks and communication systems contributed to its dissemination. The incorporation of computing devices, objects or sites for monitoring, allows a glimpse of real, physical, and enhanced with information and computing resources that can be used to facilitate human life in its various tasks (personal or social) or to improve business or organizational processes [27]. Complementarily, Banavar [27] also considers that pervasive computation can be:

(i)   the way people think about and use mobile computing devices to perform tasks;

(ii)   how applications are developed and implemented as form of support for these tasks;

(iii)   How the environment is strengthened by the emergence and ubiquity of new information and features.

Almost all the concepts mentioned above can be migrated to pervasive health care. According to Varshney [28], Pervasive HealthCare (PH) can be defined as "health for everyone, anytime and anywhere by removing restrictions such as location and time, increasing both coverage and quality of healthcare". Such approach is essentially based on information that is stored and available online [29].

One of the targets of the system is to achieve the three points described above and to deliver medical information remotely and online through ubiquitous devices [31, 32].

Such approach contributes for the mitigation of medical errors caused by the lack of correct information at the place and time that it is required e.g., misdiagnosis. Decision errors may reach 50% of the total errors [33].

Ubiquity in electronic medical records enhances the analysis of the information by authorized users, anytime and anyplace [34] promoting a real time operation.

The granularity of time, the representation and retrieval of information play an important role in the mode as the diagnosis, prognosis and treatments are performed [35].

3   The INTCare System

INTCare is an IDSS for intensive medicine that is being developed in the ICU of the Hospital Santo António (HSA) in Porto, Portugal. It makes use of intelligent agents [36-38] that are capable of autonomous actions in order to meet its goals [7], [39].

To allow a pervasive access to the IDSS some also is essential take into account some requirements and concerns like privacy and security of the users, patients and system.

3.1   System Features

In order to model the information for KDD processing, the system attends some requirements:

Online Learning - The system acts online, i.e., the DM models are induced using online data in opposition of an offline approach, where the data is gathered and processed afterwards;

Real-Time - The system actuates in real-time, for the data acquisition and storing is made immediately after the events take place to allow that decisions are taken whenever an event occurs;

Adaptability - The system has the ability to, automatically, optimize the models with new data when needed. This information is obtained from their evaluation results;

Data Mining Models - The success of IDSS depends, among others, on the acuity of the DM models, i.e., the prediction models must be reliable. These models make it possible to predict events and avert some clinical complications to the patients;

Decision Models - The achievement of the best solutions depend heavily on the decision models created. Those are based in factors like differentiation and decision that are applied on prediction models and can help the doctors to choose the better solution on the decision making process;

Optimization - The DM models are optimized over time. With this, their algorithms are in continuous training so that increasingly accurate and reliable solutions are returned, improving the models acuity;

Intelligent Agents - This type of agents makes the system work through autonomous actions that execute some essential tasks. Those tasks support some modules of the system: Data acquisition, data entry, knowledge management, inference and interface. The flexibility and efficiency of this kind of system emerges from the intelligent agents and their interaction [7].

Accuracy - The data available in the IDSS need to be accurate and reliable. The system need to have an autonomous mechanism to a pre-validation of the data. The final validation will be always done by a Human, normally by the nurse staff. This operation should be done on the ENR, moments after collection. With this, the user is sure that the data he can see online is guaranteed true.

Safety - All patient data should be safely stored in the database. The data security has to be ensured the access should be restricted. This is the one of the most critical aspects in this type of approach.

Pervasive / Ubiquitous - The system need to be prepared to work in ubiquitous devices like notebooks, PDAs and mobile phones. The internet plays an important role making the system available for users in anyplace. The ICU access policy should be available.

Privacy - There are two types of privacy: i) related to the patient and; ii) related to the health care professional. The patient identification should be always hidden to the people out of hospital. On the other hand the pieces of information recorded on this environment need to be identified and associated to one user, in order to find out responsibilities. Both types of identifications should be protected and masked.

Secure Access from Exterior - The hospital access point has to be protected from exterior connections and encrypted. A Virtual Private Network (VPN) with appropriate access protocols is a good option. Only people who have access to the ICU can see the information and operate, locally or remotely, with the IDSS. This system should implement a secure policy access and be prepared to work in a protected environment.

User Policy - The IDSS should include an inside (ICU environment) and an outside (remote connections) access policy, e.g. where and who can be consult or edit the data.

In order to accomplish the features presented above, some requirements should be considered:

• Fault tolerance capacities;

• Processing to remove null and noisy data;

• Automatic detection and processing of null patient identification;

• Automatic validation of the data taking into account the ranges (min, max) of each variable;

• Continuous data acquisition process;

• Time restrictions for the data acquisition and storage;

• Online learning mode;

• Digital data archive in order to promote the dematerialization of paper based processes (e.g., nursing records);

• Database extension to accommodate the data structures;

• Correct usage of the equipment that collects the vital signs.

4   Information Architecture

Patient management is supported by complex information systems, which brings the need for integration of the various types and sources of data [40].

In order to follow the requirements enumerated above, an information model was drawn, regarding the data acquisition module which includes three types of information sources:

• Bedside Monitors (BM);

• Lab Results (LR); and

• Electronic Nursing Records (ENR).

All sources can produce information to the system and that information can be used to develop predicting models in Intensive Care (knowledge). The development of an automated information system for ICU has to be in harmony with the whole information system and activities within the unit and the hospital [40].

The first type of sources relates to data acquisition from BM. This acquisition is in real-time, the data is received by a gateway, and it is stored on a DB table by an agent. Automatic acquisition eliminates transcription errors, improves the quality of records and allows the assembly of large electronic archives of vital sign data [40].

The second type of sources (LR) is the one that contains the less frequent observations, because the patient normally does this type of clinical analysis once or twice per day, except in extraordinary situations. With this method we can collect the data related with some clinical analysis, such as: number of blood platelets, creatinine, bilirubin, SOFA scores, partial pressure of oxygen in arterial blood and fraction of inspired oxygen.

4.1   INTCare Sub-systems Functionality

The INTCare System [6, 7] is divided into five subsystems: data acquisition, data entry, knowledge management, inference and interface. Fig 2 shows a model that is a part of INTCare system and represents an evolution of two subsystems: data acquisition and data entry. This subsystem is responsible for all activities of data acquisition and data store and will gather all required data into a data warehouse [41-43]. The evolution of this architecture is prominent. Formerly, most of the data was registered in paper format, and it was necessary to manually put it in electronic format, i.e., the information was rarely stored in computers, except the information from the BM, which was automatically collected and stored in electronic format.

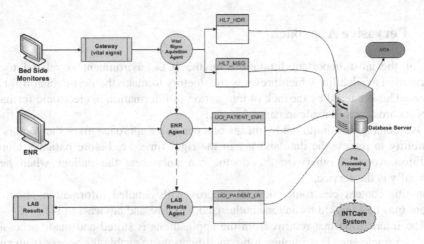

Fig. 2. The INTCare Data Acquisition Subsystem

The new architecture [41-43] contemplates the data acquisition from three sources and, regarding the information input, it is done either automatically (BM, LR) or automatically and manually (nursing records). The adjustment made to the system was the addition of one more data source and the creation of two more agents that enable storing information in the database (DB).

This modification is in course and it is the most important, because it makes possible the data acquisition in electronic and automatic mode for all data sources through multi-agent system. Whit this change, we will have all the necessary information in electronic format for the DM models and the decision support process, addressing the timing requirements of critical tasks.

4.2   How These Subsystems Work

The first type of data sources is the BM, which collects the patients' vital signs (VS). The gateway is connected to the monitors, reads the information and stores it on a DB through the data acquisition agent. This agent splits a HL7 [44] message in two, one with the header information and another with patient data.

The second source is the ENR [41]. It was developed with the objective of registering electronically the paper-based nursing records. With the ENR, the medical and nursing staff can register various types of data, like confirming if some therapeutic was performed or not, and they may consult all the present and past data about the patients. The last type of data sources is the LR, which is controlled by the clinical analyses agent that automatically stores all the LB from the patients.

All the data is stored in one DB and it can be accessed by the medical staff through a computer. The integrated data will be used by the INTCare system to create prevision and decision models.

The DM agent belongs to the sub-system knowledge management and it is in charge of retrieving the required data to feed the DM models and to train new models whenever their performance becomes unsatisfactory.

5   Pervasive Approach

One of the most important limitations in the ICU environment is related to the information availability which prevents the doctors to make the best decision for the patient. This is caused by the lack of integration of information in electronic format (a lot of records are still made in paper format).

Overcoming this limitation by means of a pervasive approach gives the doctors the possibility to review the data and act in the right time, i.e. before patient's clinical condition worsens, otherwise the doctor can only treat the patient when he is physically in the service.

In this context electronic medical records with detailed information about the patient may be analysed by anyone authorized, anytime and anywhere [34].

The information that returns from the applications is stored and made accessible from one single site. This implies some modifications to enable the access from small portable devices [29] (as defined in chapter 3).

Fig. 3 shows a pervasive perspective of the decision making process. The doctors can remotely, through a secure connection, consult the patient data or make a decision. This consult may have two purposes and can be done in real-time through the web platforms:

• ENR - shows all clinical data validated about the patient,

• INTCare - helps to make the decision based on some clinical predictions about the patient suggested by the Data Mining Engine.

The doctor's decision making process can start after the patient data is consulted or at the start, and can be made in a single fashion or discussed in a collaborative platform [45, 46].

After the decision be taken, this can be performed remotely, by a directly configuration in an ICU System or by giving some clinical instruction to the nurses. At the end, the performed decision will be used to optimize and adapt the decision and prevision models.

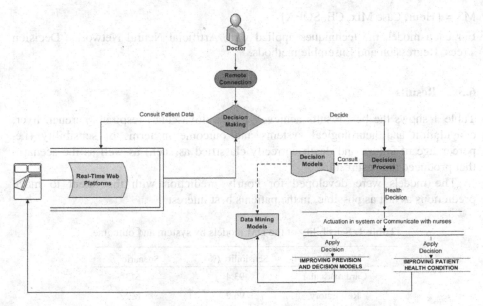

Fig. 3. Decision Support - Pervasive Approach

6   Data Mining Models

6.1   Data Description

The data used to generate the DM models originates from three distributed and heterogeneous sources: LR, BM and paper-based nursing records, presented and explained previously. Additionally, variables containing the case mix (information that remains unchanged during the patient's stay - age, admission from, admission type) were also considered. It was also included some calculated variables: Critical Events (CE), SOFA scores and a set of ratios relating the previous variables to the patients' length of stay.

The data was gathered in the ICU of HSA and it was collected in the first five days of stay of thirty two patients. The construction of the dataset was not automatically, the data from the LB and nursing records was manually registered, for the new adjustments of the system regarding the data acquisition and data entry were not developed at the time the models were generated.

6.2   Features Selection

For the prediction of the dysfunction/failure of each organic system and outcome, three scenarios were explored regarding the inclusion of the variables mentioned above – M1, M2, and M3 – where

M1 = {Hour, Case Mix, CE}
M2 = {Hour, Case Mix, CE, Ratios}

M3 = {Hour, Case Mix, CE, SOFA}.

For each model, the techniques applied were Artificial Neural Networks, Decision Trees, Regression and Ensemble methods.

6.3   Results

Table 1 shows the best results achieved for cardiovascular, respiratory, renal, liver, coagulation and neurological systems and outcome in terms of sensibility (i.e. percentage of failure and death correctly classified as such) as well as the scenario that produced the best results.

The models were developed for hourly prediction with the intent to make predictions as fast as possible, in the patients best interest.

Table 1. Sensibility of the DM models by system and outcome

System	Sensibility (%)	Scenario
Cardiovascular	93.4	M3
Respiratory	96.2	M2
Renal	98.1	M3
Coagulation	97.5	M2

7   Discussion

In this paper we presented the INTCare system, which is an IDSS for intensive medicine. It relies on the KDD process and AI algorithms to apply DM techniques for predicting outcomes that might support the course of action of doctors' decision.

Relying on intelligent agents, the system in divided into five sub-systems (data acquisition, data entry, knowledge management, inference and interface) that guarantee its functionality.

Since its beginning, INTCare has evolved towards using real-time and online clinical data so that the predictions can be as accurate and as soon as possible. As an IDSS, INTCare uses continuous data monitoring and acquisition systems that make possible for all information being available at the right time. This allows doctors to have a proactive attitude in patients' care.

The development of an ENR allows the integration of all necessary information regarding the patients' condition to be collected and integrated in just one application, which is a great gain in time and performance for the medical staff operating in the ICU. In addition to the patients' vital signs, data regarding their LR, procedures, medication, is also available by the time it is generated.

Moreover, the INTCare system is designed to address know issues of the ICU setting, such as noisy, high dimensional numerical time series data in real-time [25], as well as the data acquisition in real-time, storage, integration and rapid availability of all clinical information. The pervasive approach brought new features to the system enabling the remote access. However is necessary to reformulate the ICU environment taking into account some questions like privacy, security, connection and other concerns [34, 47-49].

8   Conclusions and Future Work

The main concern in ICU is to avoid or reverse organ failure, in order to preserve the patients' lives. The INTCare system is being developed for hourly prediction of the patients' clinical condition, i.e. the prediction of dysfunction/failure of the organ systems (cardiovascular, respiratory, renal, coagulation and liver systems) and outcome.

We believe that, with this fine grained prediction, it will be possible for the healthcare professionals to have a timely intervention and a proactive attitude, regardless the space where they are, so that worst complications for the patients may be avoided. A pervasive approach will have a strong impact. Context awareness is an important issue to adapt the applications to the current situation [45].

Further work will encompass the test of the DM models generated so far, with online and real-time data from the ICU of HSA, in order to guarantee their accuracy or, in case their performance decays, to optimize them. The models presented used data manually entered and the next step is to use them with the new adjustments of the system, i.e., online and in real-time. Prediction, optimization and adaptability are features that make INTCare an ABI system, whose maid goal it to allow the medical staff to make better decisions, at the right time and place, improving quality in health care. At the same time the pervasive approach will be tested and evaluated remotely.

The integration with the various data sources and with the rest information systems of the hospital has been supported by the development of an ENR and further related work include its test in the ICU and subsequently, its optimization.

Acknowledgements. The authors would like to thank FCT (Foundation of Science and Technology, Portugal) for the financial support through the contract PTDC/EIA/72819/2006.

References

1. Mahmoud, M.: Real-time data acquisition system for monitoring patients in intensive care unit. In: Multisensor, Multisource Information Fusion: Architectures, Algorithms and Applications, vol. 5090, pp. 320–326 (2003)
2. Pereira, M., et al.: Computer aided monitoring system of intensive care unit patients. WSEAS Transactions on Information Science and Applications 4, 78–84 (2007)
3. Gardner, R.M., et al.: Real time data acquisition: recommendations for the Medical Information Bus (MIB). International Journal of Clinical Monitoring and Computing 8, 251–258 (1991)
4. Ying, Z., et al.: Real-Time Evaluation of Patient Monitoring Algorithms for Critical Care at the Bedside. In: 29th Annual International Conference of the IEEE Engineering in Medicine and Biology Society, EMBS 2007, pp. 2783–2786 (2007)

5. Morik, K.: Data analysis and knowledge validation in intensive care monitoring (2003)
6. Santos, M.F., et al.: Intelligent decision support in Intensive Care Medicine. In: 2nd International Conference on Knowledge Engineering and Decision Support, Lisbon, Portugal, pp. 401–405 (2006)
7. Gago, P., et al.: INTCare: a knowledge discovery based intelligent decision support system for intensive care medicine. Journal of Decision Systems (2006)
8. Gago, P., Santos, M.F.: Towards an Intelligent Decision Support System for Intensive Care Units. Presented at the 18th European Conference on Artificial Intelligence, Greece (2008)
9. Gago, P., et al.: Adaptive decision support for intensive care. In: 13th Portuguese Conference on Artificial Intelligence, Guimaraes, Portugal, pp. 415–425 (2007)
10. Silva, Á., et al.: Organ failure prediction based on clinical adverse events: a cluster model approach. In: 3rd International Conference on Artificial Intelligence and Applications (2003)
11. Silva, Á., et al.: Multiple organ failure diagnosis using adverse events and neural networks. In: 6th International Conference on Enterprise Information Systems, pp. 401–408 (2004)
12. Kwon, O., et al.: UbiDSS: a proactive intelligent decision support system as an expert system deploying ubiquitous computing technologies. Expert Systems with Applications 28, 149–161 (2005)
13. Turban, E., et al.: Decision Support Systems and Intelligent Systems, 7th edn. Prentice Hall (2005)
14. Arnott, D., Pervan, G.: A critical analysis of decision support systems research. In: Conference on Decision Support Systems, Prato, Italy, pp. 67–87 (2004)
15. Michalewicz, Z., et al.: Adaptive Business Intelligence. Springer, Heidelberg (2007)
16. Negash, S., Gray, P.: Business intelligence. Handbook on Decision Support Systems 2, 175–193
17. Frawley, W.J., et al.: Knowledge Discovery in Databases: An Overview. AI Magazine 13, 57–70 (1992)
18. Fayyad, U.M., et al.: From data mining to knowledge discovery: an overview (1996)
19. Lourenco, A., Belo, O.: Promoting agent-based knowledge discovery in medical intensive care units. WSEAS Transactions on Computers 2, 403–408 (2003)
20. Silva, A.: Modelos de Inteligência Artificial na análise da monitorização de eventos clínicos adversos, Disfunção/Falência de órgãos e prognóstico do doente critico. Tese de doutoramento, Ciências médicas, Universidade do Porto (2007)
21. Hall, J.B., et al.: Principles of Critical Care: McGraw-Hill's Access Medicine (2005)
22. Ramon, J., et al.: Mining data from intensive care patients. Advanced Engineering Informatics 21, 243–256 (2007)
23. Rao, S.M., Suhasini, T.: Organization of intensive care unit and predicting outcome of critical illness. Indian J. Anaesth. 47(5), 328–337 (2003)
24. Vincent, J.L., et al.: The SOFA (Sepsis-related Organ Failure Assessment) score to describe organ dysfunction/failure. Intensive Care Medicine 22, 707–710 (1996)
25. Morik, K., et al.: Combining statistical learning with a knowledge-based approach-a case study in intensive care monitoring, pp. 268–277
26. Satyanarayanan, M.: Pervasive computing: vision and challenges. IEEE Personal Communications 8, 10–17 (2002)
27. Banavar, G., et al.: Challenges: an application model for pervasive computing, p. 274 (2000)
28. Varshney, U.: Pervasive Healthcare. Computer 36, 138–140 (2003)
29. Mikkonen, M., et al.: User and concept studies as tools in developing mobile communication services for the elderly. Personal and Ubiquitous Computing 6, 113–124 (2002)

30. Varshney, U.: Pervasive healthcare and wireless health monitoring. Mobile Networks and Applications 12, 113–127 (2007)
31. Dovey, S., Makeham, M., County, M., Kidd, M.: An international taxonomy for errors in general practice: a pilot study. The Medical Journal of Australia 177, 68–72 (2002)
32. Kohn, L.T., et al.: To Err Is Human: Building a Safer Health System. National Academy Press (2000)
33. Bergs, E.A.G., et al.: Communication during trauma resuscitation: do we know what is happening? Injury 36, 905–911 (2005)
34. Varshney, U.: Pervasive Healthcare Computing: EMR/EHR, Wireless and Health Monitoring. Springer-Verlag New York Inc. (2009)
35. Augusto, J.C.: Temporal reasoning for decision support in medicine. Artificial Intelligence in Medicine 33, 1–24 (2005)
36. Vilas-Boas, M., Portela, F., Santos, M.F., Machado, J., Abelha, A., Neves, J., Silva, A., Rua, F., Salazar, M., Quintas, C., Cabral, A.F.: Intelligent Decision Support in Intensive Care Units - Nursing Information Requirements
37. Abelha, A., et al.: Agency for Integration. Diffusion and Archive of Medical Information
38. Santos, M.F., et al.: INTCARE - Multi-agent approach for real-time Intelligent Decision Support in Intensive Medicine. Presented at the 3rd International Conference on Agents and Artificial Intelligence (ICAART), Rome, Italy (2011)
39. Jennings, N.R.: On agent-based software engineering. Artificial Intelligence 117, 277–296 (2000)
40. Fonseca, T., et al.: Vital Signs in Intensive Care: Automatic Acquisition and Consolidation into Electronic Patient Records. Journal of Medical Systems 33, 47–57 (2009)
41. Portela, F., Santos, M.F., Vilas-Boas, M., Machado, J., Abelha, A., Neves, J., Silva, A., Rua, F., Salazar, M., Quintas, C., Cabral, A.F.: Intelligent Decision Support in Intensive Care Units - Nursing Information Requirements. WSEAS Transactions on Informatics (2009)
42. Santos, M.F., Portela, F., Vilas-Boas, M., Machado, J., Abelha, A., Neves, J.: Information Architecture for Intelligent Decision Support in Intensive Medicine. Presented at the 8th WSEAS International Conference on Applied Computer & Applied Computational Science (ACACOS 2009), Hangzhou, China (2009)
43. Santos, M.F., et al.: Information Modeling for Real-Time Decision Support in Intensive Medicine. In: Chen, S.Y., Li, Q. (eds.) Proceedings of the 8th Wseas International Conference on Applied Computer and Applied Computational Science - Applied Computer and Applied Computational Science, pp. 360–365. World Scientific and Engineering Acad and Soc., Athens (2009)
44. Hooda, J.S., et al.: Health Level-7 compliant clinical patient records system, pp. 259–263
45. Villas Boas, M., et al.: Distributed and real time Data Mining in the Intensive Care Unit. Presented at the 19th European Conference on Artificial Intelligence - ECAI 2010, Lisbon, Portugal (2010)
46. Miranda, M., et al.: A group decision support system for staging of cancer. Electronic Healthcare 0001, 114–121 (2009)
47. Cook, D.J., Das, S.K.: Smart environments: technologies, protocols, and applications. Wiley-Interscience (2005)
48. Black, J.P., et al.: Pervasive Computing in Health Care: Smart Spaces and Enterprise Information Systems
49. O'Donoghue, J., Herbert, J., Sammon, D.: Patient Sensors: A Data Quality Perspective. In: Helal, S., Mitra, S., Wong, J., Chang, C.K., Mokhtari, M. (eds.) ICOST 2008. LNCS, vol. 5120, pp. 54–61. Springer, Heidelberg (2008)

Semantics and Machine Learning: A New Generation of Court Management Systems

E. Fersini[1], E. Messina[1], F. Archetti[1,2], and M. Cislaghi[2]

[1]DISCO, Università degli Studi di Milano-Bicocca, Viale Sarca 336, 20126 Milano, Italy
[2]Consorzio Milano Ricerche, Via Cicognara 7, 20129 Milano, Italy
{fersini, messina}@disco.unimib.it
{archetti, cislaghi}@milanoricerche.it

Abstract. The progressive deployment of ICT technologies in the courtroom, jointly with the requirement for paperless judicial folders pushed by e-justice plans, are quickly transforming the traditional judicial folder into an integrated multimedia folder, where documents, audio recordings and video recordings can be accessed via a web-based platform. Most of the available ICT toolsets are aimed at the deployment of case management systems and ICT equipment infrastructure at different organisational levels (court or district). In this paper we present the JUMAS system, stemmed from the homonymous EU project, that instead takes up the challenge of exploiting semantics and machine learning techniques towards a better usability of the multimedia judicial folders. JUMAS provides not only a streamlined content creation and management support for acquiring and sharing the knowledge embedded into judicial folders but also a semantic enrichment of multimedia data for advanced information retrieval tasks.

Keywords: Machine learning, Semantics, Court management systems.

1 Introduction

The use of Information and Communication Technologies (ICT) is considered one of the key elements for making judicial folders more usable and accessible to the interested parties, reducing the length of judicial proceedings and improving justice. The progressive deployment of ICT technologies in the courtroom (audio and video recording, document scanning, courtroom management systems), jointly with the requirement for paperless judicial folders pushed by e-justice plans [1], are quickly transforming the traditional judicial folder into an integrated multimedia folder, where documents, audio recordings and video recordings can be accessed usually via a web-based platform [2]. This trend is leading to a continuous increase in the number and the volume of case-related digital judicial libraries, where the full content of each single hearing is available for online consultation. A typical trial folder contains: (1) audio hearing recordings; (2) audio/video hearing recordings; (3) transcriptions of hearing recordings; (4) hearing reports; (5) attached documents (scanned text documents, photos, evidences, etc..).

The ICT container is typically a dedicated judicial content management system (court management system), usually physically separated and independent from the case management system used in the investigative phase, but interacting with it. Most of the

A. Fred et al. (Eds.): IC3K 2010, CCIS 272, pp. 382–398, 2013.

present ICT deployment has been focused on the deployment of case management systems and ICT equipment in the courtrooms, with content management systems at different organisational levels (court or district). ICT deployment in the judiciary has reached different levels in the various EU countries, but the trend toward a full e-justice is clearly in progress. Accessibility of the judicial information, both of case registries, more widely deployed, and of case e-folders, has been strongly enhanced by state-of-the-art ICT technologies. Usability of the electronic judicial folders is still affected by a traditional support toolset, being information search limited to text search, transcription of audio recordings (indispensable for text search) is still a slow and fully manual process, template filling is a manual activity, etc. Part of the information available in the trial folder is not yet directly usable, but requires a time consuming manual search. Information embedded in audio and video recordings, describing not only what was said in the courtroom, but also the way and the specific trial context in which it was said, still needs to be exploited. While the information is there, information extraction and semantically empowered judicial information retrieval still waits for proper exploitation tools. The growing amount of digital judicial information calls for the development of novel knowledge management techniques and their integration into case and court management systems. Different commercial solutions have been proposed on the market for addressing data, knowledge and e-discovery issues related to the judicial field (see [3] as survey). Most of the available products provide tools to identify, collect, process, review and produce information related to cases. The functionalities related to those tools are mainly concerned with the automatization of manual processes such as archiving trial details, acquiring scanned paper records and OCR, retrieval and consultation of digital material (as for instance minutes, manual trancriptions, pictures), management of court calendars, court event scheduler, court personnel assignment, and drafting dispositions or sentencing. The keyword that could describe these toolsets is automatization, where the role of semantics is completely discarded.

Several EU research projects have proposed actionable models of legal knowledge (as for example ESTRELLA[1] and ALIS[2]), and investigated interoperability of legal documents and the potential of text-based semantic analysis. Research about multimedia judicial trial folder and courtroom technologies in criminal trials has been addressed by e-COURT project and SecurE-Justice projects in FP5 and FP6, with the main objective of the digital trial folder and its secure accessibility. While some research projects were mainly aimed at developing standards-based platform for comprehensive legal knowledge representation (Legal Knowledge Interchange Format and Open XML Interchange Format for Legal and Legislative Resources), some others were mainly focused on managing digital trial folders from a structural point of view. Also in this case the role of semantics, especially related to the multimedia data acquired during hearings/proceedings, is completely disregarded. JUMAS project (JUdicial MAnagement by digital libraries Semantics), ended on Januray 2011 after a validation at the Court of Wroclaw (Poland) and the Court of Naples (Italy) with the support of Polish and Italian Ministries of Justice, faces the issue of a better usability of the multimedia judicial folders, including transcriptions, information extraction and semantic search, to provide to

[1] http://www.estrellaproject.org/

[2] http://www.alisproject.eu/index.php

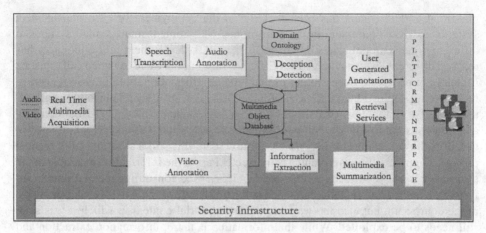

Fig. 1. Information flow in JUMAS

users a powerful toolset able to fully address the knowledge embedded in multimedia judicial folders. The JUMAS project has several scientific objectives:

- Knowledge Models and Spaces: Search directly in the audio and video sources without a verbatim transcription of the proceedings.
- Knowledge and Content Management: Exploit hidden semantics in audiovisual digital libraries in order to facilitate search and retrieval, intelligent processing and effective presentation of multimedia information. Research addresses also multiple cameras and audio sources.
- Multimedia Integration: Information fusion from multimodal sources in order to improve accuracy in automatic transcription and annotation phases.
- Effective Information Management: Streamline and Optimize the document workflow allowing the analysis of (un)structured information for document search and evidence base assessment.
- ICT Infrastructure: Service Oriented Architecture supporting a large scale, scalable, and interoperable audio/video retrieval system.

In this paper we present how the results of JUMAS can help to meet the challenges in analyzing audio and video recordings and outline the impact on ICT infrastructure in the court.

2 The JUMAS System

2.1 The JUMAS Concept

In order to explain the relevance of the JUMAS objectives we report some volume data related to the judicial domain context. Consider for instance the Italian context, where there are 167 courts, grouped in 29 districts, with about 1400 courtrooms. In a law court of medium size (10 court rooms), during a single legal year about 150 hearings per court held with an average duration of 4 hours. Considering that approximately in

40% of them only audio is recorded, in 20% both audio and video while the remaining 40% has no recording, the multimedia recording volume we are talking about is 2400 hours of audio and 1200 hours of audio/video per year. The dimensioning related to the audio and audio/video documentation starts from the hypothesis that multimedia sources must be acquired at high quality in order to obtain good performances in audio transcription and video annotation, which will in turn affect the performance connected to the retrieval functionalities. Following these requirements one can figure out a storage space of about 8.7 MB/min for audio and 39 MB/min for audio/video. The total amount of data to process in one year in a court is summarized by the following table.

Table 1. Dimension of the problem

Hypothesis of Data Amount per Year /court		
	Hearing Duration (hrs)	Required Space (TB)
Audio	2400	1,2
Audio/Video	1200	2,8

During the definition of the space dimension required on a single site, the estimation will also take into account that a trial includes some additional data: (1) textual source as for example minutes in .doc and .pdf format; (2) images; (3) other digital material. Under these hypotheses, the overall size generated by all the courts justice system only for Italy in one year is about 800 terabyte; this shows how the justice sector is a major contributor to the data deluge [4]. The typical deployment of JUMAS is at district level (about 5 courts), in which it is performed the coordination among courts. Installations in specific courts can be justified by the needs of particular trials, which involve huge quantities of very sensitive data (e.g. class actions, organized crime). The issue of scalability has to be considered for a proper dimensioning of the hardware infrastructure in which to run the system: the most computationally intensive part of JUMAS is the automatic audio transcription; the transcription engine based on an ensemble of Hidden Markov Models requires high performance computing. Also video processing has high computational needs. In order to manage such quantity of complex data, JUMAS aims to:

- Optimize the workflow of information through search, consultation and archiving procedures.
- Introduce a higher degree of knowledge through the aggregation of different heterogeneous sources.
- Speed up and improve decision processes discovering and exploiting knowledge embedded into multimedia documents, in order to consequently reduce unnecessary costs.
- Model audio-video proceedings in order to compare different instances.
- Allow traceability of the proceedings during their evolution.

2.2 System Architecture

The architecture of JUMAS is based on a set of key components: a central database, a user interface on a web portal and the integration and orchestration modules which

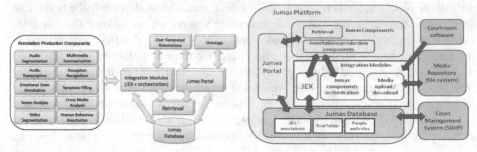

(a) JUMAS components and data flows   (b) Software architecture of JUMAS

Fig. 2. The JUMAS system

allows composing several media analysis components, as described in the following. Figure 1 presents the information flows in JUMAS: the media stream recorded in the courtroom includes both audio and video which are analyzed to extract semantic information used to populate the multimedia object database. The outputs of these processes are Annotations, which are the common name in JUMAS to identify tags that are attached to media streams and stored in the database (Oracle 11g).

The links between audio and video analysis components in Figure 2(a) show that the algorithms apply exchange information, using the annotations produced by one component as input to improve the performance of the other (e.g. through face recognition performed on video is possible to identify changes in speaker, which is used to cut audio streams to separate transcriptions). Figure 2(b) shows a higher level overview of the JUMAS components and the software architecture, which shows also the links with the external courtroom infrastructure.

The integration among modules is performed through a workflow engine and a module called JEX (JUMAS Exchange library). The workflow engine is a service application that manages all the modules for audio and video analysis (described in Sect. 3). It is written in Java; it defines an entity model for annotations with a corresponding XML schema for web services and a database schema for Hibernate persistence. JEX provides a set of services to upload and retrieve Annotations to and from the JUMAS database; these services are hosted in a J2EE application server, with SOAP support for direct usage. The integration of the JUMAS components is supported by a client library that allows manipulating objects that are stored in the DB through web services or direct database access, without having to cope with web services, SQL or XML. A client console multi-platform application exposes all the functionality of the JEX services through XML files. This integration schema allows clear and simple data exchange, making JUMAS flexible enough to be deployed in different configurations, including all or part of its components, and is also open to future improvements with the inclusion of new modules. In JUMAS it has been defined a wide set of annotations categories and events, to deal with the potentially wide set of annotations produced by the media analysis modules. The JEX client library (and the applications) can use two communication protocols (web services and database) because each one has pros and cons: web services are easier to be accessed by remote modules; moreover, this approach can be used directly in different languages/platforms. On the other side, the direct database access does not require the presence of the JEX service (i.e. of the J2EE application server). The users interact with JUMAS through the JUMAS Portal, a web application whose server runs where JUMAS is deployed. Clients can access the data through secure authentication. JUMAS actually integrates a court management system (namely SIDIP - Sistema Informativo DIbattimento Penale - (as shown in Figure 2(a)) for the creation of the trial folder, the management of lawyers and judges registries and the attachment of textual documents. The same results can be obtained by including alternative Court Management Systems (CMS). All the annotations produced by the JUMAS components and stored in the Database serve as basis for the Hyper Proceedings View described in Sect. 5.1.

3   Knowledge Extraction

3.1   Automatic Transcription

A first fundamental information source, for a digital library related to the courtroom debate context, is represented by the audio recordings of actors involved into hearings/proceedings. The automatic transcription is provided by an Automatic Speech Recognition (ASR) system [5] trained on real judicial data coming from courtrooms. Currently two languages, Italian and Polish, have been considered for inducing the models able infer the transcription given the utterance. Since it is impossible to derive a deterministic formula able to create a link between the acoustic signal of an utterance and the related sequence of associated words, the ASR system exploits a statistical-probabilistic formulation based on Hidden Markov Models [6] In particular, a combination of two probabilistic models is used: an acoustic model, which is able to represent phonetics, pronounce variability and time dynamics (co-utterance), and a language model able to represent the knowledge about word sequences. The audio acquisition chain in the courtroom has been designed specifically to improve the Word Error Rate (WER), using lossless compression such as FLAC and cross-channels analysis. This allows a good trade-off between the conflicting needs of a manageable dimension of the audio file and good quality recording. The ASR system developed was facing with several domain constraints/limitations:

- Noisy audio streams: the audio stream, recorded during a judicial trial, can be affected by different types of background environmental noises and/or by noises caused by the recording equipment.
- Spontaneous Speech: the sentences uttered by a speaker during a trial are characterized by breaks, hesitations, and false starts. Spontaneous speech - with respect to read utterance - plays a fundamental role in ASR systems, by originating higher values of WER.
- Pronounce, language and lexicon heterogeneity: the actors interacting during a trial may be different for language, lexicon and pronounce types. In particular, the judicial debates could contain many words (e.g. person names, names of Institutions or Organizations, etc.) that are not included in the dictionary of the ASR, thus increasing the number of out-of-vocabulary words and, consequently, the resulting WER.

- Variability of a vocal signal: the word sequences provided by the ASR can be influenced by different circumstance such as posture of the speaker, emotional state, conversation tones, and different microphone frequency responses. These elements introduce perturbations that are difficult to be taken into account during automatic speech transcription activities.
- Non-native speakers: the actors involved in a proceeding can be characterized by linguistic difficulties or can be non-native speaker. These linguistic distortion negatively affect the accuracy of the produced transcriptions.

If we compare error rates of speech recognition systems, it is important to distinguish between different speech recognition tasks, since acoustic conditions, degree of language and speaker variability have a big impact on word error rate. An estimation of the expected error rates is summarized in Table 2, according to the reported speech recognition tasks. Considering that the courtroom environments can best be described as meetings with distant microphones, and that the expected WER is less than 50%, the ASR modules enclosed into the JUMAS system can be considered as a really promising tool that is able to reduce the WER to 38%.

The generated automatic transcriptions represent the first contribution for populating the digital libraries behind the judicial trials. In fact, the produced transcriptions are the main information source that can be enriched by other modules and then can be consulted by the end users through the information retrieval system.

Table 2. Expected WER

	WER
Connected digits	$\leqslant 0.5\%$
Continuous dictation	$\leqslant 5\%$
Studio broadcast news	$\leqslant 10\%$
Telephone news reports	$\leqslant 20\%$
Telephone conversations	$\leqslant 30\%$
Meetings (head mounted microphone)	$\leqslant 30\%$
Meetings (distant microphone)	$\leqslant 50\%$

3.2 Emotion Recognition

Emotional states represent a bit of knowledge embedded into courtroom media streams. This kind of information represents hidden knowledge that may be used to enrich the content available in multimedia digital libraries. The possibility for the end user to consult the transcriptions, also by considering the associated semantics, represents an important achievement that allow them to retrieve an enriched written sentence instead of a flat one. This achievement radically changes the consultation process: sentences can assume different meanings according to the affective state of the speaker.

In order to address the problem of identifying emotional states embedded into courtroom events, an emotion recognition system is comprised into the JUMAS system. A set of real-world human emotions obtained from courtroom audio recordings has been gathered for training the underlying supervised learning model. In particular, this corpus encloses a set of 175 sentences uttered by different actors involved into the considered debates, i.e. judges (46 samples), witnesses (67 samples), lawyers (29 samples) and prosecutors (33 samples). The corpus contains emotional speech signals coming from 95 males and 80 females, with a duration ranging from 2 to 25 seconds. The dataset contains the following emotional states: anger, neutral, sadness and happiness. Given the emotional corpus, a features extraction step is performed in order to map the vocal signals into descriptive attributes (prosodic features, formant frequencies, energy, Mel Frequency Cepstral Coefficients, etc...). Given this representation, a supervised emotion recognition model can be trained. Into the emotion recognition component, a Multi-layer Support Vector Machines (SVMs) have been defined [7].

At the first layer a Gender Recognizer model is trained to determine the gender of the speaker, for distinguishing the "male" speakers from the "female" ones. In order to avoid overlapping with other emotional states, at the second layer gender-dependent models are trained. In particular, Male Emotion Detector and Female Emotion Detector are induced to produce a binary classification that discriminates the excited emotional states by the not excited ones (i.e. the neutral emotion). The last layer of the hierarchical classification process is aimed at recognizing different emotional state using Male Emotion Recognizer and Female Emotion Recognizer models, where only sentences uttered as excited are used to train the models for discriminating the remaining emotional states. Since SVMs are a linear learning machine able to find the optimal hyperplane separating two classes of examples (binary classification) and in our final layer we have a multi-class problem, we adopted the pairwise classification approach [8]. In this case, one binary SVMs for each pair of classes is learned to estimate the posterior probability to assign an instance to a given class label. A given instance is finally associated to the class with the highest posterior. The performance obtained by the proposed Multi-Layer Support Vector Machine has been compared to those provided by the traditional "Flat" machine learning approaches and by some recent hierarchical approaches. The comparison highlights an improvement (~6-9%) in terms of percentage of correctly classified affective states obtained by the Multi-layer Support Vector Machines with respect to the benchmark algorithms.

3.3 Human Behaviour Recognition

A further fundamental information source, for a semantic digital library into to the trial management context, is concerned with the video stream. Recognizing relevant events that characterize judicial debates have great impact as well as emotional state identification. Relevant events happening during debates trigger meaningful gestures, which emphasize and anchor the words of witnesses, highlighting that a relevant concept has been explained. For this reason, human behaviour recognition modules have been included into the JUMAS system. The human behaviour recognition modules capture relevant events that occur during the celebration of a trial in order to create semantic annotations that can be retrieved by the end users. The annotations are mainly concerned with the events related to the witness: change of posture, change of witness, hand gestures, fighting. The Human Behaviour Recognition modules are based on motion analysis able to combine localization and tracking of significant features with supervised classification approaches. In order to analyze the motions taking place in a video, the optical flow is extracted as moving points. Then active pixels are separated from the static ones for finally extracting relevant features and for recognizing relevant judicial events [9], [10]. The set of annotations produced by the human behaviour recognition modules provide useful information for the information retrieval process and for the creation of a meaningful summary of the debates (see section 5.2). The human behaviour recognition modules developed was constrained by several domain limitations:

- Quality of the input video stream: the video capturing equipment used into courtroom is mostly low-cost. This resulting in low quality input video stream has a great impact into the recognition of relevant events.
- Stationary camera hampers shot detection: cameras are usually installed in fixed positions with a background that remains nearly stationary during the whole process. Indeed, it is difficult to cut the video-stream into shots for then understanding eventual relevant events.
- Long distance shots: video is usually shot from long distance. This implies that all involved people belong to the shot and, consequently, features of peoples faces are most of the time not distinguishable. Therefore, video analysis tasks, as for instance face recognition or expression analysis, become extremely difficult or even impossible.

In order to address the mentioned challenges, the acquisition chain related to the video source has been opportunely tuned. In order to allow the video analysis procedures to use a high quality stream albeit limiting the growth of the dimension of the video into the judicial folder, a double chain has been developed. The high quality video is analyzed by the human behaviour recognition components, while the low quality video is stored into the judicial folder and synchronized with the extracted semantic annotations. Concerning the experimental investigation, although most events of interest - with correct annotations and semantics - have been detected, false alarms are also present. For instance, many small motions between the lawyers are detected, which usually do not trigger any significant event. Considering that the precision over 10 videos is close to 63%, with a recall of 94%, the false alarms could be easily reduced by tuning the detection threshold enclosed into the event recognition modules.

3.4   Deception Detection

The discrimination between truthful and deceptive assertion is one of the most important activity performed by judges, lawyers and prosecutors. In order to support their reasoning activities, aimed at corroborating/contradicting declarations (lawyers and prosecutors) and judging the accused (judges), a deception recognition module has been developed as a knowledge extraction component. The deception detection module stands at the end of the data processing chain, as it fuses the output of the ASR, Video Analysis, and Emotion Recognition modules. This module is based on the idea initially presented by Mihalcea and Strapparava [11], where a variety of machine learning algorithms have been investigated to distinguish between truth and falsehood. The deception detection module developed for the JUMAS system is based on Nave Bayes and Support Vector Machines classifiers [12] . To study the distinction between true and deceptive statements, we required a corpus with explicit labelling of the truth value associated with each statement. In order to train the model, a manual annotation of the output of the ASR module - with the help of the minutes of the transcribed sessions has been performed. The knowledge extracted for then training classification models is concerned with lies, contradictory statements, quotations and expressions of vagueness. To date, 8 sessions ($\sim$70.000 words) have been annotated, yielding 88 lies, 109 contradictory statements, and 239 expressions of vagueness. Given the training data, the Nave Bayes classifier can be induced for providing indications about future trials. In fact, the deception indications are provided only to the judges by highlighting into the text transcription, through an interactive interface, those relevant statements derived from verbal expression of witnesses, lawyers and prosecutors. In this way the identified statements may support the reasoning activities of the judicial actors involved in a trial by triggering relevant portion of the debate representing cues of vagueness and contradiction. A preliminary study on vagueness annotations for the Italian Corpus was carried out using unigrams, bigrams, and skipgrams in a Nave Bayes classifier, yielding 96% Precision and 64% Recall . Similar results have been obtained on the Polish corpus by using Support Vector Machines and a bag-of-words feature representation.

3.5 Information Extraction

The current amount of unstructured textual data available into the judicial domain, especially related to transcriptions of debates, highlights the necessity to automatically extract structured data from the unstructured ones for an efficient consultation processes. In order to address the problem of structuring data coming from the automatic speech transcription system, we defined an environment that combines regular expression, probabilistic models and relational information. The key element of the information extraction functionality is represented by the Automatic Template Filling component, which is based on a probabilistic model for labelling and segmenting sequential data by handling the correlation among features. In particular, a probabilistic framework based on Conditional Random Fields [13] for labelling a set of trial transcriptions coming from an ASR system, has been developed for JUMAS. The core elements of the information extraction module can be distinguished in:

- Data: training data represented by annotated transcriptions and domain knowledge information available into (national) relational databases. The main information sources used for producing a structured view of unstructured texts are represented by:
  - Automatic speech transcriptions that correspond to what is uttered by the actors involved into hearings/proceeding. The ASR output related to 20 sessions has been manually annotated ($\sim$165.000 words), yielding about 3.500 semantic annotations (both for Italian and Polish trials). The semantic items stated for annotating include: name of the lawyer, name of the defendant, name of the victim, name of the witness, name of cited subjects, cited date and date of the verdict.
  - Domain knowledge information, which correspond to databases containing information about National Lawyers, National Judges, Common Weapons, etc

- Model: Conditional Random Fields, which are discriminative graphical models trained with both transcriptions as training examples and domain knowledge as additional information. The traditional model has been extended in order to train from ASR features and from domain knowlegde databases.

The Information Extraction functionalities are provided to judges, lawyers, prosecutors and court clerks. In particular, the structured information is exploited in two different ways: (1) to provide additional information, to the Information Retrieval component, for an efficient storage and retrieval of proceedings; (2) to provide, through an interactive user interface, an immediate overview of trial contents for consequently speeding up the consultation process. Concerning the experimental investigation, the computational results show that Extended CRFs are able to improve both precision and the recall with respect to the traditional CRFs. In particular for the Italian case while precision is pretty close to the state of the art model, the recall shows a relative improvement around 20%. Regarding the Polish language, the relative improvement relates to both measures: ∼20% for precision and 35% for recall.

4   Knowledge Management

4.1   Information Retrieval

Currently the retrieval process of audio/video materials acquired during a trial needs the manual consultation of the entire multimedia tracks. The identification of a particular position on multimedia stream, with the aim at looking/listening at/to specific declarations, participations and testimonies, is possible either by remembering the time stamp in which the events were occurred or by watching the whole recording. In order to address this problem, an Information Retrieval system should address the following challenges for retrieving relevant multimedia objects: (1) Users tend to specify the queries by using only few keywords; (2) Search terms might be ambiguous or simply not the right ones; (3) The information retrieval system might not be able to automatically extract the information necessary, e.g., in case the search term relates to a high-level concept that cannot be understood at the machine level.

The conjunction of automatic transcriptions, semantic annotations and ontology representations (outlined in section 4.2), allow us to build a flexible retrieval environment based not only on simple textual queries, but on wide and complex concepts. In order to define an integrated platform for cross-modal access to audio, video recordings and their automatic transcriptions, a retrieval model able to perform semantic multimedia indexing and retrieval has been developed [14]. In particular, a linear combination of the following information has been developed:

- Similarity of representative frames of shots.
- Face detector output for topics involving people.
- High level feature considered relevant by text based similarity.
- Motion information extracted from videos.
- Text similarity based on ASR lattices.

The main goal of the information retrieval module is to provide the users with a flexible search system on judicial documents through the realization of the following type of services:

- Basic search: specification of single terms, list of terms, pairs of adjacent terms, prefixes or suffixes.
- Advanced search: specification of linguistic query weights associated with single terms, specification of linguistic quantifiers (most, all, at least n) to aggregate the terms, searching in specified XML sections of documents ordered by importance scores, and query translation based on bilingual dictionaries.
- Semantic search: the user may not only define simple text keyword queries, but also rely on wide and complex concepts based on multimedia content or ontology usage.

In this way, all the relevant information of a trial may be retrieved in terms of multimedia objects (audio, video and text with the associated embedded semantics) by using low level textual queries and high level semantic concepts.

4.2 Ontology as Support to Information Retrieval

An ontology is a formal representation of the knowledge, which characterizes a given domain, through a set of concepts and a set of relationships that hold among them. Into the judicial domain, ontologies represent a key element that support the retrieval process performed by the end users. Textual-based retrieval functionalities are not sufficient for finding and consulting transcriptions (and other documents) related to a given trial. A first contribution of the ontology component developed in the JUMAS system is concerned with its query expansion functionality. Query expansion aims at extending the original query specified by the end users with additional related terms for then automatically submitting the whole set of keywords to the retrieval engine. The main objective is to narrow the focus (AND query) or to increase recall (OR query). When expanding the query, new terms are enhanced with a confidence weight used by the scoring function of the retrieval component. Let's introduce an example in order to understand the aim of the query expansion functionality in the whole searching process. Suppose that a judge needs to retrieve the transcription about a weapon crime: a knife. By specifying knife in the search form the system finds only few documents containing this word. Then the retrieval module invokes the query expansion web service in order to obtain the related terms. The web service finds the knife instance in the ontology and explores the ontological relationship in order to find related terms. The query expansion module is a web service based on Jena [15] that provides a programmatic environment for OWL ontologies and includes a rule-based inference engine. A further functionality offered to the end user is related to the possibility of knowledge acquisition. In fact, the ontology component offers to the judicial users the possibility of acquiring specific domain knowledge, i.e. they have the opportunity of specifying semantic relationships among concepts available into the trial transcriptions. A final contribution of the developed ontology is related to the possibility to augment the query-based multimedia summarization toolset presented in section 5.2. This knowledge management component provides not only the possibility of more advanced and accurate search, but also the opportunity to contribute to the construction of sophisticated domain ontology without any background knowledge about the ICT and ontological modelling aspects.

4.3 User Generated Semantic Annotations

Judicial users usually tag manually some papers for highlighting (and then remembering) significant portion of the debate. An important functionality offered by the JUMAS system relates to the possibility of digitally annotating relevant arguments discussed during a debate. In this context, the user-generated annotations may help judicial users for future retrieval and reasoning processes. The user-generated annotations module enclosed into the JUMAS system allows the end-users to assign free tags to multimedia content in order to organize the trials according to their personal preferences. It also enables judges, prosecutors, lawyers and court clerks to work collaboratively on a trial, e.g. a prosecutor who is taking over a trial can build on the notes of his predecessor. The tags are analysed to suggest related tags to the user for search and to automatically find related documents that contain related terms. The user-generated semantic annotation module exposes all annotations to the common JUMAS JEX annotation infrastructure, which can be searched in by the retrieval models. To allow users of JUMAS to browse through the various available documents in a focused manner, upon viewing a specific document, a dedicated interface shows several tags, which can be used to browse the documents. The tags recommended by the module for a specific document are found by several techniques, which have been combined to a meta-recommender as proposed in [16]. In particular, Collaborative Filtering, Tag co-occurrence-based and occurrence-based recommendations have been enclosed into the meta-recommender module. As outcome, this module offers the possibility of personalizing contents according to the user preferences or working routines, providing then a better usability of multimedia contents.

5 Knowledge Visualization

5.1 Hyper Proceeding Views

As introduced in Section 2.2, the user interface of JUMAS is a web portal, in which the contents of the Database are presented in different views, to support the operations of clerks, judges and all the people involved in the trial. The basic view is the browsing of the trial archive, like in a typical court management system, to present general information (dates of hearings, name of people involved) and documents attached to each trial in the archive. JUMAS has also distinguishing features among which the automatic creation of a summary of the trial, the presentation of user generated annotations and, primarily, the Hyper Proceeding View (see Figure 3(a)), i.e. an advanced presentation of media contents and annotations that allows to perform queries on contents, jumping directly to relevant parts of media files. The Annotations are shown in the portal while media files are played and it is also possible to browse media by clicking on annotations. The contents can be browsed over several dimensions: audio, video, text annotations; the user can switch among them. This approach, typical of web applications, provides a new experience for the user that, in existing systems, can only passively watch the contents or perform time-consuming manual search.

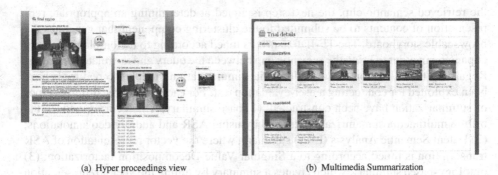

(a) Hyper proceedings view   (b) Multimedia Summarization

Fig. 3. Knowledge visualization

5.2 Multimedia Summarization

Digital videos represent a fundamental informative source of those events that occur during a trial: they can be stored, organized and retrieved in short time and with low cost. However, considering the dimension that a video source can assume during a trial recording, several requirements have been pointed out by judicial actors: fast navigation of the stream, efficient access to data inside and effective representation of relevant contents. One of the possible solutions to these requirements is represented by multimedia summarization aimed at deriving a synthetic representation of audio/video contents, characterized by a limited loss of meaningful information. In order to address the problem of defining a short and meaningful representation of a debate, a multimedia summarization environment based on an unsupervised learning approach has been developed [17].

The main goal is to create a (offline) storyboard of either a hearing or an entire proceeding, by taking into account the semantic information embedded into a courtroom recording. The storyboard construction creates a unified representation of contents. In particular, the summarization module exploits two matrices: one matrix associated with speech transcription and one matrix associated to the audio/video annotations. The first matrix represents textual transcription scoring, obtained through the TF-IDF weighting technique [18]. The second matrix, defined as binary, represents the presence or absence of a specific audio/video annotation associated to a given transcription segment. Starting from these two matrices, the multimedia summarization module may start the summary generation. The core component is based on a clustering algorithm named Induced Bisecting K-means [19]. The algorithm creates a hierarchical organization of (audio, video and textual as well automatic annotations) clips, by grouping in several clusters hearings or sub-parts of them according to a given similarity metric.

A more complex (online) multimedia summarization environment has been developed in order to tackle the problem of generating a meaningful summary according to a given user query enhanced by the domain ontology knowledge. In particular, once the end user has specified the query, it is submitted to the ontological query expansion module for deriving a set of semantically enriched keywords. The augmented query, joint with the eventually required semantic audio and video annotations, are provided to the search engine for retrieving those trials that match the required information. Given the retrieved semantic clip, the next step is aimed at determining an appropriate representation of contents to be submitted to the clustering component that generates the browseable storyboard. The TF-IDF matrix is tuned according to the ontological query expansion in order to bring the semantic gap between the query and the related concepts. Regarding the validation of the multimedia summarization modules, 6 real trials have been exploited for conducting a user-evaluation test. For each trial, three different types of summarization have been constructed: (1) base summary, which corresponds to the offline multimedia sumamrization derived by using ASR and audio/video annotations; (2) Latent Semantic Analysis (LSA) summary, where the vector representation of ASR transcription is tuned according to a Singular Value Decomposition factorization; (3) ontology-based summary, which creates a summary by tuning the vector representation of ASR text according to related concepts identified through the ontological query expansion module. The user opinion investigation highlighted a good impact (in terms of accuracy) related to the summary generated by the ontology-based approach ($\sim$65%), a medium significance related to the base summary ($\sim$40%) and a low accuracy for the LSA summary ($\sim$30%). The resulting storyboard is presented to judge, prosecutor, lawyer and court clerk as shown in Figure 3(b).

6   Potentiality of the JUMAS System

The JUMAS project and the related JUMAS system have been presented to several audiences both in Italy and Poland, with particular focus on Criminal Justice and Public administration targets. Some of the potentiality emerged during this demonstration phase are the following:

1. The JUMAS portal has been perceived as a straightforward tool for supporting common daily activity such as retrieval of transcriptions, consultation of multimedia streaming and annotation of relevant contents.
2. JUMAS could be used for linking different trials and therefore used as a learning tool exhibiting to young magistrates how a similar topic has been dealt with in different processes.
3. Environmental recordings, wired taps and recordings law enforcement interviews acquired during the investigative phase could be analysed by using the JUMAS functionalities for extracting relevant semantic information . This information could be used as confirmatory or drive towards the formulation of alternative hypothesis.
4. The tags associated to a video content, automatically suggested by the system or manually provided by the end-user, represent useful and appropriate information for the corresponding media stream. This functionality is appreciated for helping the end user to find and retrieve relevant cases and for cooperation/sharing purposes.
5. the JUMAS core could be exploited for addressing other important challenges in the judicial domain. For example it could be customised to be used as an E-learning platform for prosecutors and judges or an Analysis and visualization platform for the investigative/pre-trial phase or again as a cooperation platform for judges, prosecutors, lawyers and low enforcement (this is particularly relevant according to the Treaty of Lisbon).

7 Conclusions

The main objective of JUMAS is to support the establishment of procedures for acquiring and sharing the knowledge embedded into a corpus of multimedia documents. This objective is obtained through innovative techniques both in audio and video processing jointly with information retrieval services which improve usability, accessibility, scalability and cost effectiveness. Automatic template filling, semantic enrichment of the judicial folder through audio and video processing, enhanced transcription process, help judges, prosecutors not only to save time, but also to enhance the quality of their judicial decisions and actions. These improvements are mainly due to the possibility to search not only text, but also events that occurred in the courtroom. The demonstration and validation have provided valuable recommendation for the next generation of ICT-empowered courtrooms. Recording systems and ICT infrastructure in the courtroom that are actually under deployment can be designed in order to support audio and video processing capabilities, while information retrieval relies on state-of-the art ICT infrastructure. JUMAS is providing not only tools for a better usability of the trial folder, being the audio-related tools the closest to a first deployment, but also inputs for a more future oriented specification of e-justice systems in different countries.

Acknowledgements. This work has been supported by the European Community FP-7 under the JUMAS Project (ref.: 214306) and by "Dote ricercatori" - FSE, Regione Lombardia.

References

1. Council of the European Union: Multi-annual european e-justice action plan 2009-2013. The Official Journal of the European Union (OJEU) 75(1) (2009)
2. Velicogna, M.: Use of information and communication technologies (ict) in european judicial systems. In: European Commiission for the Efficiency of Justice Reports, Council of Europe (2008)
3. Kershaw, A., Howie, J.: Ediscovery institute survey on predictive coding. Technical report, Electronic Discovery Institute (2010)
4. The Economist: Data, data everywhere. The Economist (February 25, 2010)
5. Falavigna, D., Giuliani, D., Gretter, R., Loof, J., Gollan, C., Schlueter, R., Ney, H.: Automatic transcription of courtroom recordings in the jumas project. In: ICT Solutions for Justice (2009)
6. Rabiner, L., Juang, B.H.: Fundamentals of Speech Recognition. Prentice-Hall Inc. (1993)
7. Fersini, E., Messina, E., Arosio, G., Archetti, F.: Audio-Based Emotion Recognition in Judicial Domain: A Multilayer Support Vector Machines Approach. In: Perner, P. (ed.) MLDM 2009. LNCS, vol. 5632, pp. 594–602. Springer, Heidelberg (2009)
8. Hastie, T., Tibshirani, R.: Classification by pairwise coupling. In: Proc. of the 1997 Conference on Advances in Neural Information Processing Systems, pp. 507–513 (1998)
9. Briassouli, A., Tsiminaki, V., Kompatsiaris, I.: Human motion analysis via statistical motion processing and sequential change detection. EURASIP Journal on Image and Video Processing (2009)
10. Kovács, L., Utasi, Á., Szirányi, T.: VISRET – A Content Based Annotation, Retrieval and Visualization Toolchain. In: Blanc-Talon, J., Philips, W., Popescu, D., Scheunders, P. (eds.) ACIVS 2009. LNCS, vol. 5807, pp. 265–276. Springer, Heidelberg (2009)

11. Mihalcea, R., Strapparava, C.: The lie detector: explorations in the automatic recognition of deceptive language. In: Proc. of the ACL-IJCNLP 2009 Conference, pp. 309–312 (2009)
12. Ganter, V., Strube, M.: Finding hedges by chasing weasels: hedge detection using wikipedia tags and shallow linguistic features. In: Proc. of the ACL-IJCNLP 2009 Conference, pp. 173–176 (2009)
13. Lafferty, J.D., McCallum, A., Pereira, F.: Conditional random fields: Probabilistic models for segmenting and labeling sequence data. In: Proc. of the 18th International Conference on Machine Learning, pp. 282–289 (2001)
14. Daróczy, B., Nemeskey, D., Petrás, I., Benczúr, A.A., Kiss, T.: Sztaki @ trecvid 2009 (2009)
15. Carroll, J., Dickinson, I., Dollin, C., Reynolds, C., Seaborne, A., Wilkinson, K.: Implementing the semantic web recommendations. Technical Report HPL-2003-146, Hewlett Packard (2003)
16. Jäschke, R., Eisterlehner, F., Hotho, A., Stumme, G.: Testing and evaluating tag recommenders in a live system. In: Proc. of the 3rd ACM Conference on Recommender Systems, pp. 369–372 (2009)
17. Fersini, E., Messina, E., Archetti, F.: Multimedia summarization in law courts: A clustering-based environment for browsing and consulting judicial folders. In: Proc. of 10th Industrial Conference on Data Mining (2010)
18. Salton, G., Buckley, C.: Term-weighting approaches in automatic text retrieval. Information Processing and Management 24, 513–523 (1998)
19. Archetti, F., Campanelli, P., Fersini, E., Messina, E.: A Hierarchical Document Clustering Environment Based on the Induced Bisecting k-Means. In: Larsen, H.L., Pasi, G., Ortiz-Arroyo, D., Andreasen, T., Christiansen, H. (eds.) FQAS 2006. LNCS (LNAI), vol. 4027, pp. 257–269. Springer, Heidelberg (2006)

A Workspace to Manage Tacit Knowledge Contributing to Railway Product Development during Tendering

Diana Penciuc, Marie-Hélène Abel, and Didier Van Den Abeele

Alstom Transport, 93482 Saint-Ouen Cedex , France
Heudiasyc CNRS UMR 6599, University of Technology of Compiègne
BP 20529 , 60 205, Compiègne, France
{diana.penciuc,didier.van-den-abeele}@transport.alstom.com
marie-helene.abel@utc.fr

Abstract. Railway products development during the Tendering phase is very challenging and tacit knowledge of experts remains the key of its success. Therefore, there is a need to capture and preserve tacit knowledge used during Tendering in order to improve, in time, railway products development. In this context, we think that tacit knowledge management and organizational learning need to be intertwined and supported by a dedicated workspace. In this paper, we propose the model and the implementation of such a workspace.

Keywords: Tacit knowledge, Knowledge management, Organizational learning, Organizational memory, Railway products development, Product-line development.

1   Introduction

Modern large-scale companies are more and more dynamic and rise to the increasing customer demands by continually coming up with innovatory products. It becomes important to minimize the production cost of new products by shortening the product life cycle. One method of achieving this is reusability of not only the existing formalized knowledge (e.g. architectures), but also the knowledge that employees gained after years of experience.

This paper focuses on finding ways to discover, store and reuse this knowledge, referred to as tacit knowledge.

Product-line engineering is a method of creating products in such a way that it is possible to reuse product components and apply planned variability to generate new products. This yields to a set of different products sharing common features [3]. These common features are summed up in a core or a "reference platform" and used to engineer each of the products in the product-line. A few examples of commonalities (core assets) are: architecture, software components, documents etc. The advantage of the product-line engineering lies in the reusability of this "reference platform", which leads to significant gains, such as engineering work reduction, time-to-market and costs reduction, or improved quality. This paper addresses the case of an international company providing railway transport systems. Railway market is characterized by a great diversity and dynamics of demands guided by the customer's background (e.g. habits,

A. Fred et al. (Eds.): IC3K 2010, CCIS 272, pp. 399–412, 2013.

historical reasons), evolving technologies, competitors etc. In the context of this hetero-geneity, having a "reference platform" brings considerable improvements, but it cannot perfectly match each request. Thus, the problem of adapting it appears.

When responding to a customer's demand the core has to be properly adapted in order to map the specific needs of a customer. Adapting the reference platform relies not only on its explicit definition or adaptation rules, but also on the tacit knowledge [15] of the employees, on non-formalized practices and exchanges between employees etc. As this knowledge is volatile, it can be easily forgotten, put aside or ignored; contrary to explicit knowledge it cannot be stored and reused, and therefore cannot constitute a lasting capital for the firm [4]. Past decisions, lessons learnt, solutions to specific problems are a few examples of knowledge that may be lost. Consequences of this fact are: spending time to reinvent past or existing solutions, repeating mistakes from the past. Once codified, this knowledge could be a source for improving existing practices, avoid re-iteration and reinforce the "re-use" strategy.

Our goal is to find a solution for managing tacit knowledge involved in the process of adapting the "reference platform". A set of 40 employees was interviewed, and data gathered from their answers. Data revealed the need for tacit knowledge management. An analysis of this need led to a set of requirements for a future knowledge management platform and the proposal of first solutions.

The paper is organized as follows. First, we provide the context of our work. Next, we present the analysis of needs we carried out throughout interviews and emphasize the need of a workspace dedicated to managing tacit knowledge and supporting organizational learning. In order to justify our modelling and implementation choices of the workspace, we first discuss background information on Enterprise 2.0 technologies and the MEMORAe approach to organizational learning. In the final part we present the workspace model and its implementation. We conclude with future directions.

2   Context

It is a fact that one of the most critical processes of product development is "Tendering". This process is triggered as a consequence of a request for proposal (RFP) announce-ment and its purpose is to build a formal offer that will be submitted to the customer. It is an early stage of product development with a significant influence on contract es-tablishment and success of the final product. Furthermore, decisions taken here often cannot be changed and time allocated to this process can be very short (up to six month).

For these reasons, our research is focused on analyzing the "Tendering" process and providing knowledge management solutions adapted to this process needs.

2.1   The Tendering Process

The study was carried out in the company A. Its mission is to provide solutions for railway transport management. In the following, the stakeholders and the workflow of the Tendering process are described. Stakeholders are the customer and the "ad-hoc teams" of companies competing to obtain the contract. In general, in a railway context, the term "customer" may stand for:

- the demander (a society demanding a railway product, local authorities)
- the consultant (the demander can employ a consultant to write its demand)
- the operators of the railway system
- the final users (voyagers)

Throughout this paper, the term "customer" will be used to refer to parties involved in specifying the request for proposal (RFP), which encompass the first three above mentioned categories. "Ad-hoc teams" are built according to the specificity of each offer. Functions of the members are related to: technical, commercial, planning, and support to the Tendering process. Figure 1 presents a simplified workflow of the "Tendering" process within the studied Company A, competing with the Company B to obtain the contract of a customer.

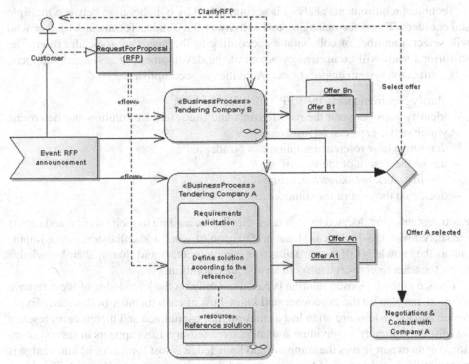

Fig. 1. Tendering process in the company A competing with the company B

In the company A, comparing the demand with the reference solution leads to the building of a customer solution reflected in the offer. Understanding of the RFP and customer's need is an iterative activity which results in several successive proposals coming from the two companies *(OfferA1, OfferB1;...;OfferAn, OfferBn)*. The customer then chooses the company proposing the most convenient offer and the Tendering process ends with negotiations and contract with the chosen company.

Given the complexity of this process and the diversity of specific knowledge coming from the various disciplines, we have decided to limit our study to the technical aspects of the "Tendering process". These are presented in more details in the following section.

2.2   The Technical Analysis

The purpose of the technical analysis is to understand the technical requirements of the customer and to propose an appropriate solution/system to answer the demand.

The entry to this activity is the technical part of the RFP and knowledge acquired during a previously-made rough analysis of the RFP. The RFP consists of any type of resources used by the customer to specify his need. Typical resources are one or more paper or digital documents containing text and graphics or images, but they can be videos as well.

The result of the activity is a "Technical Answer" of the offered system comprising mainly the architecture of the system, the specification of the requirements as understood by the team, the compliance with the demand and the cost of the solution.

Technical requirements analysis is accomplished by collaboration between multiple stakeholders. The actor managing this activity is the Tender Technical Manager who will select a number of collaborators according to the needs of a specific offer. The so-formed team will comprise key actors of the development of a system: suppliers, subcontractors, system engineers, etc. Activities to accomplish are:

- clarify requirements in the RFP
- identify gaps between the requirements and the reference solution and determine which of the gaps can be resolved
- determine how reference solution can be adapted
- define the design of the system
- identify work breakdown structure
- decide on the cost of the solution

Requirements are allocated to each team member according to their mission and results are shared with the Technical Manager. Division of work and collaboration are important as the knowledge of the participants is complementary and sharing their knowledge helps the team to converge faster to an adequate solution.

Choice of the proposed solution is based not only on the knowledge of the reference solution, but also on the experience and know-how of each member of this team. Experience and know-how are often lost as they are not capitalized and therefore represent a loss for the company's individual and collective memory. Loss appears in several situations: a) an expert leaves the company and knowledge is lost forever or b) knowledge is stored in the inactive part of the memory and therefore is not actively used. This latter is due to the unshared knowledge (individual written/unwritten knowledge), to difficulties to locate existing knowledge, or to ignorance. On the contrary, once capitalized, knowledge would serve as a means to speed-up the Tendering process, to improve practices for future offers, to trigger new knowledge and to support apprenticeship. Benefits are two-folded: boost individual knowledge as well as collective knowledge resulted not only from the sum of individual knowledge, but also from the added-value of the collaboration between individuals.

To accomplish this, members should be helped not only on their individual tasks but also on the collaborative ones. In order to understand how this could be accomplished, a set of needs were identified through a better understanding of the tendering process and particularly of the Technical analysis.

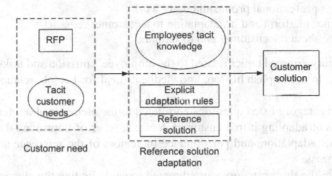

Fig. 2. Role of explicit and tacit knowledge in finding a good customer solution

These needs were further used to determine how knowledge management could respond to the problems previously mentioned.

3   Analysis of Needs

As stated before, success of a customer solution is due to: a good adaptation of the reference solution and learning and understanding customer's need. Figure 2 details the contribution of these elements by highlighting the explicit (represented by squares) and the tacit ones (represented by ellipses).

The adaptation of the reference solution is based not only on explicit rules but also on tacit rules dictated by employees' tacit knowledge. Knowing customer's need implies understanding the need expressed explicitly in the RFP but also the un-written, un-said need which may be learnt by asking questions, observation etc. To summarize, three factors are determinant for the success of a customer solution: 1) the employees, 2) the reference solution and its adaptation to a demand and 3) the knowledge about a customer. In order to better understand how these three factors impact the "Tendering" process and the technical analysis, a study of the internal environment has been undertaken using interviews. Details about interviews data and results are exposed in the next paragraphs.

3.1   Interviews Data

From the sample of 80 employees selected, 40 of them agreed to collaborate. Employees have been selected from different divisions and with different functions, which allowed to have a diversity of opinions and a much deeper understanding of the business. Divisions considered were Tender group and its links to other divisions contributing to carry out activities during tender: finances, platform group etc. The subjects were selected based on the suggestions coming from the management and from the already interviewed employees.

Therefore, we selected a set of 15 questions directed to find out more about:

- employee's professional profile and daily work
- the reference platform and its adaptation to a customer request
- knowledge about the customer and its demand

The first category grouped topics related to the employee's mission and tasks, tools and resources used to accomplish his job, knowledge critical for his job, relations to other people.

The second category listed questions correlated to the elements of a reference solution, problems of adapting it to a customer demand, ways of improving the reference solution and its adaptation, and performance indicators of the reference solution and adaptation process.

Questions in the third category were directed towards finding the characteristics of a request for proposal, revealing problems related to the understanding of the RFP, and determining the elements impacting the adaptation of the reference solution to the request stated in the RFP.

3.2 Interviews Results

The results of our interviews are two-folded:

- a number of activities impacted by prevalent tacit knowledge use were identified
- specific needs were identified

In what the first matter concerns, the selection criteria consisted in the degree of human contribution in accomplishing the activities. By way of illustration, we cite: "capture customer needs", "estimate risks", "identify gaps between the reference solution and the demand of the customer".

As regard to the technical analysis, results were grouped into three categories which are: employee's professional profile and daily work, the reference platform and its adaptation to a customer request, and knowledge about the customer and its demand. According to these categories, needs identified were: a common vocabulary inside the company and between the company and the customer, advanced tools to capitalize key knowledge (past experience, expertise etc.), adequate solutions to better handle the complexity of the reference solution, and to capture more knowledge about the customer. Detailed results of the interviews are presented in [14].

The study revealed that a heterogeneity of tools/methods are used to accomplish technical analysis, which does not allow a rigorous capitalization of knowledge. Informal knowledge coming from individual work and informal exchange (e.g. meetings) are lost, as they are not registered.

Observations showed that a common workspace supporting individual and collective work during the technical analysis is needed, although it does not exist today. This will sustain formal and informal exchange of knowledge. Existing technologies were studied in order to establish the basis for the future workspace.

4 Related Work

The preliminary study of the Technical analysis revealed the importance of a workspace enabling knowledge capitalization, sharing and new knowledge creation. Moreover, it clearly showed the importance of tacit knowledge and collaboration in accomplishing tasks during Technical analysis.

The way collaborators are chosen, decisions are taken, priorities are identified, anticipation is accomplished, are all issues of tacit knowledge. Nonaka [11] considers that through socialization, the barriers of tacit knowledge can be overcome and knowledge transmitted to others and thus become a collective good. For all these reasons we can affirm that we are facing a problem of organizational learning and therefore we are seeking for appropriate means to challenge it. Indeed, according to [18], an organizational learning process is 3-folded and concerns three processes: individual learning, social learning (allowing collaboration between individuals) and knowledge management.

Considering this, we chose to examine Enterprise 2.0 technologies and MEMORAe [1], an approach to support organizational learning.

4.1 MEMORAe

The aim, within the MEMORAe approach, is to put into practice organizational learning. In order to do so, this approach associates: knowledge management to support capitalization, semantic Web to support sharing and interoperability and Web 2.0 to support the social process [9]. The underlying concept for the knowledge management is the organizational memory, which Dieng [6] defines as an: "explicit, disembodied, persistent representation of knowledge and information in an organization, in order to facilitate its access and reuse by members of the organization, for their tasks". By adapting this concept to the learning process, the concept of Learning Organizational Memory was proposed and its implementation uses ontologies that index learning resources. Social processes in exchange, are facilitated by Web 2.0 technologies.

The advantage of the MEMORAe approach over using Enterprise 2.0 technologies is the integration of the workbench, learning and socialization support into the same platform. From the knowledge management perspective, this allows to accomplish a directed knowledge management in such a way that knowledge creation and sharing are guided by the learning process, which avoids knowledge overabundance and favors innovation [2]. The implementation of this approach is the eMEMORAe2.0 environment. As shown in Figure 3, the interface contains the following elements: the ontology tree in the middle, a menu at the top of the page from which the user may select some elements of general interest (agenda, content annotation, etc.), and some individual boxes (windows) displaying different categories of interest for the user: individual, group or organizational content, a list of entry points (concepts) that allow to locate concepts of interest in the ontology.

4.2 Enterprise 2.0

Enterprise 2.0 is a term first used by McAffee in 2006 [10] to name digital "platforms that companies can buy or build in order to make visible the practices and outputs of their knowledge workers". These platforms are the equivalent for enterprise intranet of the popular "Web 2.0" technologies on the Internet, bringing to the light the benefits of socialization and collaboration.

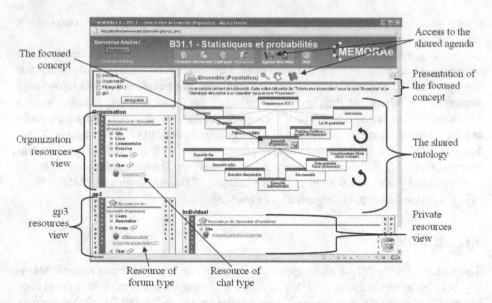

Fig. 3. eMEMORAe2.0 interface (in French)

We considered for our study the following technologies enabling knowledge capture, organizing, storing and sharing: RSS feed, wiki, blog, microblog, forum, social networks, folksonomies. RSS feed can be used for real-time capture of knowledge from the sources one is interested with. This technology is not Web2.0 specific but it could be suitable when combined with collaborative tools, given the short time available for the team to accomplish its mission.

Microblogs would be appropriate to communicate short pieces of news and guide users to other sources more complex of information (in the way Twitter does it). A comparison of 19 enterprise microshraing tools is given in [7]. Blogs could be employed by each team member to write their own thoughts and notes regarding assumptions made or notes about requirements etc. Forum could be employed to allow exchange of questions/answers in order to clarify requirements not well understood.

A social network could keep in contact the members of the team with collaborators sharing the same interests (maybe a collaborator closer to the customer, or subcontractors) or help them detect experts. Generally, we can affirm that a social network can relate between them different communities of practice [8].

A wiki could be provided to increasingly "build" knowledge on a specific customer. A study presented by [16] showed that, in order to provide concrete results, corporate wikis have to solve a clearly specified problem crucial for the business and the work practices of employees.

An analysis of commercial and open source Enterprise 2.0 tools according to the services they provide is presented in [5].

We have shown how Enterprise 2.0 technologies may contribute to the socialization and collaboration processes. Nevertheless, these technologies should be combined in order to provide efficient support. Moreover, a simple combination of tools is not enough and therefore, further knowledge management support has to be added to them in order to increase their capability.

5 Workspace Modelling

Based on the previous observations, it was decided that a workspace was needed to support organizational learning in the way MEMORAe approach does it.

In order to explain how this workspace was modelled, we first expose in the next sections several scenarios revealing different use cases.

5.1 Workspace Use Cases

We consider the case of a RFP containing text documents and we take as example the following statement contained in the RFP: "The system shall provide an emergency power production system synchronous with the public supply". Typically, a reader reinterprets this statement according to his experience: "Power supply should be continuous even in the case of an incident".

Case 1. The reader may choose to make an annotation to the initial statement in order to remember easily its meaning. The expression "emergency power production system" is not usual, but due to his experience he understands that the customer is talking about a source of energy which may be a battery or an electric generator.

Case 2. There is no other sentence in the RFP that can clarify customer need. The reader decides, taking into account the context of the demand what to offer to the customer.

Case 3. There is no other sentence in the RFP that can clarify customer need. The reader decides he has not sufficient information to take a decision. He may: a) consult his collaborators or b) propose to send the question to the customer or c) consult resources he may consider relevant (e.g. previous demands of the same costumer).

Case 4. When customer consulted, the following situations may appear: a) requirement is reformulated, as to be clear for all the parties; b) a new requirement will be added, to specify missing points.

Case 5. The reader finds another statement specifying that "The emergency power production system should be able to function at least 2 days". The reader infers that the customer needs an electric generator, given that a battery could not provide supply for such a long period.

Case 6. It was decided the customer needs an electric generator, but the reference solution does not support this component, so the requirement is considered a gap. Two cases are possible: a) if the gap can be solved, statement is marked compliant; b) if the gap cannot be solved, statement is marked non-compliant.

5.2 Workspace Requirements

The use cases led us to the decision that the workspace will have two main components: a workbench and a communication space. The workbench will be dedicated to individual learning while the communication space will be dedicated to the collaborative learning. In addition, a knowledge management component will support the capitalization of knowledge emerging from the two spaces and will allow users to query and search for existing knowledge. The requirements were then grouped in: requirements for the workbench and requirements for the communication space.

Requirements for the Workbench. The workbench should satisfy at least the requirements:

- The workbench shall contain all necessary resources for one to do his job. Resources consist of RFP resources and personal resources which one may need in order to accomplish his tasks (like stated in Case 2c)
- The system shall provide to the user a means to visualize the requirements to be analyzed or any other resource related to a customer or his requirements
- User shall be able to edit his notes (as free-text annotations) while reading a statement contained in the RFP
- The system should allow new requirements registration for the case new needs, which are not already specified, are identified (like in Case 3b)
- The system should allow multiple annotations on a requirement
- Each user will decide on the visibility of each of his annotations to other stakeholders

Requirements for the Communication Space. The communication space should satisfy at least the requirements:

- The communication space shall allow user to collaborate with other stakeholders and to obtain/transmit real-time knowledge about a topic he is interested in, or news
- System shall allow authorized users to create ad-hoc communities (e.g. tender technical manager should be allowed to create a tender technical team)
- The system should allow easy communication between the team members on an annotation of a requirement (Case 2, a),b)).
- The system should provide a way for a user to find and stay in contact with any other person/community that could help him in his work (e.g. with a community dedicated to the customer owner of the current RFP)
- A user will be informed each time news appear (e.g. new instructions are given from the management) or a task was allocated to him
- System should allow a user to locate knowledge about a topic

5.3 Workspace Model

Once requirements defined, we proceeded to the modelling of the workspace. A simplified model is presented in Figure 4, which highlights workbench and communication space elements as well as the way the two are intertwined through one or more topics.

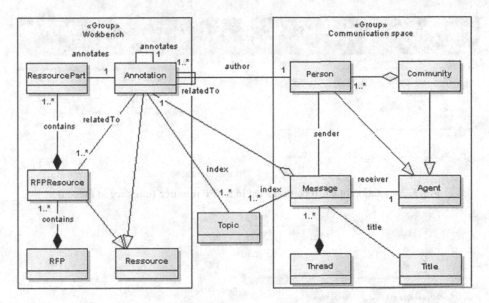

Fig. 4. Workspace model

One or more topics will be chosen by a user to index an annotation made on a part of a resource. Users will contribute with their knowledge externalized trough annotations, messages exchanged or other resources and stored in a knowledge base. Once stored, users will have the possibility to look-up for knowledge by locating it trough a direct search or by locating a potential owner of this knowledge via the communication space.

The workbench main elements are the RFP resources and the annotations one can make on existing resources. An annotation may annotate a part of a resource (e.g. a text representing a requirement) or another annotation (e.g. an annotation which is the response to another annotation edited by a stakeholder). Annotations may be related one to another in two cases: 1) they correspond to the same topic or 2) the user decides to make a direct link between them via the "relatedTo" relation. This latter can also be used to relate the annotation to a resource (e.g. the user decides to attach a drawing to complete the annotation).

The author of an annotation is a Person (e.g. a member of the technical team) which may choose to send it as part of a message to an Agent (e.g. a colleague or to the whole community: the technical team).

The communication space is represented by persons/communities and the threads tracking messages exchanged between users. A Message contains an annotation and corresponds to one or more topics.

We note that topics allow not only to index annotations but also the messages exchanged in the network, allowing therefore to capture and store informal content and link it to formal content which is the RFP content. Topics are concepts of an ontology of the reference solution and its adaptation. As stated in [17], the role of the ontology is to: 1) assist communication between people and organizations, 2) achieve interoperability between systems, 3) improve system engineering.

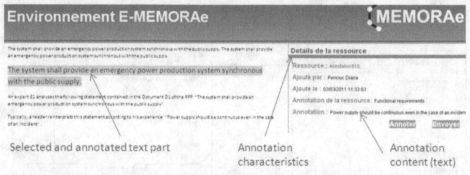

Fig. 5. Annotation of a text part selected in a RFP resource (interface in French)

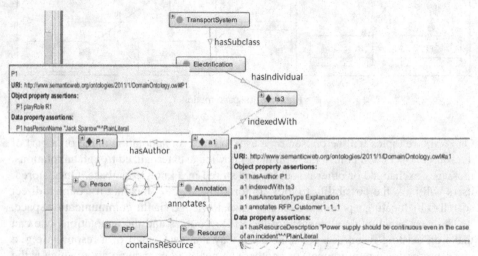

Fig. 6. View of the domain and application ontology after Person P1 annotates the selected text part with the annotation a1

The ontology will provide the common vocabulary, allowing understanding between not only the members of the technical team, but also between the technical team and the customer. Furthermore, once created, its consistency can be checked (when represented in an appropriate formalism) and reused.

Topics, along with the information about the members of the technical community and the information specific for a Tender (e.g. about the customer, the strategy, etc.) will provide a context to any knowledge captured during the Technical analysis. Contextual information can then be used to locate and retrieve needed knowledge. Further details about the workspace modelling are given in [12] and [13].

6   Workspace Implementation

A first implementation of the workspace was realized using the eMEMORAe2.0 environment presented earlier. For the purpose of this paper, we only explain how knowledge management is supported by the use of ontologies and how tacit knowledge is captured through annotation of the RFP content. Two ontologies were developed in OWL: a domain ontology that comprise general knowledge about the organization (persons, groups, type of resources etc.) and an application ontology that corresponds to the reference solutions available in the company. This second ontology is used when indexing content available in the stored resources.

Annotation management is supported by a module that allows user to visualize the content of the RFP resources and select the text to analyze. Figure 5 exemplifies the case "Case 1": an expert "Jack Sparrow" analyzes the sentence "The system shall provide an emergency power production system synchronous with the public supply" and writes an annotation to remember its meaning: "Power supply should be continuous even in the case of an incident". Next, he indexes the annotation with the concept "Electrification" of the ontology. We note that "Electrification" and "Power supply" are equivalent terms. This observation led us to a future objective for our research which is to define vocabularies which would give the alternative terms for concepts defined. Corresponding to this case, Figure 6 shows the domain and application ontologies. The domain ontology contains the concepts Person, Resource, Annotation, RFP and TransportSystem with the subclass Electrification. The application ontology contains a detailed description of "Electrification" and the individuals P1 (a Person) and a1 (the annotation created by P1 and which is annotated with an instance of the "Electrification" concept).

7   Conclusions

In this paper we have shown the contribution of tacit knowledge in accomplishing tasks during railway product development. We think this knowledge can be made productive trough a workspace favoring organizational learning. We concluded that this workspace has to sustain three processes: an individual learning process, a social process for collaborative learning and a knowledge management process. Based on the analysis of needs, we deduced that these processes will be supported by: 1) the workbench helping employees to accomplish their analysis 2) the communication component allowing socialization and collaboration, 3) the knowledge management component unifying the previous two and allowing to capitalize knowledge by classifying it into topics. Our future work will consist in the implementation of supplementary functionalities to the workspace and the generalization of the method for the subsequent phases of the product lifecycle.

References

1. Abel, M.-H., Leblanc, A.: Knowledge Sharing via the E-MEMORAe2.0 Platform. In: Proceedings of 6th International Conference on Intellectual Capital, Knowledge Management & Organizational Learning, Montreal Canada, pp. 10–19 (2009)
2. Abel, M.-H., Leblanc, A.: A web plaform for innovation process facilitation. In: IC3K 2009 International Joint Conference on Knowledge Discovery, Knowledge Engineering and Knowledge Management, Madeira Portugal, pp. 141–146. ACM (October 2009)
3. Birk, A., Heller, G., John, I., Schmid, K., von der Maßen, T., Müller, K.: Product Line Engineering: The State of the Practice. IEEE Software, 52–60 (November/December 2003)

4. Boughzala, I., Ermine, J.-L.: Trends in Enterprise Knowledge Management. ISTE, USA (2006)
5. Büchner, T., Matthes, F., Neubert, C.: A Concept and Service Based Analysis of Commercial and Open Source Enterprise 2.0 Tools. In: Proceedings of the 1st International Conference on Knowledge Management and Information Sharing, Madeira, Portugal, pp. 37–45 (2009)
6. Dieng, R., Corby, O., Giboin, A., Ribière, M.: Methods and Tools for Corporate Knowledge Management. In: Proceedings of the Eleventh Workshop on Knowledge Acquisition, Modeling and Management (KAW 1998), Banff, Alberta, Canada, pp. 17–23 (1998)
7. Fitton, L.: Enterprise Microsharing Tools Comparison (2008), http://pistachioconsulting.com/wpcontent/uploads/2008/11/enterprise-microsharing-tools-comparison-110320081.pdf (retrieved May 19, 2010)
8. Garrot-Lavoué, E.: Interconnection of Communities of Practice: A Web Platform for Knowledge Management. In: International Conference on Knowledge Management and Information Sharing (KMIS 2009), Madeira, Portugal, October 6-8, pp. 13–20 (2009)
9. Leblanc, A., Abel, M.: Linking Semantic Web and Web 2.0 for Learning Resources Management. In: Proceedings of the 2nd World Summit on the Knowledge Society: Visioning and Engineering the Knowledge Society. A Web Science Perspective, Chania, Crete, Greece, September 16-18 (2009)
10. McAfee, A.P.: Enterprise 2.0: The Dawn of Emergent Collaboration. Sloan Management Review 47(3), 21–28 (2006)
11. Nonaka, I., Takeuchi, H.: The knowledge-creating company: How Japanese companies create the dynamics of innovation. Oxford University Press, New York (1995)
12. Penciuc, D., Abel, M.-H., Van Den Abeele, D.: Modelling and architecture of a workspace for tacit knowledge management in railway transport. In: Proceedings of the Twenty-Second International Conference on Tools with Artificial Intelligence, ICTAI 2010, Arras France, vol. 2, pp. 356–357 (2010)
13. Penciuc, D., Abel, M.-H., Van Den Abeele, D.: Requirements and modelling of a workspace for tacit knowledge management in railway product development. In: Proceedings of the International Conference on Knowledge Management and Information Sharing, KMIS 2010, Valence Spain, pp. 61–70 (2010)
14. Penciuc, D., Abel, M.-H., Van Den Abeele, D.: From intangibles identification to Requirements for Intangibles Management. In: Proceedings of the 7th International Conference on Intellectual Capital, Knowledge Management and Organizational Learning, Hong Kong China, pp. 628–634 (2010)
15. Polanyi, M.: The Tacit Dimension. Routledge (1966)
16. Stocker, A., Tochtermann, K.: Exploring the value of enterprise wikis - A Multiple-Case Study. In: IC3K 2009 International Joint Conference on Knowledge Discovery, Knowledge Engineering and Knowledge Management, Madeira Portugal, pp. 5–12. ACM (October 2009)
17. Uschold, M., Gruninger, M.: Ontologies: principles, methods and applications. Knowledge Engineering Review 11(2) (June 1996)
18. Zhang, R., Zhang, Y.: Systems requirements for organizational learning. Communications of the ACM 46(12), 73–78 (2003)

Semantics for Enhanced Email Collaboration

Simon Scerri

Digital Enterprise Research Institute, NUIG, IDA Business Park
Lower Dangan, Galway, Ireland
simon.scerri@deri.org

Abstract. Digital means of communications such as email and IM have become a crucial tool for collaboration. Taking advantage of the fact that information exchanged over these media can be made persistent, a lot of research has strived to make sense of the ongoing communication processes in order to support the participants with their management. In this Chapter we pursue a workflow-oriented approach to demonstrate how, coupled with appropriate information extraction techniques, robust knowledge models and intuitive user interfaces; semantic technology can provide support for email-based collaborative work. While eliciting as much knowledge as possible, our design concept imposes little to no changes, and/or restrictions, to the conventional use of email.

Keywords: Email visualisation, Semantic email, Email workflows.

1 Introduction

Despite the rise of competing technologies, email remains a crucial business communication tool and an important source of enterprise information. Email's successes are attributed to a very simple but effective protocol, whose asynchronousity frees the participants from the constraints of time and space. However, the use of email for functions which go beyond its intended design results in an email overload problem, which in turn induces widespread (inter)personal information management problems. These problems are especially affecting users that thoroughly depend on email to carry out their daily work.

Email serves as a virtual extension to the user's workplace, within which they collaborate, generating and sharing new personal information in the process. From this perspective, email overload can be considered as a workflow management problem where, users become overwhelmed with the increasing amount and complexity of co-executing workflows. Although ad-hoc in nature, these workflows are conceptually well-formed. We think that the source of email overload lies partly in the lack of structure imposed by the email model, and partly in the fragmented way in which these workflows are represented on the user's desktop. Our approach investigates whether by providing automated support for structured email workflows, email overload and the ensuing information management hardships can be reduced.

A. Fred et al. (Eds.): IC3K 2010, CCIS 272, pp. 413–427, 2013.

2   Email Overload

Many research efforts have targeted email overload [1] by enabling machines to support the users with better managing their email data. Some have advocated a direct approach, e.g., through automatic email classification [2][3][4], enhanced search and retrieval [5]; whereas others have taken less direct approaches to solving the problem, e.g., by supporting embedded activities [6] or facilitating visualisation [7], to name but a few. Most of these efforts however, offer a somewhat superficial solution that does not target the source of the problem – which lies in email technology being utilised not only as a simple communication means, but also to effectively perform collaborative work. From this perspective, we feel that the source of this problem lies partly in the lack of structure imposed by the email model, and partly in the fragmented way in which email collaborations, or workflows, are 'represented' on the user's conventional desktop. Our approach to this problem is to identify and place patterns of email communication into a structured form, without changing the email experience for the end-user. We start by considering *Action Items* embedded in email content (e.g. Task Assignment, Meeting Proposal). Sequences of related action items exchanged in email messages are then treated as implicit but well-defined *Ad-hoc Email Workflows* (e.g. Task Delegation, Meeting Scheduling) [8]. The nature of these workflows is such that they occur spontaneously and evolve dynamically over time. Besides their lack of support, the way these workflows (or rather, their artefacts) are represented on the user's conventional desktop system is too different than the way the user would visualise them through their mind's eye.

We will explain this situation via the example in Fig. 1, which illustrates how Martin conceives an email conversation (workflow) in his mind and how he can see the corresponding fragmented information physically on his desktop. At time t1, Martin writes an email (1) to Dirk and Claudia, which amongst other things contains a Meeting Proposal action item asking about their availability for a group meeting. This initiates an implicit Meeting Scheduling email workflow, which splits in two co-executing paths at time t2 – control of which is passed to Dirk and Claudia individually. Dirk reacts to the meeting proposal immediately by sending an email (2) with his feedback (Deliver Feedback action item) back to Martin. Claudia instead, is not sure about the purpose of the meeting and thus sends an email back to Martin (3) with her inquiry (Information Request action item). This is considered a sub-workflow of the currently executing workflow. Martin deals with this sub-workflow at time t3 by replying with an Information Delivery in Email 4. Martin's answer to Claudia's query terminates this sub-workflow, upon which Claudia can get back to the initial workflow. At time t4, she also sends her feedback back to Martin (Email 5). At this point Martin has all the required information for the meeting proposal he sent in Email 1. Thus at time t5 the two parallel workflow paths to merge back together and Martin is passed back its control. He decides on a specific date and time for the meeting right away and sends another email (6) with an Event Notification to both Dirk and Claudia. Upon sending the email, an event involving Dirk, Martin and Claudia has been generated for Martin. After both Dirk and Claudia have acknowledged the Event Announcement action item at time t6, the same shared event has been generated for all of them.

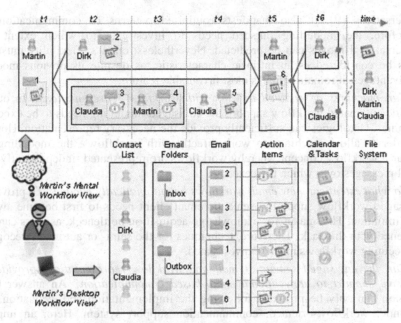

Fig. 1. A user's mental and desktop views of a workflow

Unfortunately for Martin, the email workflow knowledge as presented above is in no way similar to what he can visually gather through conventional email clients on his desktop. There is no straightforward way for users to quickly get an overview of past email conversations, especially when there are many such conversations running at the same time, within tens or hundreds of email messages. Fig. 1 shows the fragmented physical view of the same workflow with which Martin would have to contend. The main workflow components are scattered within a number of separate, largely unconnected, data 'islands'. The action items making up the workflow are obscurely strewn across a number of usually (physically) unrelated email messages, belonging to different email folders. People in the contact list are only associated with these emails, and their roles in the contained workflows remain unspecified. The workflow artefact generated at the end of the example is stored in the Calendar, with little or no connection to the email or the email thread wherein it was generated. Workflow artefacts can also be dispersed in additional data islands, such as generated tasks which end up in a separate task list or having attached documents propagated onto the file system without keeping any connection to their source emails.

3   Research Challenges

Our workflow-oriented approach to email overload faced the following challenges:

1. *How can one model and represent email workflows?* Given the ad-hoc nature of email workflows, each conversational move – or update to the workflow – is largely unpredictable. However, being more similar to spoken rather than written

conversations [9], it also manifests routinised patterns of communication [10]. Therefore, the modelling pursued needs to investigate to which extent email conversational moves can be predicted. Nevertheless, email's flexibility must at all times be considered as an intrinsic characteristic, as despite its obvious modelling disadvantages, it remains email's most favourable feature.

2. *Can machines be enabled to work with workflow representations?* In order to provide automated workflow support, the envisaged modelling needs to be exposed to machines. Ontologies can sufficiently provide the necessary representation. However, in order to allow machines to work directly with workflows, the modelling must favour a workflow concept whereby workflows are represented independently of the email messages over which they execute.

3. *To what extent can new email workflows be automatically elicited?* To provide the envisaged workflow support, a semantic email client needs to first become aware of their initiation. This makes for a knowledge-acquisition bottleneck, as users cannot be burdened with this task. Therefore, techniques for the semi- or automatic recognition of executing workflows must be investigated.

4. *Can the envisaged semantic email and workflow representations provide non-intrusive support to conventional email-based communication?* An answer to this question can only be provided following the implementation and evaluation, of a semantic, workflow-oriented, communication support system. Here, an important design criterion is that additional workflow support is provided alongside conventional emailing practices, which must change minimally, if at all. This calls for a seamless integration of semantic technology within the existing technical landscape. An intelligent user interface (UI) can play a major role in mediating between the introduced semantic technology and the conventional email user, who must never be directly exposed the former.

Our solutions to the first three challenges, and corresponding evaluations, are covered in Section 4. The fourth challenge is tackled separately in section 5, which introduces Semanta: a semantic application that extends existing email clients with workflow-supportive features, including their elicitation, handling, integration and visualisation.

4   Semantic E-mail

The provision of the said features required a robust knowledge representation. We will next review the models enabling *Semantic E-mail*, that is, email enhanced with machine-processable metadata about the underlying workflow knowledge. Here we intend only to provide an overview of this research[1], as the details have been extensively covered in separate publications [11][12][13].

The first milestone was the design of a concise but expressive model by which various email action items could be represented. The model is based on aspects of speech act theory [14], which states that every explicit utterance (or in this case, written text) has one or more associated explicit or implicit acts. This idea is not new, and speech acts were considered in some of the earlier-mentioned approaches [4][6]. In the context of this research, the more user-friendly 'action item' is interchangeable

---

[1] http://smile.deri.ie/projects/smail/

with the more technical 'speech act' term. To start with, our speech act-based approach at structuring email required an appropriate system of categorisation. An earlier speech act taxonomy, also designed for the purpose of email classification [15] was first considered. However, it was deemed unsatisfactory as its design did not focus on speech act sequencing. A number of other ways to improve the flexibility of this model were also identified, resulting in a custom speech act model that is more expressive yet more concise. The model and the results of its evaluation, are introduced in the first subsection below.

The next step was to investigate the nature and strength of relationships between successive speech act instances in email threads. A study in email speech act sequentiality led to the design of a speech act-based workflow model, introduced in the second subsection. In view of the second research challenge, the speech act and workflow models were exposed to machines via a semantic email ontology. Based on this ontology, an Ontology-based Information Extraction (OBIE) technique was then implemented to enable the recognition of action items and the consecutive workflows in email. An overview of this technique is provided in the third subsection.

4.1   A Speech Act Model

We first define a Speech Act as the triple (a,o,s), where a denotes an *action* i.e., what is being performed e.g. a request, delivery or assignment; o the *object* of the action e.g. a request for a meeting; and s the *subject* or agent of the action e.g. meeting participant(s). Actions can have varying *roles* (Fig.2), e.g. *initiative* for actions used to initiate discourse and *continuative* otherwise. Actions may serve particular roles in different situations (e.g. Deliver can be *responsive* as a response to a request, and *informative* otherwise). The speech act object represents instances of *nouns*, categorised under *data nouns* - representing something which can be represented within email (Information, Resource, Feedback); and *activities* - representing external actions occurring outside of email (Task, Event). The speech act subject is applicable to speech acts with an activity as their object, and it represents who is involved in that activity – e.g. for a meeting, the subject can be the *sender* ("Can I attend?"), the *recipient* ("You have to write the document") or *both* ("Let's meet tomorrow").

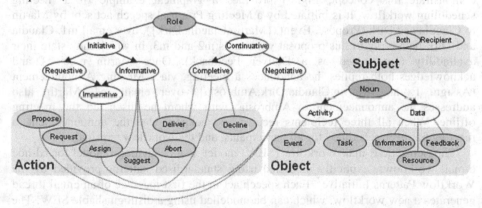

Fig. 2. The Speech Act Action, Object and Subject

The model enables speech act instances, e.g. a request for permission to attend an event is represented as (Request, Event, Sender), a notification of a joint task as (Assign, Task, Both), and a request for information as (Request, Information, Ø). An experiment to measure the human inter-annotator agreement was carried out to evaluate the model's adequacy [12]. The two annotators had the task of identifying instances of 22 valid combinations of speech act parameters in 174 messages from the Enron email corpus. The resulting agreement - 0.811, fares better than the 0.756 observed for the earlier model [15].

Each speech act is assigned a semantics, in the form of two properties – the sender's expected action (SEA) on sending a speech act and the recipient(s)'s expected reaction (RER) on receiving it. The available re/actions are: *await* (a reply from recipient), *do nothing*, *acknowledge* (on receipt), *reply* (to sender) and *attend* (e.g. attend to a task or to a joint event after committing or approving it). These two characteristics define a speech act's conditions of satisfaction and determine the state of an exchanged speech act, which can theoretically be either pending, or completed. For example, a speech act with an SEA of await will remain pending until a reply is sent back. The speech act state is a crucial property in view of our approach – which considers an exchanged speech act as either the initiation, or the resumption of an executing workflow.

4.2   A Behavioural Model for Semantic E-mail

Related sequences of speech acts in email threads can be interpreted as independent but concurrently executing e-mail workflows. By breaking down email into speech acts, each of which executes within a formal workflow, the email process can be given a semantics. The inter-annotator experiment was extended to study speech act sequencing in email threads. Confirming the view withheld by some linguists, we observed that although unpredictable by nature, a "significant percentage of conversational language is highly routinised into prefabricated utterances" [10], or in this case, speech act sequences which carry out e-mail ad-hoc workflows. To demonstrate this concept, Fig. 3 provides a graphical example of a meeting scheduling workflow. It is initiated by a Meeting Proposal speech act sent by Martin to Claudia and Dirk (Propose, Event, {Martin,Claudia,Dirk}), over email m1. Claudia and Dirk both reply to this proposal via emails m2 and m3, in which they state their availability (represented as a Deliver Feedback). Once Martin receives and acknowledges both replies, he announces a meeting, via a Meeting Announcement (Assign, Event, {Martin,Claudia,Dirk,Ambrosia}) over email m4. Martin also addresses the announcement to Ambrosia, with whom he discussed the meeting offline. When all three recipients receive and acknowledge the announcement, a meeting has been scheduled for all participants, and the workflow terminates.

The Semantic E-mail Workflow (SEW) model is a behavioural model for ad-hoc e-mail workflows, explicitly modelled using standardised patterns provided by the Workflow Patterns Initiative[2]. Each speech act in the first message of an email thread generates a new workflow, which can be modelled using a distinguishable SEW. The

---

[2] http://www.workflowpatterns.com/

SEW models frequently occurring email workflows, e.g. the one in Fig. 3, while simultaneously providing support for their ad-hoc nature. Spontaneous courses of action in the workflows are not considered as deviances, but as an intrinsic part of the workflow model. To demonstrate its flexibility, the earlier workflow example is revisited, this time represented by a relevant snapshot (Fig. 4) from the entire workflow model [11], which is expressed using UML 2.0 Activity Diagram notation[3]. A vertical swimlane distinguishes between the workflow initiator (Martin, left-hand side) and each other workflow participant (right-hand side). The workflow initiates when Martin sends a proposal for an activity (an event, i.e. the meeting) to the two participants. Activity Proposals are just one of seven categories of speech acts which Martin can initiate a workflow with. When each participant (right-hand side) receives the proposal, the only default option provided by the SEW is to send a Feedback Delivery back to provide their time/date preferences. However, just as at any other point in a workflow, all participants can pursue any other option, e.g. send a related Information Request (e.g. 'Why are you proposing this meeting?') before providing their availability. These possibilities are represented by a "[...]", and would result in the initiation of a subworkflow of the executing workflow.

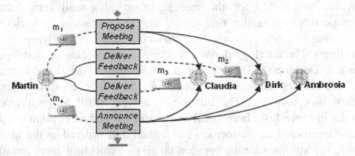

Fig. 3. Email ad-hoc workflow example – scheduling a meeting over email

The 'Collect Feedback' component is a special component which features a persistant trigger workflow pattern that continuously checks whether all participants have provided their feedback (refer to corresponding publication for details). When all participants have provided their availability, back on the initiator's side the SEW provides Martin with two default options for resuming the workflow. On the event that there is no common availability, Martin can amend and resend the proposal. In the example, this is not necessary and instead, Martin replies to all participants (including Ambrosia) with a meeting announcement at the chosen date and time, represented as an Assign Activity (event) speech act. As shown in Fig. 4, since Martin is also implied in this event (as one of the speech act subjects) the parallel split requires him to manage the activity. Similarly, on acknowledgment of the meeting announcement on the right-hand side, all participants are expected to do the same, as shown with the equivalent parallel split. Typically this requires exporting the generated activity to an appropriate manager, e.g. a calendar in this case. The participants however, have the option of delegating the meeting to their subordinates

---

[3] http://www.omg.org/spec/UML/2.0/

via the Delegate path, which splits in two: first to notify Martin of the delegation via a Deliver Data (Information) speech act, and second to perform the actual delegation to a chosen 3rd party.

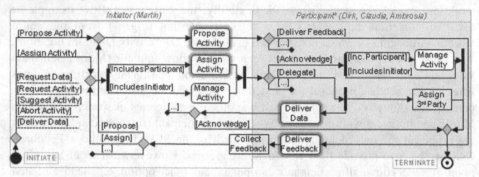

Fig. 4. Email Ad-hoc Workflow Model – A Snapshot for the given example

The two above models are exposed to machines by the semantic email ontology[4], which provides vocabulary for the representation of e-mail (nmo:Email) and 'standard' properties (e.g. sender, recipient, date, etc.), instances for various kinds of embedded speech acts (smail:SpeechAct), and underlying workflows (smail:Workflow). The ontology derives concepts from ontologies developed for the Social Semantic Desktop (SSD) [16], e.g. the messaging (NMO) and contact (NCO) ontologies. This enables integration of data on the SSD from email to other domains, e.g. calendars, task lists, etc. The ontology's design fulfill its requirements, and, although at the instance level both email and workflow instances point to speech acts (unlinked and unordered in the former, and linked and ordered in the latter), in the ontology there is a strict separation between email and workflow representations. This enables workflows to be managed and visualised separately, as demonstrated shortly.

4.3   Semi-automatic Action Item Elicitation

In theory, the SEW can assist on-the-fly workflow knowledge elicitation in communication support systems through user interaction. However, for this to be enabled, they are required to be aware of the executing workflows in the first place. Thus, speech acts that are not already bound to an existent workflow (i.e. sequence of speech acts) need to be correctly recognised. For the purpose, a rule-based classification model for the classification of communicative textual clauses into speech acts, was developed. Based on this model [13], which takes into consideration a number of linguistic features, such as sentence form, tense, modality and the semantic roles of verbs, an OBIE technique is employed to elicit instances of speech acts from email content. A GATE [17] implementation consists of standard GATE components (e.g. tokeniser, POS tagger), a gazetteer lookup augmented with customised keywords/phrases, and an own set of handcoded JAPE [18] grammars.

---

[4] http://ontologies.smile.deri.ie/smail#

The grammars constitute a cascade of finite state transducers over patterns of annotations, such that the output of one transducer becomes the input of the next. Each transducer consists of a collection of phases containing pattern/action rules. The most important is the Speech Act Transducer, which matches combinations of intermediate annotations to one of the speech act instances in the ontology. It alone consists of 58 rules within 14 different phases, such that textual clauses matched in the initial phases are excluded from later ones.

In an evaluation of this technology, 12 people were employed to review annotations generated for over 100 emails. Positive ratings, representing correct classifications, amounted to 41%, negative ratings (representing false positives) to 31%, and 28% were missing speech acts manually highlighted by the evaluators. The resulting F-measure of 0.58 needs to be interpreted in the light of the result obtained earlier for the inter-annotator agreement experiment (0.81), which indicated the difficulty of the classification task even when performed by humans. Also, both results are well within the reasonable performance range described in related literature for information extraction tasks of a similar complexity [19]. Although a reliability of 58% is not suitable for full automation, the technique is suitable for the provision of semi-automatic speech act annotation, in the form of user-reviewable suggestions.

5   Semanta

We now demonstrate how, equipped with the presented models and technology and an intelligent UI providing innovative features, Semanta serves as both a semantic communication support system, and an email-based workflow management tool.

5.1   Implementation

Fig. 5 depicts Semanta's architecture. Knowledge expressed in the semantic email models at the conceptual level is exploited within the knowledge representation (KR) level, by means of the semantic email ontology. The services in the service level provide for all the business logic of the system. The text analytics service performs email speech act classification whereas the semantic email service is responsible for the generation, retrieval and querying of all metadata. Expressed in machine-processable RDF format[5], the metadata is stored in a personal RDF store (instance data, KR level). The semantic email service acts as an intermediary between the KR level and the enhanced semantic UI. The complete separation between business logic and the UI allowed extensions to two popular email clients – Microsoft Outlook and Mozilla Thunderbird. The shaded levels in Fig. 5 stress the fact that Semanta relies on existing mail user agents and email transfer technology. As a result, aside from the additional functionality provided by Semanta, the user's email experience remains relatively unchanged.

---

[5] http://www.w3.org/TR/rdf-primer/

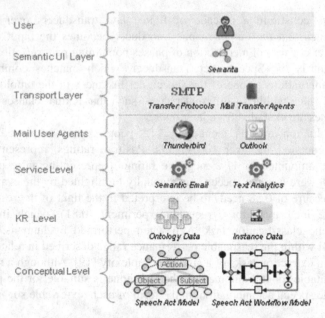

Fig. 5. Semanta's architecture: Separation between

Semanta is one of the semantic applications developed for the SSD. Although Semanta still functions on a conventional desktop, the SSD has the added benefit of desktop data integration – whereby machine-processable data generated by semantic applications can be shared across multiple applications. Thus, the objects shown in Fig. 1 could be linked and related to other physical and abstract personal concepts, e.g. to a concept representing a project to which the scheduled meeting is related. Additionally, the social aspect of the SSD allows all those involved in the meeting to actually share the same instance of the meeting across their desktops. The merits of Semanta as one of many interoperable SSD applications have been discussed in [11]. In this chapter we instead focus on the personal information management support provided by Semanta as a stand-alone semantic application.

5.2   Eliciting Workflow Knowledge

The meeting scheduling example provided earlier will now serve as a use-case to show how Semanta provides on-the-fly support to email communication, and on-request email workflow visualisation to ease their management. To start with, before Martin sends his email, the results of automatic classification are highlighted in the content and presented for review. If required, an intuitive wizard supports the user with the easy creation or modification of action item annotations, hiding the complexity of the speech act model beneath a simple 3 to 4 step procedure (Fig. 6a). The resulting action item annotations together with other harvested metadata are then encoded in RDF within the email headers and transported alongside the email. Once Semanta is aware of the initial action items in a workflow the email workflow model is employed keep track of subsequent action items (updates to the workflow) and to support with their management on both the sender and the recipient(s) side.

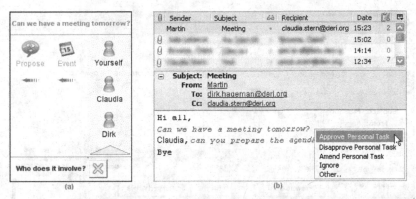

Fig. 6. Email Ad-hoc Workflow Model – A Snapshot for the given example

As shown in Fig. 6b, when a semantic email reaches Claudia and Dirk's inbox, Semanta displays the number of pending action items alongside other conventional properties in the inbox, e.g. date (right-hand column). This number adjusts dynamically as action items are taken care of. Items are also highlighted in displayed messages, enabling users to interact with them directly. Depending on the semantics of the speech acts, a number of relevant options are provided for each item as outlined by the SEW. For example, whereas upon receiving the Meeting Announcement at the end of the example, Claudia, Dirk and Ambrosia are required to simply acknowledge it, the Task Request shown in Fig. 6b requires some form of reply. As shown by the context menu, Claudia can approve, reject or amend the request. However, action items can also be ignored (and unignored) indefinitely, and more importantly, users are allowed to react in additional ways ('Other..' option). Here, the user can type in free text which is again automatically classified into a speech act for review. Whenever the user's interaction results in a new speech act to be sent in an email reply, e.g. a Task approval, Semanta notes this as the next workflow update.

5.3   Integrating Workflow Artefacts

Workflow metadata enables Semanta to detect events/tasks generated over email. For example, on approving a task, the user is prompted to export it directly to the associated Tasklist. Fig. 7 shows how the stored links between three workflow artefacts are exploited. Since the Task Request in email 1 (Fig. 7-1) was answered with a Task Approval in email 2 (Fig. 7-2), the two are linked via the 'Previous/Next Email' buttons. The task generated at the end of the workflow (Fig. 7-3) is linked from the related emails by the 'Related Activity' button. Furthermore, Semanta extends the display of task/calendar items by a 'Conversation' pane which shows the workflow history leading to that item, in terms of text associated with the underlying speech acts. The user can then jump from these items to the email messages wherein they were generated, i.e., back to email 1 and 2 (Fig. 7-1,2).

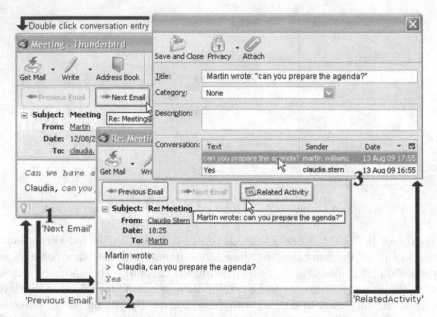

Fig. 7. Linking E-mail Workflow Artefacts: Speech Acts, Messages, Events, etc

5.4   Visualising Email Workflows

Semanta exploits the gathered knowledge to visualise workflows, enabling users to navigate from a workflow to the individual messages over which it executes. A novel 'Workflow Treeview' (Fig.8-S) is available *alongside* Thunderbird's default email treeview (Fig. 8-T). It provides three views, each of which enables the UI components on the right-hand side. The main view ('All') displays a list of workflows that have taken place or are still running/pending (displayed in bold) in the 'Workflow List' (Fig. 8-1). When one is selected (e.g. the Meeting Proposal underlying "Can we have a meeting tomorrow"), the speech acts it constitutes are shown, in order of succession, in the 'Workflow Details' below (Fig. 8-2). In turn, when one of these is selected Semanta retrieves the email within which it has been exchanged and displays it in the 'Email Message' component below (Fig. 8-3). The user can thus quickly identify those action items workflows which are still pending and require attention.

This workflow visualisation is more akin to Martin's mental recollection of workflows, such as that shown in Fig. 1. Alongside the main view, the workflow treeview provides two other useful views: 'Pending Incoming' shows all incoming speech acts (e.g. requests, assignments, suggestions) which the user still needs to address; and 'Pending Outgoing' shows all outgoing action items (e.g. requests) for which the user is still awaiting a reply. These enable the user to keep track of email action items, resume stalled workflows, or send reminders urging others to do so.

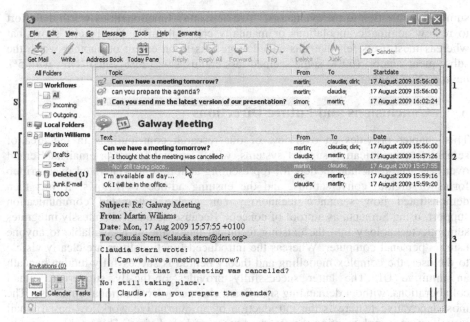

Fig. 8. Semanta's alternative workflow-based e-mail visualisation

5.5   Evaluation

Semanta's evaluation consisted of a formative stage – where the initial system prototype was improved following a controlled study; and a summative stage – where users tried the improved prototype in their actual day-to-day email work. Given its platform independency, this stage was based on Semanta's Thunderbird add-on. The results of the formative and summative evaluations were published in detail in [20] and [21] respectively. All material used for the evaluation, including the full results, is available online[6]. The main results of the summative evaluation are reproduced below. As our main hypothesis was that the use of Semanta improves the email experience over the use of a standard non-semantic email client, Semanta was compared to the standard Thunderbird, with no add-ons. Here, a total of 18 users, collaborating in subgroups of between 2 and 6 people were instructed to use Semanta for at least 10 days.

To interpret the ratings that the evaluators gave to Semanta in comparison to Thunderbird, we constructed a one-sample t-test (two-tailed, 99% confidence interval). Summarising these results, it was observed that the use of Semanta does improve the user's email experience, helping them keeping track of email action items, retrieving additional workflow information from previously physically unrelated items, and aggregating content and messages to visualise entire workflows. However, an increase in the email writing time was also reported, mainly due to the required annotation reviewing stage. The evaluation exercise also posed the following

---

[6] http://www.smile.deri.ie/projects/semanta/evaluation/

summative question to the evaluators: "Are Semanta's functionalities worth the effort to review automatic annotations or manually create them?". The results show that whereas most users felt that this added stage was to an extent or another worth the subsequent support (69%) some seemed less enthusiastic (25% voted 'No', with 5% having 'No Opinion').

6   Conclusions and Future Directions

This chapter provides an overview of our research contributions in the area of semantic communication support systems, with a special focus on email and email overload. These include models for representing action items generated over written forms of digitial conversations, and the ensuing ad-hoc workflows. It was then demonstrated how semantic technology can enable automated communication support, using Semanta as a proof of concept. Because Semanta seamlessly integrates semantic technology into the existing technical landscape, it is available to anyone using a personal computer. Whereas the introduced functionalities are clearly visible to the user, the complex modelling and theory behind is successfully hidden beneath an intuitive UI. The latter successfully provides on-the-fly support to email collaborations, without demanding significant changes to conventional email use. The most innovative feature is the introduction of an alternative workflow-based email visualization, supplementing the standard message-based view. This enables users to look at the 'big picture', rather than fragmented information, when managing their email collaborations. Our experience shows that semantic technology can provide high-level support to day-to-day computer work, without exposing users to the complexity of the underlying models. In order to provide reliable support, users are still required to oversee a semantic annotation process. However, provided that the resulting benefits are of multiple orders to the added cognitive costs, most users feel that the ensuing support is worth the annotation reviewing stage.

To further reduce this extra cost, future work will investigate machine-learning and case-based reasoning techniques to improve automatic annotation. An extension to the text classification grammars to enable the recognition of more information, e.g. matching person names in text to the user's email contacts, recognition of dates and times related to upcoming events or task deadlines, etc., is also in the pipeline. Architecture-wise, Semanta will be extended to also mine action items from non-semantic email, when the corresponding users are not using Semanta. Finally, the workflow views will be extended to incorporate any resulting events/tasks. The status of tasks can then also be dynamically updated when the responsible participant(s) update it as such. As a final note, although the presented contributions focus specifically on email, the semantic technology underlying Semanta is doman-independent, and can easily and without much effort be applied to other digital communication media such as IM, text messaging, customer services or online fora.

Acknowledgements. The work presented in this paper was supported (in part) by the Lion project supported by Science Foundation Ireland under Grant No. SFI/02/CE1/I131 and (in part) by the European project NEPOMUK No FP6-027705. In addition, I would like to thank all people contributing to this research over the years, especially Dr. Siegfried Handschuh, Mr. Brian Davis, and Mr. Gerhard Gossen.

References

1. Whittaker, S., Sidner, C.: Email overload: exploring personal information management of email. In: CHI 1996, pp. 276–283. ACM, New York (1996)
2. Chan, J., Koprinska, I., Poon, J.: Co-training with a single natural feature set applied to email classification. In: Web Intelligence, pp. 586–589. IEEE Computer Society (2004)
3. Dredze, M., Wallach, H.M., Puller, D., Pereira, F.: Generating summary keywords for emails using topics. In: IUI 2008, pp. 199–206. ACM (2008)
4. Goldstein, J., Sabin, R.E.: Using Speech Acts to Categorize Email and Identify Email Genres. In: Proc. System Sciences, HICSS (2006)
5. Dumais, S., Cutrell, E., Cadiz, J.J., Jancke, G., Sarin, R., Robbins, D.C.: Stuff I've seen: a system for personal information retrieval and re-use. In: SIGIR 2003, pp. 72–79. ACM (2003)
6. Khoussainov, R., Kushmerick, N.: Email task management: An iterative relational learning approach. In: CEAS (2005)
7. Rohall, S.L., Gruen, D., Moody, P., Wattenberg, M., Stern, M., Kerr, B., Stachel, B., Dave, K., Armes, R., Wilcox, E.: Remail: a reinvented email prototype. In: CHI 2004. ACM (2004)
8. Voorhoeve, M., Van der Aalst, W.: Ad-hoc workflow: problems and solutions. In: 8th International Workshop on Database and Expert Systems Applications (1997)
9. Khosravi, H., Wilks, Y.: Routing email automatically by purpose not topic. Nat. Lang. Eng. 5(3), 237–250 (1999)
10. Stubbs, M.: Discourse Analysis. Blackwell (1983)
11. Scerri, S., Handschuh, S., Decker, S.: Semantic Email as a Communication Medium for the Social Semantic Desktop. In: Bechhofer, S., Hauswirth, M., Hoffmann, J., Koubarakis, M. (eds.) ESWC 2008. LNCS, vol. 5021, pp. 124–138. Springer, Heidelberg (2008)
12. Scerri, S., Mencke, M., Davis, B., Handschuh, S.: Evaluating the Ontology underlying sMail - the Conceptual Framework for Semantic Email Communication. In: LREC 2008 (2008)
13. Scerri, S., Gossen, G., Davis, B., Handschuh, S.: Classifying action items for semantic email. In: LREC 2010. European Language Resources Association, ELRA (2010)
14. Searle, J.: Speech Acts. Cambridge University Press (1969)
15. Carvalho, V.R., Cohen, W.W.: Improving email speech acts analysis via n-gram selection. In: ACTS 2009 Workshop, pp. 35–41. ACL (2006)
16. Reif, G., Groza, T., Scerri, S., Handschuh, S.: Final NEPOMUK Architecture Deliverable D6.2.B., NEPOMUK Consortium (2008)
17. Cunningham, H.: Gate, a general architecture for text engineering. Computers and the Humanities 36(2), 223–254 (2002)
18. Cunningham, H.: JAPE: a Java Annotation Patterns Engine. Research memorandum, Department of Computer Science, University of Sheffield (1999)
19. Cunningham, H.: Information Extraction, Automatic. Encyclopedia of Language and Linguistics (2005)
20. Scerri, S., Davis, B., Handschuh, S., Hauswirth, M.: Semanta – Semantic Email Made Easy. In: Aroyo, L., Traverso, P., Ciravegna, F., Cimiano, P., Heath, T., Hyvönen, E., Mizoguchi, R., Oren, E., Sabou, M., Simperl, E. (eds.) ESWC 2009. LNCS, vol. 5554, pp. 36–50. Springer, Heidelberg (2009)
21. Scerri, S., Gossen, G., Handschuh, S.: Supporting digital collaborative work through semantic technology. In: KMIS 2010, Valencia, Spain (2010)

Author Index